NEW TECHNOLOGIES FOR HUMAN RIGHTS LAW AND PRACTICE

New technological innovations offer significant opportunities to promote and protect human rights. At the same time, they also pose undeniable risks. In some areas, they may even be changing what we mean by human rights. The fact that new technologies are often privately controlled raises further questions about accountability and transparency and the role of human rights in regulating these actors. This volume – edited by Molly K. Land and Jay D. Aronson – provides an essential roadmap for understanding the relationship between technology and human rights law and practice. It offers cutting-edge analysis and practical strategies in contexts as diverse as autonomous lethal weapons, climate change technology, the Internet and social media, and water meters. This title is also available as Open Access.

MOLLY K. LAND is Professor of Law and Human Rights at the University of Connecticut School of Law and Associate Director of the University of Connecticut's Human Rights Institute. Her research focuses on the intersection of human rights, science, technology, and innovation.

JAY D. ARONSON is the founder and director of the Center for Human Rights Science and Associate Professor of Science, Technology, and Society in the History Department at Carnegie Mellon University.

New Technologies for Human Rights Law and Practice

Edited by

MOLLY K. LAND

University of Connecticut School of Law

JAY D. ARONSON

Carnegie Mellon University Center for Human Rights Science

CAMBRIDGE
UNIVERSITY PRESS

University Printing House, Cambridge CB2 8BS, United Kingdom

One Liberty Plaza, 20th Floor, New York, NY 10006, USA

477 Williamstown Road, Port Melbourne, VIC 3207, Australia

314–321, 3rd Floor, Plot 3, Splendor Forum, Jasola District Centre, New Delhi – 110025, India

79 Anson Road, #06–04/06, Singapore 079906

Cambridge University Press is part of the University of Cambridge.

It furthers the University's mission by disseminating knowledge in the pursuit of education, learning, and research at the highest international levels of excellence.

www.cambridge.org
Information on this title: www.cambridge.org/9781107179639
DOI: 10.1017/9781316838952

© Cambridge University Press 2018

This work is in copyright. It is subject to statutory exceptions and to the provisions of relevant licensing agreements; with the exception of the Creative Commons version the link for which is provided below, no reproduction of any part of this work may take place without the written permission of Cambridge University Press.

An online version of this work is published at 10.1017/9781316838952 under a Creative Commons Open Access license CC-BY-NC-ND 4.0 which permits re-use, distribution and reproduction in any medium for non-commercial purposes providing appropriate credit to the original work is given. You may not distribute derivative works without permission. To view a copy of this license, visit https://creativecommons.org/licenses/by-nc-nd/4.0

All versions of this work may contain content reproduced under license from third parties. Permission to reproduce this third-party content must be obtained from these third-parties directly.

When citing this work, please include a reference to the DOI 10.1017/9781316838952

First published 2018

Printed in the United States of America by Sheridan Books, Inc.

A catalogue record for this publication is available from the British Library.

Library of Congress Cataloging-in-Publication Data
NAMES: Land, Molly K., editor. | Aronson, Jay D., editor.
TITLE: New technologies for human rights law and practice / Edited by Molly K. Land, University of Connecticut School of Law, Jay D. Aronson, Carnegie Mellon Center for Human Rights Science.
DESCRIPTION: New York : Cambridge University Press, 2018. | Includes bibliographical references and index.
IDENTIFIERS: LCCN 2017045843 | ISBN 9781107179639 (Hardback : alk. paper) | ISBN 9781316631416 (pbk. : alk. paper)
SUBJECTS: LCSH: Human rights. | Technological innovations–Law and legislation.
CLASSIFICATION: LCC K3240 .N497 2017 | DDC 342.08/5–dc23
LC record available at https://lccn.loc.gov/2017045843

ISBN 978-1-107-17963-9 Hardback

Cambridge University Press has no responsibility for the persistence or accuracy of URLs for external or third-party internet websites referred to in this publication and does not guarantee that any content on such websites is, or will remain, accurate or appropriate.

Contents

Notes on Contributors		page vii
Acknowledgements		xiii
1	The Promise and Peril of Human Rights Technology *Molly K. Land and Jay D. Aronson*	1
	PART I NORMATIVE APPROACHES TO TECHNOLOGY AND HUMAN RIGHTS	21
2	Safeguarding Human Rights from Problematic Technologies *Lea Shaver*	25
3	Climate Change, Human Rights, and Technology Transfer: Normative Challenges and Technical Opportunities *Dalindyebo Shabalala*	46
4	Judging Bioethics and Human Rights *Thérèse Murphy*	71
5	Drones, Automated Weapons, and Private Military Contractors: Challenges to Domestic and International Legal Regimes Governing Armed Conflict *Laura A. Dickinson*	93
	PART II TECHNOLOGY AND HUMAN RIGHTS ENFORCEMENT	125
6	The Utility of User-Generated Content in Human Rights Investigations *Jay D. Aronson*	129

7	Big Data Analytics and Human Rights: Privacy Considerations in Context *Mark Latonero*	149
8	The Challenging Power of Data Visualization for Human Rights Advocacy *John Emerson, Margaret L. Satterthwaite, and Anshul Vikram Pandey*	162
9	Risk and the Pluralism of Digital Human Rights Fact-Finding and Advocacy *Ella McPherson*	188

PART III BEYOND PUBLIC/PRIVATE: STATES, COMPANIES, AND CITIZENS 215

10	Digital Communications and the Evolving Right to Privacy *Lisl Brunner*	217
11	Human Rights and Private Actors in the Online Domain *Rikke Frank Jørgensen*	243
12	Technology, Self-Inflicted Vulnerability, and Human Rights *G. Alex Sinha*	270
13	The Future of Human Rights Technology: A Practitioner's View *Enrique Piracés*	289

Index 309

Contributors

Jay D. Aronson is the founder and director of the Center for Human Rights Science and Associate Professor of Science, Technology, and Society in the History Department at Carnegie Mellon University. He currently conducts research on the acquisition and analysis of video evidence in human rights investigations. His recent book, *Who Owns the Dead? The Science and Politics of Death at Ground Zero* (2016), analyzes the recovery, identification, and memorialization of the victims of the 9/11 World Trade Center attacks. It represents the culmination of more than a decade of work on forensic identification in criminal justice and humanitarian contexts. Aronson received his PhD in the history of science and technology from the University of Minnesota and was both a pre- and post-doctoral fellow at Harvard University's John F. Kennedy School of Government. From 2012 to 2018, he served as a member of the Committee on Scientific Freedom and Responsibility with the American Association for the Advancement of Science.

Lisl Brunner is an international human rights lawyer. As a staff attorney at the Inter-American Commission on Human Rights, she worked on cases and urgent measures dealing with counterterrorism and the protection of journalists, as well as projects to define Inter-American standards on freedom of expression and the Internet. She also served as Policy Director and facilitator for the Telecommunications Industry Dialogue at the Global Network Initiative, an international multistakeholder initiative that seeks to protect and advance freedom of expression and privacy in the ICT sector. Brunner recently moved to AT&T, where she focuses on privacy issues in her role as Director of Global Public Policy. Brunner received her JD from the University of Pittsburgh and a bachelor's degree from the College of William & Mary.

Laura A. Dickinson is the Oswald Symister Colclough Research Professor of Law at The George Washington University Law School. Her work focuses on national security, human rights, foreign affairs privatization, and qualitative empirical approaches to international law. Dickinson's book, *Outsourcing War and Peace*

(2011), examines the increasing privatization of military, security, and foreign aid functions of government, considers the impact of this trend on core public values, and outlines mechanisms for protecting these values in an era of privatization. In addition to her scholarly activities, Dickinson has served as special counsel to the general counsel of the Department of Defense and as a senior policy adviser to Harold Hongju Koh, Assistant Secretary of State for Democracy, Human Rights, and Labor at the US Department of State. She is a former law clerk to US Supreme Court Justices Harry A. Blackmun and Stephen G. Breyer, and to Judge Dorothy Nelson of the Ninth US Circuit Court of Appeals. She received her JD from Yale University and her AB from Harvard.

John Emerson is a research scholar at the Center for Human Rights and Global Justice at the New York University School of Law. As a creative technologist, he works at the intersection of digital design, data, and social change. Based in New York City, he has designed web sites, printed materials, and data visualizations for leading media companies as well as local and international human rights organizations including Amnesty International, Human Rights Watch, the Committee to Protect Journalists, and the United Nations. His writing about graphic design has been published in *Communication Arts* and *Print* and featured in *Metropolis* and *The Wall Street Journal*. Since 2002, he has published Social Design Notes, a weblog on design and activism at http://backspace.com. He received his BFA from the Cooper Union.

Rikke Frank Jørgensen is a senior researcher at the Danish Institute for Human Rights in Copenhagen. Her research focuses on the interface between human rights and technology, the role of private actors in the online domain, social media platforms, and Internet governance (gaps). Besides her scholarly activities, she has served as an adviser to the Danish government, participated in Council of Europe's Committee on Human Rights for Internet Users, and been closely involved in civil society networks such as European Digital Rights. Her most recent book, *Framing the Net: The Internet and Human Rights* (2013), examines how internet-related metaphors instruct policy. She holds a PhD in communication and information technology from Roskilde University and a master's degree from Aarhus University.

Molly K. Land is Professor of Law and Human Rights at the University of Connecticut School of Law and Associate Director of the University of Connecticut's Human Rights Institute. Her research focuses on the intersection of human rights, science, technology, and innovation. Her most recent work considers the duties of Internet companies to promote and protect rights online as well as the effects of new technologies on human rights fact-finding, advocacy, and enforcement. Land has authored several human rights reports, including a report for the World Bank on the role of new technologies in promoting human rights. She is currently a member of the Committee on Scientific Freedom and Responsibility with the American

Association for the Advancement of Science. Land received her JD from Yale Law School and was a Fulbright Scholar at the University of Bonn in Germany.

Mark Latonero leads the human rights program at Data & Society, a nonprofit research institute in New York. He is a senior fellow at both USC Annenberg School's Center on Communication Leadership & Policy and Leiden University's Institute for Security and Global Affairs. In addition, he serves as innovation consultant to the UN Office of the High Commissioner for Human Rights. Previously he was a research professor and research director at USC where he led the Annenberg Technology and Human Trafficking Initiative. Latonero examines the social risks, harms, and benefits of emerging technology, particularly in human rights and humanitarian contexts where vulnerable populations are concerned. He has published on the role of digital technologies in human trafficking, child exploitation, migration, and refugee crises. He has led field research in Bangladesh, Cambodia, Greece, Haiti, Indonesia, Pakistan, Philippines, Serbia, Sri Lanka, and Thailand. Latonero completed his PhD at the University of Southern California and was a postdoctoral scholar at the London School of Economics.

Ella McPherson is the University of Cambridge's Lecturer in the Sociology of New Media and Digital Technology and the Anthony L. Lyster Fellow in Sociology at Queens' College, Cambridge. She is also Co-Director of Cambridge's Centre of Governance and Human Rights. McPherson's research focuses on symbolic struggles surrounding the media in times of transition. Her current research examines the potential of using social media by human rights NGOs to generate governmental accountability. This involves understanding the methodological and reputational implications of using social media and related networks as data sources and dissemination tools, as well as social media's effects on pluralism in human rights discourse. McPherson's previous research, drawing on her media ethnography of human rights reporting at Mexican newspapers, identified the contest for public credibility between state, media, and human rights actors as a significant driver of human rights coverage. She also leads The Whistle, a human rights reporting and verification app in development at the University of Cambridge. McPherson earned her PhD from the Department of Sociology at Cambridge, her MPhil in Latin American studies from Cambridge, and her BA from the Woodrow Wilson School of Public and International Affairs at Princeton University.

Thérèse Murphy is Professor of Law and Director of the Health and Human Rights Unit at Queen's University Belfast. Her work focuses on health and human rights, including the human right to science, new health technologies and human rights, and the rights of incarcerated prisoners. She is also interested in methodological issues in human rights research. Her books include *Health and Human Rights* and the edited collections *The United Nations Special Procedures System*, *European Law and New Health Technologies*, and *New Technologies and Human*

Rights. She is a member of the editorial board of the *Human Rights Law Review* and co-editor of the book series Law and Health. Murphy studied law at University College Dublin and Cambridge University.

Anshul Vikram Pandey is the co-founder and chief technology officer of Accern, a data design company that develops predictive analytics for institutional investors using traditional and alternative datasets. His primary research interests are in big data analytics and sensemaking, natural language processing, and human computer interaction. He received a BE in electrical and electronics engineering from Birla Institute of Technology and Science (Pilani, India) in 2012, and a PhD in computer science from New York University in 2017. He has published numerous articles in the areas of data visualization, text analytics, and human-computer interaction.

Enrique Piracés manages the technology program at the Carnegie Mellon University Center for Human Rights Science. His primary responsibilities are to serve as a liaison between the human rights community and researchers at Carnegie Mellon, and to ensure that the methods and tools developed at CMU can be effectively integrated into the workflows of human rights practitioners. He has been working at the intersection of human rights, science, and technological innovation for over a decade, previously as vice president of the human rights program at Benetech and as a technology specialist at Human Rights Watch. His focus has been both the implications of the use of technology in the context of human rights and the opportunities that new scientific and technological developments open for NGOs and practitioners. He is an advocate for the use of open source technology and strong crypto in human rights documentation and journalistic work. His experience ranges from fact-finding and evidence gathering to data science and digital security. Piracés began his education in Mexico and Chile, and ultimately received a BA from Hofstra University.

Margaret L. Satterthwaite is Professor of Clinical Law, Faculty Director of the Robert L. Bernstein Institute for Human Rights, and Co-Chair of the Center for Human Rights and Global Justice at the New York University School of Law. Her research interests include economic and social rights, human rights and counterterrorism, methodological innovation in human rights, and vicarious trauma among human rights workers. Before joining the academy, she worked for a number of human rights organizations, including Amnesty International, Human Rights First, and the Commission Nationale de Verité et de Justice in Haiti. She has authored or co-authored more than a dozen human rights reports and dozens of scholarly articles and book chapters. She has worked as a consultant to numerous UN agencies and special rapporteurs and has served on the boards of several human rights organizations. She received her JD from the New York University School of Law.

Dalindyebo Shabalala is Assistant Professor of Law at the University of Dayton Law School. His research focuses on the interaction of intellectual property law, especially patent law, with the rights of indigenous peoples and climate change law. He conducts research on the rights of indigenous peoples and traditional communities to their traditional knowledge and culture and the role of international intellectual property treaties in enabling or preventing the realization of those rights. Shabalala also conducts research on the interaction of patent law with climate change, focusing on the role of technology licensing and transfer in enabling the technology goals of the climate change convention (UNFCCC). He is a member of the Climate Action Network Technology Working Group and serves as the Environmental NGO representative to the UNFCCC Technology Executive Committee's Task Force on Innovation, Research, Development, and Demonstration. Shabalala received his PhD in law from Maastricht University, his JD from the University of Minnesota School of Law, and his BA from Vassar College.

Lea Shaver is Professor of Law at Indiana University's Robert H. McKinney School of Law. Her work focuses on the intersection of intellectual property and human rights law and informs the concept of "the right to science and culture" recently embraced at the United Nations. Her work applies a social justice perspective to the study of copyright law. By recognizing that copyright's incentive system creates both winners and losers, she argues, copyright scholars can help identify ways to adapt copyright law and cultural policy to better serve all of society. Shaver is currently writing a book exploring solutions to the profound shortage of mother-tongue reading material in most of the world's languages, a problem affecting more than one billion children worldwide. Shaver holds a JD from Yale Law School and a BA and MA from the University of Chicago. She was a Fulbright Scholar in South Africa.

G. Alex Sinha is currently a litigation associate at Arnold & Porter Kaye Scholer LLP in New York, and will be joining the faculty of the Quinnipiac University School of Law as an assistant professor in the fall of 2018. His research interests include privacy, national security, and human rights. He has published on topics such as surveillance, the human right to privacy, and children's rights under humanitarian law. He was formerly the Aryeh Neier Fellow at Human Rights Watch and the American Civil Liberties Union. In that role, he litigated national security cases and authored a report on the chilling effects of large-scale US government surveillance on journalists and lawyers. Sinha holds a PhD in philosophy from the University of Toronto and a JD from the New York University School of Law.

Acknowledgements

We are grateful to the numerous people and institutions that made this book possible. First and foremost, we would like to thank the contributors, who first gathered in October 2015 at the University of Connecticut's Human Rights Institute to share their ideas and lay the groundwork for the theoretical approaches to understanding technology and human rights that are introduced in this volume. We are deeply appreciative of the financial support provided by the University of Connecticut's Human Rights Institute for that initial workshop, as well as crucial logistical and administrative support provided by Rachel Jackson and Lyndsay Nalbandian.

We are fortunate that our home institutions (University of Connecticut School of Law and Carnegie Mellon University) were willing and able to provide the financial support that allows us to offer this book under a Gold Open Access license. This license allows anyone anywhere around the world to read, download, and utilize the material in this volume directly from Cambridge University Press under a Creative Commons license. We are proud of this arrangement and hope that it becomes the norm for all scholarly work that relates to human rights and human flourishing.

Jay Aronson would like to acknowledge Humanity United, MacArthur Foundation, and Oak Foundation for their generous support of his scholarship, and the work of the Center for Human Rights Science at Carnegie Mellon, over the past several years. Molly Land would like to thank the School of Law for summer fellowship support for her research on these issues.

We also thank Tatyana Marugg, Katherine Graichen, Magdalena Narozniak, and Sarah Hamilton at the University of Connecticut School of Law for their outstanding editing and research assistance.

It almost goes without saying that a project like this demands sacrifice from family members at critical points of the editorial process. Our deepest thanks to our families for their love, patience, and support throughout the production of this volume.

Finally, we would like to express our greatest debt to the practitioners and front-line defenders who continue to inspire us and motivate our efforts to understand the profound impacts that technology is having on human rights advocacy and accountability.

1

The Promise and Peril of Human Rights Technology

Molly K. Land and Jay D. Aronson

The first two decades of the twenty-first century have seen a simultaneous proliferation of new technological threats to and opportunities for international human rights. New advances – not only the Internet, social media, and artificial intelligence but also novel techniques for controlling reproduction or dealing with climate change – make clear that scientific and technological innovations bring both risks and benefits to human rights. Efforts to protect and promote human rights have to take seriously the ways in which these technologies, and the forms of knowledge creation, production, and dissemination they enable, can create harms and be exploited to violate rights. At the same time, human rights practitioners must continue to seek creative ways to make use of new technologies to improve the human condition. This dichotomy is the central tension that animates both this volume and the emerging field of human rights technology.

The overriding purpose of the volume, and of the University of Connecticut workshop that launched it, is to encourage human rights institutions, experts, and practitioners to take seriously the risks and opportunities of technology for the promotion and protection of human rights. The volume uses diverse case studies to examine how the dynamic of intertwined threat and opportunity plays out in a range of contexts. Case studies focus on assisted reproductive technologies, autonomous lethal weapons, climate change technology, the Internet and social media, and water meters. Considering the relationship between technology and human rights across these diverse areas reveals areas of both continuity and discontinuity in terms of how technology affects the enjoyment of human rights.

We begin by laying out the principles that animate the project. These principles have been derived chiefly from international human rights law and practice, and also draw on the scholarly study of science, technology, and the law. Based on these principles, we define a "human rights" approach to the study of technology. Finally, we identify and analyze the cross-cutting themes that unite the book – power and justice, accountability, and the role of private authority – to chart a road map for further study of the relationship between technology and human rights.

I DEFINING A "HUMAN RIGHTS" APPROACH TO TECHNOLOGY

This collection goes beyond analyzing the risks and opportunities of technology to articulate a human rights-based approach to understanding the impact of technological change on human rights. A human rights-based approach to technology in this context is defined by two elements: a reliance on international human rights law as a source of normative commitments; and a focus on accountability strategies derived from human rights practice. In order to examine human rights law and practice as they intersect with technology, the book also makes use of ideas and concepts from cyberlaw and science and technology studies.

Although human rights is clearly not the only lens through which we can view technological change, it is an essential one. Understanding how human rights law and practice intersect with technology offers a global baseline for addressing the cross-border impacts of technology, and it also provides guidance for human rights advocates who are deploying new techniques in their work and responding to the impacts of new technologies.

A *Human Rights Law*

To say that technology presents both opportunities and challenges for human rights is not to suggest it is neutral. To the contrary, the book is motivated by the recognition that the design of technology reflects and influences societal values and norms.[1] Technology matters for human rights not only because it can be used in ways that have negative or positive consequences for the enjoyment of human rights, but also because its very design can make those consequences more or less likely. The well-known maxim "code is law" is shorthand for the idea that design can encourage or discourage particular activities by making them more or less costly, and thus promote particular outcomes.[2]

The ability of technological design to steer outcomes necessarily means that decisions about design will reflect preexisting normative commitments. Those commitments can be derived from a variety of sources, including community values,

---

[1] See generally L. DeNardis, *Protocol Politics: The Globalization of Internet Governance* (Cambridge, MA: MIT Press, 2009); M. Flanagan, D. C. Howe, and H. Nissenbaum, "Embodying Values in Technology: Theory and Practice," in Jeroen van den Hoven and John Weckert (eds.), *Information Technology and Moral Philosophy* (Cambridge: Cambridge University Press, 2008).

[2] L. Lessig, *Code and Other Laws of Cyberspace* (New York: Basic Books, 1999), p. 6.

constitutional precepts, or individual morals. This volume uses international human rights law to orient its discussion of technological design and implementation. This includes not only international human rights law but also a range of specific commitments that characterize human rights practice, including commitments to participation in decision-making and an emphasis on the needs of the most vulnerable. In this sense, the volume uses "human rights" in the specific rather than the general sense – not as a general proxy for "social good," but rather as a set of internationally recognized legal norms and established practices. In fact, these norms and practices are increasingly characterized as a "human rights based-approach" in a variety of social justice contexts.[3]

In reorienting discussions about human rights and technology on the core values of international human rights, the contributors to this volume disavow two tropes that generally dominate such analysis. The first is reductionist thinking about technology, which focuses on innovation and technology as silver bullets or even as goals in and of themselves, rather than as tools that embody both opportunities and risks. The second trope is reductionist thinking about human rights, which tends to reflect unrealistic assumptions about the effectiveness and functioning of international institutions or emphasizes legal accountability over other methods of responding to human rights violations.

We want to move the conversation away from these well-worn paths and reorient it on the fundamental values of a human rights-based approach, which emphasizes universality/inalienability, indivisibility, interdependence/interrelatedness, equality and nondiscrimination, participation/inclusion, and accountability/rule of law.[4] Each of these principles yields insights for understanding the contribution of a human rights-based approach to technology.

Equality and nondiscrimination require attending to the situation of the most vulnerable and demand that all people, regardless of their position in society, have access to the tools and knowledge needed to make their lives better. Accountability means that people must also have access to institutional spaces and mechanisms that allow them to make rights claims and to seek redress from accountable parties, whether governmental or non-state, when these tools and innovations have negative impacts on their lives or when they lack access to the benefits of these tools. Participation means that users and others affected by technological innovation must be meaningfully involved in, not just consulted on, the development and design of technology. Universality and inalienability require us to look beyond the ostensible neutrality of technology to recognize the power and privilege that are embedded in technological systems. Finally, indivisibility, interdependence, and interrelatedness

[3] See A. E. Yamin, *Power, Suffering and the Struggle for Dignity: Human Rights Frameworks for Health and Why They Matter* (Philadelphia: University of Pennsylvania Press, 2016), p. 5.
[4] HRBA Portal, "The Human Rights Based Approach to Development Cooperation: Towards a Common Understanding among UN Agencies," http://hrbaportal.org/the-human-rights-based-approach-to-development-cooperation-towards-a-common-understanding-among-un-agencies.

mandate attention to the effects of technology not only on civil and political rights, such as freedom of expression and privacy, but also on rights to water, health, and education, among others. Focusing on these core values of the human rights-based approach can help cut through some of the deterministic thinking that technology engenders and provide the foundation for an approach to technology and innovation that centers on people, not things or institutions.

B *Human Rights Practices*

Technology can also play a central role in human rights accountability practices. Human rights practitioners have developed a set of accountability strategies over the past several decades that have emerged from the peculiarities of international human rights law. Human rights are protected by international treaties that create binding legal commitments for the states that ratify them. Almost always, though, these international treaties are paired with extremely weak enforcement mechanisms. The result is that human rights practitioners have had to rely on indirect compliance strategies, most notably "naming and shaming."[5] While shame can be a component of domestic law enforcement as well,[6] it has over time become a primary strategy for holding states accountable for violations of international human rights law. This feature of human rights practice has important consequences when considering the effects of technology on the promotion and protection of rights. As we discuss below, although new technologies allow greater participation by ordinary citizens in accountability processes and offer new methods for preserving evidence, the use of technology by state actors also fragments state authority and thus makes accountability efforts more challenging.

In focusing on the effects of technology on compliance and enforcement, this volume also contributes to ongoing debates about the effectiveness of international human rights law. In the absence of a centralized authority to enforce rights, international human rights law and institutions may seem far more toothless than would be expected given the importance of the values they claim to protect.[7] Yet the power of human rights is located not in its coerciveness, but in its ability to serve as a

---

[5] See, e.g., A. Chayes and A. Handler Chayes, *The New Sovereignty: Compliance with International Regulatory Agreements* (Cambridge, MA: Harvard University Press, 1998); T. M. Franck, *Fairness in International Law and Institutions* (Oxford: Oxford University Press, 1995); R. Goodman and D. Jinks, "How to Influence States: Socialization and International Human Rights Law" (2004) 54 *Duke Law Journal* 621–703; A. Guzman, "A Compliance-Based Theory of International Law" (2002) 90 *California Law Review* 1823–87; H. H. Koh, "Why Do Nations Obey International Law?" (1997) 106 *Yale Law Journal* 2599–659.

[6] See L. C. Porter, "Trying Something Old: The Impact of Shame Sanctioning on Drunk Driving and Alcohol-Related Traffic Safety" (2013) 38 *Law & Social Inquiry* 863–91; S. Gopalan, "Shame Sanctions and Excessive CEO Pay" (2007) 32 *Delaware Journal Corporate Law* 757–97.

[7] O. Hathaway, "Between Power and Principle: An Integrated Theory of International Law," (2005) 72 *University of Chicago Law Review* 469–521 at 490.

vehicle for the assertion of political demands. Thus, the book envisions a multidimensional model of social change – with human rights operating bottom-up as well as top-down and in which a variety of actors engage both horizontally and vertically in an iterative process of incremental change.[8]

The volume also examines how technology affects this process. Can choices about technological design strengthen efforts to protect rights by encoding human rights values directly into the structures in which communication, knowledge creation, and reproduction take place? Or will technological innovation disproportionately serve the interests of the powerful because of disparities in the knowledge and resources needed to use, deploy, and interrogate it critically? In some cases, might the use of technology slow down processes of social change by rendering invisible deliberate choices that have been made to restrict rights? As discussed at the outset of this introduction, technology is not an either/or proposition; the same technology may do all of these things and more. A goal of the contributions in this volume is to tease out when, where, and under what conditions technology can strengthen and protect rights.

C Cyberlaw and STS

The volume also adopts an interdisciplinary approach aimed at bringing international human rights law into conversation with two scholarly disciplines that examine the intersection of law and technology: cyberlaw and science and technology studies (STS). These disciplines share a common commitment to better understanding the technical, social, political, legal, and cultural dimensions of the development of new technologies and the new social arrangements they both accompany and foster.

STS, for example, recognizes that there is a certain amount of experimentation involved in the introduction of a new technology into society. STS scholars explore the intentions of those who deploy new technologies in society, examine the unexpected consequences of the introduction of new technologies, and analyze how societies respond to and shape these new tools, methods, and domains of knowledge.[9]

This conversation promises to be generative and challenging for both human rights and STS. For example, the idea that the introduction of new technologies is an "experiment" seems at first to legitimize the idea of human experimentation,

---

[8] See, e.g., K. Sikkink, "Patterns of Dynamic Multi-Level Governance and the Insider-Outsider Coalition," in D. della Porta and S. Tarrow (eds.), *Transnational Protest and Global Activism* (Oxford: Rowman & Littlefield Publishers, 2005), p. 156 (dynamic multilevel governance); Yamin, *Power, Suffering, and the Struggle for Dignity*, p. 64 ("rights constitute social practices that create spaces for vital social deliberation on how to arrange social institutions to meet population needs, especially of the most disadvantaged").

[9] W. E. Bijker et al. (eds.), *The Social Construction of Technological Systems: New Directions in the Sociology and History of Technology*, 2nd ed. (Cambridge, MA: MIT Press, 2012).

which is a violation of international human rights. On the other hand, the *concept* of experimentation provides a foundation for questioning the circumstances and effects of decisions associated with the introduction of a new technology, and for integrating greater human rights protections in the process. For example, as Lea Shaver notes in Chapter 2, vulnerable populations are often chosen as initial targets for the introduction of new technology when the impacts of the new technology are unknown, even when, and sometimes precisely when, negative outcomes are directly anticipated.[10] A human rights-based approach to technology informed by the insights of STS might recognize and accept that the introduction of new technology is inevitably experimental, but require that in the process, vulnerable populations be protected from the accompanying risks and share in the potential benefits.

Cyberlaw scholarship has also been a highly generative frame for thinking about how human rights law is affected by, and should respond to, new technological innovations. As legal scholar Lawrence Lessig notes, law and technology are two "modalities of regulation" that can serve to undermine, strengthen, narrow, expand, or displace one another by making regulation invisible, ensuring precision in the delivery of essential goods and services, or fragmenting decision-making.[11] Human rights law provides a unique case study for testing the regulatory effects of technology. While much of cyberlaw focuses on how technology regulates individual behavior, human rights is interested in how technology might constrain or enable regulation by the state. Although it is essential from a human rights perspective to understand how technology can be used by states to affect rights, we are equally concerned with the use of technology to promote human rights within domestic and international law.

II CROSS-CUTTING THEMES

The chapters in this volume, which are described and analyzed in brief at the beginning of each section, highlight three common themes associated with interactions between human rights and technology: the relationship between technology and power, the effect of technological innovation on accountability, and the shifting boundary between public and private.

A *Technology, Power, and Justice*

One of the clearest and most important themes running through all of the contributions to this volume is the relationship between technology and power, and the

---

[10] K. Sunder Rajan, *Biocapital: The Constitution of Postgenomic Life* (Durham, NC: Duke University Press, 2006); R. Rottenberg, "Social and Public Experiments and New Figurations of Science and Politics in Postcolonial Africa" (2009) 12 *Postcolonial Studies* 423–40.
[11] L. Lessig, "The Law of the Horse: What Cyberlaw Might Teach" (1999) 113 *Harvard Law Review* 501–46 at 506.

effect of this relationship on the achievement of social justice and human rights. Although often heralded as a means to decentralize and destabilize power relationships, technology also reinforces and exacerbates inequality. Part of the value of combining a human rights approach with STS is to reveal the linkages between technology and power and examine how resources are distributed.

Technology is often seen as a means to shift power to the powerless. For example, mobile phones, social media, and the Internet can decrease the cost of communication, thereby making it more accessible to the public. In theory, this ought to shift power to ordinary individuals to participate in social, cultural, economic, and political life, and to take part in efforts to seek accountability for human rights violations. These shifts in power thereby destabilize and reconfigure the human rights domain. As Jay Aronson notes in Chapter 6, changes in how information is produced can alter the role and authority of human rights researchers and can give a voice to those affected by human rights violations.[12] Technology can also increase the delivery of essential services to remote areas, thus enabling the fulfillment of economic and social rights, or make possible choices about family formation that were not previously accessible to many.

At the same time, these shifts occur against the backdrop of unevenly distributed resources – and in many instances may exacerbate that unevenness. In Chapter 9, Ella McPherson illustrates how the deployment of these new techniques creates risk that not all human rights organizations are equally equipped to handle. Further, as John Emerson, Margaret Satterthwaite, and Anshul Vikram Pandey discuss in Chapter 8, technology may enable human rights defenders to convincingly articulate their demands for justice and restitution, but promoting this technology may be harmful if the technology does not come with the resources needed to enable organizations to collect, manage, and use information safely. The use of remote sensing, big data, data-visualization techniques, or even quantitative analysis may require knowledge and expertise outside the reach of most human rights advocacy organizations. Dalindyebo Shabalala (Chapter 3) illustrates global inequities in climate change technologies, which, like many emerging technologies, are often developed in well-resourced settings and only later diffused, if at all, to other parts of the world.[13] G. Alex Sinha (Chapter 12) similarly emphasizes the difficulty that even technologically savvy and well-resourced individuals face in protecting their own security and privacy online. For human rights defenders and organizations operating on a shoestring budget, with little aid directed to general operating expenses, it is nearly impossible.

---

[12] *See also* M. Beutz Land, "Peer Producing Human Rights" (2009) 46(4) *Alberta Law Review* 1115–39 at 1116.
[13] S. Cozzens and S. Thakur, "Problems and Concepts," in S. Cozzens and S. Thakur (eds.), *Innovation and Inequality: Emerging Technologies in an Unequal World* (Cheltenham: Edward Elgar, 2014), p. 5.

Further, technological innovations may be fundamentally skewed *toward* inequality. STS literature, for example, has long emphasized that "conventional science and innovation policies increase inequalities, unless they are designed specifically to do otherwise."[14] Systems of innovation that reward innovation through the market, for example, are structurally biased to produce goods that benefit those who are already well-off.[15] Fewer resources are invested in the development of technology that benefits poor individuals. Moreover, as Lea Shaver demonstrates in Chapter 2, when technology is deployed in poor areas, it can have the effect of limiting rights rather than protecting them. The water meters in Soweto, South Africa that anchor her analysis did not shift power to the poor, but rather consolidated control and authority in the state and the affiliated entity installing and running the meters. McPherson makes a similar point in her contribution. Although ICTs do enable human rights communication, the associated risk they engender means that they might only amplify the voices of the largest and most powerful organizations, leaving little room for the opening up of global audiences for smaller and less well-resourced groups.

Climate mitigation technologies also exacerbate global inequality by pitting the interests of the powerful against the less powerful.[16] Shabalala, in Chapter 3, examines the way in which those most affected by climate change and in need of mitigation technology are precisely those in the least position of power to bargain for that technology. There is an assumption that simply allowing information to be "free" will have positive social justice returns. In reality, who has access to this information is determined by existing power dynamics that depend on capital, investment, know-how, and intellectual property rights. Without changes to those underlying conditions, technology will at best make only marginally positive contributions to addressing inequalities at the national, regional, or global level. More likely, it will serve to reinforce those inequalities.[17]

For technology to serve the interests of the less powerful, accessibility is not enough. Technologies do not work in a vacuum, but rather depend upon complex networks of expertise, maintenance, and governance that embody structural inequalities. Efforts to

---

[14] Ibid., p. 8.
[15] *See* A. Kapczynksi, "The Cost of Price: Why and How to Get Beyond Intellectual Property Internalism" (2012) 59 *UCLA Law Review* 970–1026 at 978.
[16] Communication technologies have long been critiqued as exacerbating rather than alleviating global power inequalities. *See* R. F. Jørgensen, *Framing the Net: The Internet and Human Rights* (Cheltenham: Edward Elgar, 2013), pp. 43–44 (discussing the call for a New World Information and Communication Order that would enable countries of the Global South to participate more fully in global communication networks).
[17] S. D. Gatchair, I. Bortagaray, and L. A. Pace, "Strong Champions, Strong Regulations: The Unexpected Boundaries of Genetically Modified Corn," in Cozzens and Thakur, *Innovation and Inequality: Emerging Technologies in an Unequal World*, p. 116 (noting that Argentina has profited more from genetically modified crops because it has more large commercial farms than other developing countries, and thus can benefit more from the improvements that these products enable).

introduce new technologies to human rights problems must begin by asking a series of important questions about power, including who stands to benefit from any changes that the technology makes to the status quo – and to understand that these benefits are not equally distributed. It is essential to guard against the intentional bias built into technologies and their implementation, as well as unintentional negative consequences. We also need to think much more carefully about the state obligation not just to promote technological innovation and access to technology, but also the obligation to promote technological innovation in a way that supports rather than hinders the enjoyment of human rights.

B *The Challenge of Accountability*

This book also examines the impact of new technologies on efforts to promote accountability for human rights violations. As demonstrated in Part II, new technologies are often seen as having the capacity to revolutionize accountability efforts, providing opportunities for predicting, preventing, and mitigating atrocity crimes[18] as well as holding human rights abusers accountable for those violations.[19] The case for technology as a crucial new accountability tool has several dimensions:

- The falling cost of documentation technologies means that ordinary individuals now often possess the tools they need to capture and share information about violations. Rather than having to rely on trained researchers, documentation opportunities now exist wherever there is someone with a smartphone and an Internet connection. Citizen video generated in this way has been instrumental in identifying human rights abuses in many recent cases, such as in Israel's attacks in Gaza in 2014.[20]
- Many technologies offer opportunities to gather information in remote or even inaccessible areas. Mobile phones can be distributed to isolated communities, thus enabling them to gather and transmit information

---

[18] See, e.g., S. E. Kreps, "Social Networks and Technology in the Prevention of Crimes against Humanity," in R. I. Rotberg (ed.), *Mass Atrocity Crimes: Preventing Future Outrages* (Washington, DC: World Peace Foundation, 2010), p. 175; C. Tuckwood, "The State of the Field: Technology for Atrocity Response" (2014) 8 *Genocide Studies and Prevention: An International Journal* 81–86 at 81; C. Hargreaves and S. Hattotuwa, *ICTs for the Prevention of Mass Atrocity Crimes* (ICT for Peace Foundation, October 2010), http://ict4peace.org/wp-content/uploads/2010/11/ICTs-for-the-Prevention-of-Mass-Atrocity-Crimes1.pdf.
[19] S. Livingston and G. Walter-Drop, "Conclusions," in S. Livingston and G. Walter-Drop (eds.), *Bits and Atoms: Information and Communication Technology in Areas of Limited Statehood* (Oxford: Oxford University Press 2014), p. 169.
[20] Amnesty International, "Launch of Innovative Digital Tool to Help Expose Patterns of Israeli Violations in Gaza," July 8, 2015, www.amnesty.org/en/latest/news/2015/07/launch-of-innovative-digital-tool-gaza/.

about violations.[21] Satellite images can be used to collect information about violations occurring in places off-limits to researchers.[22]
- Digitization may contribute to accountability efforts. Because digital evidence is easy to share and tends to be difficult to destroy once widely distributed or preserved, it may be harder for states to keep evidence of human rights violations from reaching the hands of interested constituencies.
- Social networking technology supports the formation of groups, which can augment social movements designed to promote rights. Although scholars and advocates contest the existence and extent of the impact of technology on social mobilization, new technologies present at least the opportunity for mobilization around human rights documentation, advocacy, and capacity building.[23]

To be clear, technology is not a silver bullet.[24] At the same time, it is a critical element in present and future efforts to hold human rights abusers accountable.[25]

Behind the push to incorporate new technologies into human rights accountability efforts is an assumption about technology that is fundamentally incorrect: that because new information and communication technologies can be used to collect, analyze, and disseminate information, they are automatically biased toward greater disclosure and transparency. Technology, however, can be used just as easily to disguise, hinder, and obscure responsibility.

This is not intended as a tired recitation of the truism that technology is a tool that can be directed to both good and bad ends. Clearly, the use of technology by states and other duty bearers can undermine accountability efforts. And reliance on technology by human rights organizations can divert those organizations and their resources away from other activities that may have more of an impact on rights protection and promotion. But technology is not just *used* in good and bad ways; rather, it is in many contexts actually biased *against* disclosure and accountability. For example, the use of technology by states can obscure and fragment authority and thus disable the mechanisms that human rights advocates use to promote

---

[21] M. K. Land et al., *#ICT4HR: Information and Communication Technologies for Human Rights* (Paris: World Bank, 2012), pp. 8–9.

[22] See, e.g., American Association for the Advancement of Science, "High-Resolution Satellite Imagery and Housing Destruction in Ulu, Sudan," www.aaas.org/page/high-resolution-satellite-imagery-and-housing-destruction-ulu-sudan.

[23] See generally Z. Tufekci, *Twitter and Tear Gas: The Power and Fragility of Networked Protest* (New Haven, CT: Yale University Press, 2017).

[24] See Tuckwood, "The State of the Field," p. 82 ("Very few observers still believe that simply introducing an unspecified category of tools labeled 'technology' will be the panacea to defend human rights and save lives."); see also Kreps, "Social Networks and Technology," p. 175.

[25] L. Diamond, "Liberation Technology," in L. Diamond and M. F. Plattner (eds.), *Liberation Technology: Social Media and the Struggle for Democracy* (Baltimore: John Hopkins University Press, 2012), pp. 10–12.

accountability. Because of the absence of coercive mechanisms for enforcing human rights, human rights advocates have traditionally relied on the deployment of shame – exposing and publicizing human rights abuses – to put pressure on violators to change their behavior. This methodology functions most effectively when those exposing the abuse are able to tell a story that points to a specific "violation" attributable to the actions or decisions of a specific "violator" and for which there is a potential remedy.[26] Simply publicizing harms without explaining who is at fault or how the harms can be remedied may not motivate either the public or those who have the power to respond.

It can be more difficult to use shame to challenge activities that impact rights when those activities are mediated by technology. Technology disables shaming as a modality of enforcement because it obscures agency.[27] Activities that are accomplished through automation, for example, appear to be the inevitable result of predetermined processes set in motion by an invisible hand – even when those processes are the product of decisions that reflect and embody value judgments. In Shaver's chapter, for example, the introduction of water meters that deprived residents of adequate water was not immediately seen as a human rights violation because the act of cutting off water was done via automatic shutoff valve, not a human being. As Laura Dickinson (Chapter 5) illuminates, the automation of decision-making in armed conflict also creates an open question about who is actually making the decision that results in a violation, and thus who can and should be held accountable. Over time, technology may make rights-impacting decisions seem inevitable rather than the product of human agency, thus further complicating efforts to attribute these decisions to particular actors.

Technology also obscures agency because it interrupts the relationship between actor and effect. The chapters by Dickinson on drones, by Rikke Frank Jørgensen on the Internet (Chapter 11), and by Mark Latonero on big data (Chapter 7) demonstrate that these questions become particularly complicated and challenging when automation is involved. Who, if anyone, is responsible for a violation of humanitarian law when the targeting decision is made by an automated or semiautomated system? If a computer program systematically identifies individuals from a minority group as suspicious and those individuals are targeted by law enforcement, who is responsible for this discriminatory treatment?[28] Even the very complexity of new

---

[26] K. Roth, "Defending Economic, Social and Cultural Rights: Practical Issues Faced by an International Human Rights Organization" (2004) 26 *Human Rights Quarterly* 63–73 at 67–68.

[27] Lessig, *Code and Other Laws*, pp. 96, 238.

[28] On the topic of discrimination and algorithms, see generally Executive Office of the President, Big Data: A Report on Algorithmic Systems, Opportunity, and Civil Rights (May 2016), https://obamawhitehouse.archives.gov/sites/default/files/microsites/ostp/2016_0504_data_discrimination.pdf; S. Barocas and A. Selbst, "Big Data's Disparate Impact" (2016) 104 *California Law Review* 671–732; N. Diakopoulos, *Algorithmic Accountability: On the Investigation of Black Boxes* (Tow Center for Digital Journalism, December 3, 2014), http://towcenter.org/research/algorithmic-accountability-on-the-investigation-of-black-boxes-2/; Algorithmic Fairness,

technologies can interrupt the relationship between actor and effect. Technological artifacts can require inputs from a variety of different actors along the supply chain, any of whom might have contributed to the harm. If a decision to use lethal force is made based on a range of inputs analyzed according to predetermined algorithms programmed by a team of engineers, who is responsible for any resulting violation of humanitarian law? Even if new fact-finding technologies can be used to bring to light previously hidden information, this may not be sufficient if the deployment of technology obscures and attenuates the relationship between these violations and those responsible for them.

More attenuation in the relationship between actor and violation may also undermine other mechanisms that exist for promoting compliance with human rights law. Human rights law relies not just on shame, but also on processes of acculturation to achieve respect for rights. Duty bearers comply with human rights norms not only for fear of sanction, but also because of social and cognitive pressures that encourage conformity.[29] Yet the mechanisms of acculturation may be less effective when a state deploys technology in ways that affect rights. Actors who have merely set the technology in motion may feel less responsible for any resulting violations because they were not themselves the proximate cause of the harm, and thus may be more immune to the pressures of acculturation. This is exacerbated when technology also introduces distance between actor and violation, such as drone strikes – the subject of Chapter 5, by Dickinson – that are now piloted by individuals located far outside the theater of war.

The use of technology by human rights actors may also undermine accountability by diverting attention from what is fundamentally a political struggle. New technologies do indeed facilitate the collection of previously inaccessible information, but the primary obstacle to seeking accountability for human rights violations is not in most instances a lack of information. As Jay Aronson points out in Chapter 6, efforts to seek accountability for human rights violations often fail because powerful actors – those who have the ability to exert pressure on human rights violators or hold them directly accountable – do not have the political will to act on the information that they receive. This is not a reason to stop collecting and preserving evidence of violations, of course. And perhaps more or better quality information, or information displayed in more powerful ways, might help nudge those actors toward action. Information, of course, is connected to and influences politics.

Our concern is that the current focus on technology and, by extension, on the collection and preservation of information, poses a much more fundamental risk – namely, the risk that those who fund and carry out human rights advocacy will focus their already limited energy and resources on developing technological rather than

---

http://fairness.haverford.edu/; Centre for Internet and Human Rights, Ethics of Algorithms, https://cihr.eu/ethics-of-algorithms/.

[29] See, e.g., Goodman and Jinks, "How to Influence States," p. 626.

political solutions. This is in part a version of what Evgeny Morozov calls "technological solutionism" – "a fancy way of saying that for someone with a hammer, everything looks like a nail."[30] We seize on technocratic solutions because we think those problems might actually be capable of being solved – and for human rights, where the enforcement mechanisms are quite weak but the harms so grave, a solvable problem is like a siren song. But pouring energy into problems that can be solved requires us then to emphasize information collection as the problem, so we can justify these choices. If we are to use new technologies in human rights accountability efforts, it is imperative that we resist the impulse to frame problems in terms of the available solutions and thus divert resources from addressing even more pressing challenges. Investing in new technologies to improve evidence collection is important, but we should not neglect the more traditional advocacy and grassroots mobilization strategies that are necessary to generate the political will required for social change.

This is not, however, just a case of technological solutionism. Focusing on technological responses also risks depoliticizing human rights debates and thus depriving human rights rhetoric of the source of its power. Human rights frames are powerful because they are moral claims backed up by legal obligations. Fundamentally, rights claims are about challenging existing power structures by vesting the ability to make political claims in those who are affected by political decisions. Technological responses to human rights problems risk transforming this discourse from one that is fundamentally about power to one about technocratic solutions. Others have explored the way in which integrating human rights risk assessment into the procedures and policies of institutions and businesses may transform human rights from a claim on the powerful to a box to be checked.[31] The question is whether the attempt to develop technological solutions to human rights violations might similarly depoliticize human rights advocacy.

C *Technology and Private Authority*

The final cross-cutting theme of the volume addresses how human rights law can and should respond to the growth in private authority that results from the introduction of new technologies. For historical, economic, and political reasons, new technological developments and innovations often involve significant roles for the

---

[30] E. Morozov, *To Save Everything, Click Here: The Folly of Technological Solutionism* (New York: Public Affairs, 2013), p. 6.
[31] G. A. Sarfaty, *Values in Translation: Human Rights and the Culture of the World Bank* (Stanford, CA: Stanford University Press, 2012), p. 134. Even the very process of "translating" human rights in ways that have local resonance raises this tension. S. Engle Merry, *Human Rights & Gender Violence: Translating International Law into Local Justice* (Chicago and London: University of Chicago Press, 2006), p. 5 ("Rights need to be presented in local cultural terms in order to be persuasive, but they must challenge existing relations of power in order to be effective.").

private sector, albeit with considerable support and intervention from the state.[32] Intermediaries are extraordinarily powerful gatekeepers for information and communication; they exert control over our expressive activity, our associations with others, and our access to information.[33] Non-state actors play central roles in developing and implementing new technology outside of the information and communication technology sector as well – including in the fields of water technology, reproductive technologies, and autonomous weapons, among others.

As Jørgensen explores in Chapter 11, human rights law is at a disadvantage in responding to the impact that non-state actors can have on human rights because it creates few direct legal obligations for these actors. Under principles endorsed by the United Nations in the *Guiding Principles on Business and Human Rights*, non-state actors typically have only a moral – not a legal – obligation to respect human rights.[34] Human rights law attempts to address harms from non-state actors by imposing legal obligations on the state to protect individuals from such harm and also to provide remedies when rights have been violated.[35] Thus, in most instances, the activities of non-state actors do not constitute human rights violations unless the state has failed to protect, punish, and remedy the violation.

The challenge of responding to human rights harms by non-state actors is not a new issue,[36] but it has particular significance in the context of human rights and technology. Because technology is often owned and operated by private actors, its use shifts decision-making authority into the private sphere and outside of public mechanisms of accountability. This effect is compounded by the practice of outsourcing, which is prevalent not only in the context of automated weapons, as Dickinson discusses, but also in the provision of services, as noted by Shaver (Chapter 2), among other areas. Private actors are also increasingly at the forefront of efforts to respond to and remedy human rights violations by others, particularly in the case of information and communication technologies. It is unclear whether human rights law will be up to the task of responding to this multidimensional growth in private authority over human rights.

Perhaps most fundamentally, the essays in this volume also raise questions about what constitutes a human rights violation. When a state fails to control a security company that abuses individuals in a local community, this failure to protect is clearly a violation of the state's international obligations. The abuses themselves are

---

[32] See generally M. Mazzucato, *The Entrepreneurial State: Debunking Public vs. Private Sector Myths* (London: Anthem Press, 2014).
[33] E. B. Laidlaw, Regulating Speech in Cyberspace, Gatekeepers, Human Rights and Corporate Responsibility (Cambridge: Cambridge University Press, 2015), pp. 46–56.
[34] UN Office of the High Commissioner for Human Rights, *Guiding Principles on Business and Human Rights: Implementing the United Nations "Protect, Respect and Remedy" Framework* (New York and Geneva: United Nations, 2011), p. 13.
[35] Ibid., pp. 3, 27.
[36] See, e.g., A. Clapham, *The Human Rights Obligations of Non-State Actors* (Oxford: Oxford University Press, 2006).

also clearly human rights harms. Can we apply the same reasoning when an Internet company systematically disadvantages a particular political viewpoint?[37] Or when that same company curates controversial videos, allowing some but removing others?[38] Does the state's failure to ensure that encryption technologies are both available and easy to use breach its obligation to create an enabling environment for the fulfillment of rights?[39] A focus on the role of non-state actors in the design and implementation of technology that affects human rights brings into sharp focus not only the state action problem inherent in all of human rights, but also new questions arising from automation and algorithmic decision-making.

Contributions to this volume ultimately illuminate four important and related challenges that human rights law will need to address to effectively respond to human rights harms by private actors with respect to the introduction and use of new technologies. First, the contributions illustrate the importance of finding better solutions to regulating the conduct of non-state actors when their activities have impacts on human rights. Relying on the state to regulate these actors is often not effective, given the lack of political will, as well as state interest and even complicity in many of these rights violations. Self-regulation by companies is also unlikely to constrain abuses in the long run.

Second, the volume illustrates how important it is to understand the application of the *Guiding Principles on Business and Human Rights* in context. Industries vary widely in terms of how private companies affect rights and what changes need to occur to better protect rights. In the water meter case, private contractors were acting at the behest of public authority, but the local government initially did not intervene once it became clear that the meters were resulting in harms to rights. In the context of information and communication technologies, many of the relevant harms seem to emanate from an excess of public authority, such as state efforts to monitor private communications or remove particular content from the Internet. At the same time, the state is also failing to take the positive measures needed to create an enabling environment that allows individuals to protect their own privacy. More work needs to be done to understand the nature of the human rights harms, and the application of the *Guiding Principles* to those harms, in these very fact-specific contexts.

---

[37] M. Nunez, "Former Facebook Workers: We Routinely Suppressed Conservative News," *Gizmodo*, May 9, 2016, http://gizmodo.com/former-facebook-workers-we-routinely-suppressed-conser-1775461006.

[38] J. Concha, "Graphic Videos Spark Questions for Facebook, Journalism," *The Hill*, July 10, 2016, http://thehill.com/homenews/287166-graphic-videos-spark-questions-for-facebook-journalism; "Facebook Decides Which Killings We're Allowed to See," *Slashdot*, July 7, 2016, https://tech.slashdot.org/story/16/07/07/1652224/facebook-decides-which-killings-were-allowed-to-see.

[39] Emphasizing their importance for freedom of expression, the Special Rapporteur on Freedom of Expression has said that "States should promote strong encryption and anonymity." *Report of the Special Rapporteur on the Promotion and Protection of the Right to Freedom of Opinion and Expression*, David Kaye, U.N. Doc. A/HRC/29/32 (May 22, 2015), ¶ 59.

Third, the volume as a whole also points to the challenge of regulating the growing role of non-state actors in governance activities. Non-state actors are involved in a broad range of regulatory activities previously thought to be the exclusive province of the state. Business entities are engaged in cooperative relationships with the state to provide essential services, from water to health care. In the information and communication technology sector, non-state actors are increasingly engaged in regulating the speech of others – removing defamatory statements and other forms of problematic expression – often at the behest of the states in which they do business. Private companies now routinely decide whether evidence of human rights violations uploaded to a private platform will be publicly available. They also build and market technologies that both enable and prevent surveillance, cooperate – or refuse to cooperate – with government requests to monitor activists or political dissidents, and create algorithms and weapons systems that may determine whether an individual lives or dies. As Jørgensen's contribution makes clear, these decisions are often motivated by commercial interests rather than concerns about the public good or human rights.

Clearly, human rights law has long sought to understand the nature of the public and private obligations associated with privatization of essential services.[40] Nonetheless, the current framework for addressing human rights harms inflicted by business entities is built on the distinction between public authority (the responsibility of the state to protect) and private authority (the duty of the company to respect). As a result, it applies less well to activities that blur this distinction. When non-state actors are providing essential services or engaging in speech regulation, it is not clear that they have, or should have, only a moral duty to respect rights. What are the duties of private actors operating in these grey areas between public and private?

The distinction between public and private is further muddied by the fact that many of these governance activities are done at the behest of, or under the compulsion of, governmental authority, such as when states compel Internet providers to police speech online. Indeed, a good case can be made that in such instances, the activity is not actually "private" and thus gives rise to direct state responsibility.[41] In other cases, the line between public and private may not be very clear as a factual matter. The exponential growth of public-private partnerships in a range of industries may, over time, render the *Guiding Principles*, with their clear division of public and private, less and less relevant.

Industry "self-regulation" will not likely provide an answer. The European Commission, for example, recently negotiated a "Code of Conduct" with Facebook,

---

[40] See, e.g., K. De Feyter and F. Gómez Isa (eds.), *Privatisation and Human Rights in the Age of Globalisation* (Antwerp, Oxford: Intersentia, 2005).
[41] M. Land, "Regulating Private Harms Online," in R. F. Jørgensen (ed.), *Private Actors and Human Rights Online* (Cambridge: MIT Press, forthcoming).

Microsoft, Twitter, and YouTube for the purpose of combatting illegal hate speech online. This is a pledge the EU extracted from dominant market players to regulate the expression of those who use their platforms, without any regulation or oversight.[42] Even if private regulation may in some instances be consistent with human rights, there are still few mechanisms of accountability that govern private actors engaged in this kind of regulation. Although state control is often deeply problematic, private control lacks even the trappings of accountability and transparency that usually accompany governmental regulation. For example, the Code of Conduct has been followed by regulation in Europe (such as Germany's new social media law) imposing heavy fines against social media companies that fail to quickly remove harmful online content.[43] Although the German law also poses risks for freedom of expression online, it was at least enacted pursuant to transparent and democratic processes designed to consider the public interest. (The more troubling part of the law, of course, is that it delegates much of the actual policing of online content to private social media platforms.)[44]

Finally, the contributions also illustrate that human rights law must focus on the particular duties of the non-state actors who build, design, and program technology. What are the obligations of programmers and software engineers? Who is responsible for errors? Do we need to think differently about human rights accountability in sectors heavily driven by technological innovation? Technology embodies values, and those who design the technology can make respect for human rights more or less costly, efficient, or easy to accomplish. Routers can be built to permit surveillance, or not. Do those who build technology have moral or legal obligations to respect rights? If so, how do they integrate this commitment into their work?

III CONCLUSION: THE ROLE OF LAW

The contributions to this volume seek to raise awareness about the very real opportunities and costs of technology for the protection and promotion of human rights. Human rights actors seeking to deploy technology in pursuit of human rights must be aware of its strengths and weaknesses, and they must be prepared to have these very tools turned against them. Moreover, they should guard against the impulse to allow solutions to obscure problems or ignore the way in which

---

[42] European Commission (Press Release), "European Commission and IT Companies Announce Code of Conduct on Illegal Online Hate Speech" (May 31, 2016), http://europa.eu/rapid/press-release_IP-16-1937_en.htm.

[43] See, e.g., N. Lomas, "Germany's Social Media Hate Speech Law Is Now in Effect," *TechCrunch*, Oct. 2, 2017, https://techcrunch.com/2017/10/02/germanys-social-media-hate-speech-law-is-now-in-effect/; J. Kastrenakes, "EU Says It'll Pass Online Hate Speech Laws if Facebook, Google, and Others Don't Crack Down," *The Verge*, Sept. 28, 2017, www.theverge.com/2017/9/28/16380526/eu-hate-speech-laws-google-facebook-twitter.

[44] Land, "Regulating Private Harms Online."

technology might reinforce rather than dismantle power disparities, not only between individuals and the state, but also between large and small human rights organizations.

The range of topics covered in the book makes clear that an important goal of those interested in human rights technology must be to promote capacity. There is a need for greater technical expertise – indeed, even simply greater comfort in learning about and engaging with technological innovation – within the human rights community. Moreover, existing expertise is far from evenly distributed, and there must be significantly more attention paid to building the technological capacity of small human rights defenders around the world. Conversely, there is also a need among technology entrepreneurs and innovators for an understanding of what human rights law is and the opportunities and limits it presents. An important aim of this book is to help to help launch conversations between technologists and human rights practitioners, with the intention of promoting these critical linkages.

Human rights law itself also has an important role to play in maximizing the benefits and minimizing the risks of new technologies. At the very least, a human rights-based approach to technology should reorient decisions about technology to individuals and the impact these decisions have on their rights. For example, human rights law might be used to advocate against efforts to introduce new technologies using utilitarian rationales that neglect important sectors of society. The risks of new technologies should be assessed prior to their introduction. Decisions about technology should not just consider its overall benefits to a society or its impact on development, but also actively prioritize the needs of the most vulnerable members of that society. This applies equally well to the design of technology as to its introduction. Programmers and engineers must view the design of technology and the creation of technological standards as value-based decisions that need to support rather than hinder the enjoyment of international human rights.

This not to say that human rights law must, or should, hinder technological innovation. As this volume makes clear, new technologies also have important benefits for human rights. When technological innovation is oriented toward human rights enjoyment, it can serve as an important tool for the promotion of human rights around the world. To the extent that technological development results in limitations on rights in order to protect the rights of others or achieve important public policy objectives, such limitations should meet the tests of legality, necessity, and proportionality – the limits must be provided by law, be directed toward a legitimate purpose, and be narrowly tailored to achieve that purpose.[45]

In the modern world, technology does much more than simply limit or protect rights. New technological developments are also putting pressure on the many fissures, ambiguities, and discontinuities that already exist within human rights

---

[45] *Report of the Special Rapporteur on the Promotion and Protection of the Right to Freedom of Opinion and Expression*, Frank La Rue, U.N. Doc. A/HRC/17/27 (May 16, 2011) ¶ 24.

law. Human rights law is generally focused on principles and not technologies, and there is no reason to expect that existing law will be unable to keep pace with technological change.[46] The introduction of technology does, however, illuminate situations in which existing human rights law is not sufficient, or sufficiently developed, to protect rights. For example:

- Some areas of human rights law may rely implicitly on slippage in enforcement of domestic law to ensure that a right is adequately met. Perfect enforcement enabled by technology[47] – such as the introduction of water meters that prevent households from taking more than their allotted amount of water – can expose the inadequacies of existing standards.
- New technological developments may also create opportunities for violations that did not previously exist. Prior to digitization and the Internet, individuals might have generally relied on the fact that information disclosed about them in one context would be unlikely to find its way to another, or that in most cases, information, once disclosed, would eventually fade from public scrutiny. Today, however, information is perpetually available and infinitely sharable. Information about us that is disclosed in one context can now follow us forever[48] or be combined with other data and used in ways we could not have foreseen.[49] What should the international human right to privacy mean in the digital world, and how can we reconcile an expansion of this right with the right to free expression?[50]
- Technology may also reveal ambiguities in our understanding of particular terms that had previously seemed natural and unproblematic. As Thérèse Murphy emphasized in the workshop organized around this volume, the meaning of the term "parent," which previously seemed to have a fixed reference and definition, has changed in light of new reproductive technologies. Although "parent" has long had multiple

---

[46] M. K. Land, "Toward an International Law of the Internet" (2013) 54 *Harvard International Law Journal* 393–458 at 408.
[47] Lessig, *Code and Other Laws*, p. 6; J. Grimmelmann, Note, "Regulation by Software" (2005) 114 *Yale Law Journal* 1719–58 at 1723–24.
[48] *Google Spain SL v. AEPD*, Case C-131/12, 2014 EUR-Lex 62012CJ0131 (May 13, 2014), ¶ 92 (interpreting the European Union Data Protection Direction to require search engines to delist search results if they are "inaccurate, irrelevant or excessive").
[49] D. G. Johnson, P. M. Regan, and K. Wayland, "Campaign Disclosure, Privacy and Transparency" (2011) 19 *William & Mary Bill of Rights Journal* 959–82 at 969 (describing bouncing, shading, and highlighting).
[50] As Sinha notes in this volume, US law takes a fairly bright-line approach to this question, ostensibly removing protection from any information that has been disclosed to a third party. On the reconciliation of privacy and free expression, *see, e.g.*, A. Chander and U. P. Lê, "Free Speech" (2015) *Iowa Law Review* 501–49 at 539–42.

dimensions, ambiguities in the term did not have practical consequences until advances in reproductive technology made possible new familial formations and roles.
- Technological developments can also precipitate changes in the law itself. As Dickinson notes in Chapter 5, the combination of increasing automation and the use of contractors has reduced the likelihood of US casualties in foreign interventions to the point that new legal arguments can be made supporting an expanded executive role governing use of force in US law. At the same time, automation and the increased use of private contractors may be undermining the ability of international humanitarian law to provide a basis for international accountability.

Finally, human rights law must also grapple with the fact that technological innovations seem to be putting pressure on areas in which human rights law is weakest. One such area includes the positive state duty to fulfill rights. States are obligated under human rights law to respect, protect, and fulfill rights. What does this obligation to fulfill look like with respect to technological innovation? What does it mean when individuals must ensure their own digital security but lack access to appropriate expertise and affordable, easy-to-use tools for doing so? Technology is also challenging human rights law in the area of non-state actors. Should human rights law regulate the companies that create and build technologies and, if so, how? What obligations might human rights law impose on companies that not only themselves affect rights, but also serve as the gatekeepers for expressive activity that violates the rights of others?

As Laura Dickinson noted in the workshop, understanding the relationship between human rights law and technology may ultimately require a pluralistic approach. Technology constrains and influences behavior, of both individuals and states, in a variety of ways. Human rights law can no more control these effects than it can dictate the course of economic activity. The focus of a study on the intersection of human rights and technology must instead be to understand how technology interacts with human rights law to produce particular results – both the ways in which technology provides opportunities and risks for human rights enjoyment, and how the norms and practices of human rights advocacy are affected by new technological developments. The aim of this book is to begin that conversation.

PART I

Normative Approaches to Technology and Human Rights

The essays in Part I focus on the relationship between human rights law and technological developments – specifically what human rights law requires when new technologies are introduced and disseminated in society, and how new technologies have the potential to fragment the very legal authority needed to address the negative impacts of technology.

The first two chapters consider how human rights law applies to technology. In Chapter 2, "Safeguarding Human Rights from Problematic Technologies," Lea Shaver explores what the emerging right to science requires of states when new technologies are introduced in a way that significantly impacts human rights. Using South African litigation over restrictive water meters as her lens, she argues that human rights activists and institutions need to be conscious of the technological element of human rights violations. She proposes looking to human subjects protection as a framework for legal accountability for what may amount to involuntary and harmful technological experiments. In Chapter 3, "Climate Change, Human Rights, and Technology Transfer: Normative Challenges and Technical Opportunities," Dalindyebo Shabalala focuses on state obligations to engage in technology transfer through a human rights lens. Shabalala argues that human rights provides a stronger basis than development approaches for making claims to climate change technology transfer. Human rights emphasizes the needs of vulnerable populations within countries and provides a basis for differentiating between and prioritizing particular technologies.

The next two chapters consider the introduction of new technologies through the frames of ethics and the law of war. In Chapter 4, "Judging Bioethics and Human Rights," Thérèse Murphy considers the intersection of human rights law and bioethics, which addresses ethical issues associated with medical and biological technology. She examines leading cases in the jurisprudence of the European Court of Human Rights on the regulation of reproductive technology in order to bring bioethics and human rights into deeper conversation. In Chapter 5, "Drones,

Automated Weapons, and Private Military Contractors: Challenges to Domestic and International Legal Regimes Governing Armed Conflict," Laura Dickinson considers how the automation[1] and privatization[2] of war interact with each other to exacerbate the effects on human rights. Dickinson argues that these developments undermine domestic limits on the power of the US president to declare war, and also obscure and fragment decision-making authority on the use of deadly force in ways that diminish existing mechanisms of accountability under international law.

One of the clearest contributions of the chapters in Part I is their dramatic illustration of the variety of ways in which technology can affect rights. New technological innovations have significant consequences for human rights, in terms of both the opportunities they offer for the fulfillment of rights and the harms they can cause. States and non-state actors can use technology to limit rights in unanticipated and often invisible ways, as Shaver's and Dickinson's chapters make clear. These consequences are not limited to civil and political rights. Part I reveals that technology is just as central to the enjoyment of economic and social rights – health, water, and the environment – as it is to the rights to association, privacy, family, and expression.

The application of technology alters cultural understandings around concepts like privacy and family in ways that affect the application of international human rights norms to these problems. As technology expands the possibilities available for forming and extending families, it also puts tension on what it means to have a right to found a family – how far that right extends and the role that human rights law should play in reconciling competing claims regarding reproductive decisions.[3] These essays demonstrate that the relationship between human rights law and technology is not unilateral, but mutually constitutive. Just as international human rights law is transformed by the introduction of new technologies, technology is also affected by international human rights law. As legal regimes generate new rights and transform others, they also shape the path of technological development.[4] As Shabalala's chapter (Chapter 3) illustrates, for example, the international rules

---

[1] Concerns about the impact of automation have been raised recently in contexts as diverse as self-driving cars and the law of war. *See, e.g.*, A. Etzioni and O. Etzioni, "Keeping AI Legal" (2016) 19 *Vanderbilt Journal of Entertainment & Technology Law* 133–46; M. Wagner, "The Dehumanization of International Humanitarian Law: Legal, Ethical, and Political Implications of Autonomous Weapons Systems" (2014) 47 *Vanderbilt Journal of Transnational Law* 1371–1424; *see generally* E. Parasidis, "Emerging Military Technologies: Balancing Medical Ethics and National Security" (2015) 47 *Case Western Reserve Journal of International Law* 167–183.

[2] *See, e.g.*, M. Minow, *Partners, Not Rivals: Privatization and the Public Good* (Boston: Beacon Press, 2002); D. Barak-Erez, "The Private Prison Controversy and the Privatization Continuum" (2011) 5 *Law & Ethics of Human Rights* 138–157.

[3] *See, e.g., Artavia Murillo et al.* ("In Vitro Fertilization") v. Costa Rica, Judgement, Inter-Am. Ct. H.R. (ser. C) No. 257 (November 28, 2012).

[4] S. Jasanoff, "Introduction: Rewriting Life, Reframing Rights," in S. Jasanoff (ed.), *Reframing Rights: Bioconstitutionalism in the Genetic Age* (Cambridge, MA: MIT Press, 2011), p. 3.

regarding intellectual property affect the path of green technology development and transfer in developing countries.

The chapters in Part I also consider the obligations that human rights law puts on the regulation of technology itself. Human rights law is technologically neutral on its face, not anchored to any particular form of technology or system of knowledge production. At the same time, the human rights corpus, especially the right to science, requires states to ensure that the introduction of new technology does not harm rights and to create an enabling environment that facilitates rights promotion in the face of new technology. States also have obligations to individuals in other countries to promote the transfer of knowledge when this knowledge is necessary for those individuals to enjoy their fundamental human rights. This duty becomes ever more important as technological development exacerbates global inequalities and technology plays an increasingly prevalent role in fulfilling economic and social rights.

The essays in Part I also grapple with the question of how international human rights law ought to respond to technological change, including how technology can better incorporate human rights into its design. Two aspects of international human rights law help it to be robust enough to respond to the challenges presented by new technologies. First, as a product of political compromise and an attempt to articulate rules that apply across widely diverging national systems, human rights law is, in general, ambiguous and underdeveloped. Although this can be a source of frustration for many, it also helps ensure that human rights law can be interpreted in ways that meet new challenges.[5] For example, Shaver's chapter (Chapter 2) demonstrates how human rights law might be interpreted and adapted to respond to the threat some technologies pose to the enjoyment of human rights.

Second, human rights treaties generally allow, either directly or indirectly, a fair measure of discretion for states in terms of how they implement their treaty obligations. Although human rights law does at times prohibit particular conduct on the part of the state, states have leeway to determine how best to achieve certain outcomes.[6] This discretion also enables human rights law to evolve in ways that meet technological challenges, including challenges that require the state to undertake new initiatives to meet their international obligations.

At the same time, while the broad discretion and ambiguity that characterize human rights law allow it to evolve with technology, new interpretations must retain

---

[5] Established principles of treaty interpretation require that terms be read in light of current conditions to account for new technological developments or other circumstances the drafters did not or could not have considered, in order to ensure that the treaty remains effective in realizing its object and purpose. R. K. Gardiner, *Treaty Interpretation*, 2nd ed. (Oxford: Oxford University Press, 2015), p. 254.

[6] State progress in meeting international human rights obligations can be measured in terms of both conduct and results. M. Green, "What We Talk About When We Talk About Indicators: Current Approaches to Human Rights Measurement" (2001) 23 *Human Rights Quarterly* 1062–97 at 1075.

a focus on the core values of international human rights. This is achieved not by "updating" the law, but rather by consciously engaging with questions of risk, harm, and social disruption that inevitably accompany the introduction of new technology. Thus, a core value of a human rights-based approach to technology is attention to the consequences of technological innovation. Further, this attention must be focused on distribution of power and resources. As the chapters by Shaver (Chapter 2) and Shabalala (Chapter 3) make clear, the effects of technological innovation on human rights are not experienced equally along lines of race, class, gender, or other status. Instead, technology is deployed in uneven and unequal ways that often have a negative impact on the most vulnerable. Moreover, the inequitable distribution of technology along national, regional, and socioeconomic lines also has consequences that can reinforce power imbalances, particularly global power imbalances.

2

Safeguarding Human Rights from Problematic Technologies

*Lea Shaver**

"Water is life. Life without water is not life. One cannot speak of a dignified human existence if one is denied access to water."

High Court Decision (2008), Para. 124

"[T]he right of access to sufficient water ... does not require the state ... to provide every person with sufficient water ... [R]ather it requires the state to take reasonable legislative and other measures progressively to realise the achievement of the right of access to sufficient water, within available resources."

Constitutional Court Decision (2009), Para. 50

Beginning in 2004, impoverished black residents of Phiri township, in South Africa's Soweto area, began to encounter a previously unknown technology: the prepaid water meter. Phiri residents had previously enjoyed a standard piped water supply billed at a flat monthly rate. The majority of these desperately poor households could not afford to pay their water bills. Because it is illegal under South African law to disconnect water services as a penalty for debt, however, the water supply had long been, in effect, free. The new technology was installed on each home's water line in order to restrict this free water supply. Each household was allotted a monthly ration of free water. Additional amounts could be released only by purchasing tokens to insert into the meter.

This "demand-management" technology was hailed by the City of Johannesburg's water utility as an ideal technical solution to a persistent financial problem. Large amounts of water were being consumed yet never paid for in very poor neighborhoods. The water meters achieved the intended goals of conserving water and exacting greater payment. They also led to intense hardship for desperately poor

---

* The author assisted the South African legal team representing the *Mazibuko* applicants during 2006 and 2007 while a Fulbright Scholar at the Centre for Applied Legal Studies at the University of Witwatersrand, Johannesburg.

residents, who had to redirect already scarce resources or go without water for days or weeks at a time. (As an example of the extreme poverty prevalent in Phiri, one resident, Lindiwe Mazibuko, testified that her household of fourteen subsisted on a combined income of under $100 per month.)[1] The prepaid meters quickly became a detested symbol of material deprivation, political marginalization, and the long shadow of apartheid.

Popular frustration with the prepaid water initiative led to community demonstrations and an organized social movement, led by the populist left-wing Anti-Privatization Forum (APF). Massive resistance initially delayed installation of prepaid meters in Phiri. Johannesburg Water obtained court orders forbidding residents from interfering with the work, and a campaign of arrests and punitive water disconnections followed. Organized resistance then weakened, and installations continued. At this point, human rights lawyers heard about the situation and recommended constitutional litigation. With the support of the APF and a mass meeting of Phiri residents, the lawyers agreed to bring suit on behalf of the residents, challenging the installation of the water meters as unconstitutional.

Throughout the resulting litigation, the Phiri prepaid meter conflict was evaluated primarily in terms of the right to water. This right is explicitly recognized in the South African Constitution.[2] *Mazibuko and Others* v. *City of Johannesburg and Others* led to a victory for the human rights plaintiffs at the trial court. That result was largely upheld at an intermediate appellate level. The rights-favorable outcome was dramatically reversed, however, upon final appeal to the nation's highest court. South Africa's Constitutional Court approached the *Mazibuko* case through the lens of its increasing skepticism of judicial enforcement of social and economic rights. Emphasizing the difficulties of adjudicating rights subject to progressive realization, the Constitutional Court declined to find a violation of the right to water.

Since that time, however, human rights scholars and United Nations bodies have developed a fuller understanding of another human right implicated by prepaid meters: the right to science. Both the Universal Declaration of Human Rights and the International Covenant on Economic, Social, and Cultural Rights assert that everyone has a right to share in the benefits of scientific progress and its applications.[3] This has been referred to as "the right to enjoy the benefits of scientific progress" or, even more simply, "the right to science." The right to science entails both a *positive* right of *access to* technologies essential for a life with dignity, and a

---

[1] Founding Affidavit, *Mazibuko and Others* v. *City of Johannesburg and Others* (06/13865) [2008] ¶¶ 68–73.

[2] Constitution of the Republic of South Africa, art. 27(1) ("Everyone has the right to have access to ... sufficient food and water.").

[3] Universal Declaration of Human Rights, G.A. Res. 217A, December 10, 1948, art. 27(1) ("Everyone has the right ... to share in scientific advancement and its benefits."); International Covenant on Economic, Social and Cultural Rights, G.A. Res. 2200A (XXI) December 16, 1966, art. 15(1)(b) ("the right of everyone ... to enjoy the benefits of scientific progress and its applications").

negative right to *protection from* the imposition of technology in ways detrimental to human rights and human dignity.

This chapter revisits the *Mazibuko* case to consider how the conflict over prepaid meters might be analyzed through the lens of the right to science. I suggest that we should understand the rollout of prepaid water meters as a vast technological experiment that was conducted on vulnerable people, at significant risk to their health and well-being. Characterized in this way, the crucial question becomes whether the experiment was conducted with adequate human rights safeguards. Following widely accepted principles of ethical research, such safeguards should include free and informed consent, careful risk-benefit assessments, and appropriate selection of subjects. This new perspective on the prepaid meter controversy offers several forward-looking contributions:

First, this chapter offers a detailed case study of how a seemingly innocuous or even beneficial technology can operate in a context of unequal power to seriously endanger human rights. Various policy and design decisions made around the prepaid water meter technology created a situation that resulted in serious harm. These decisions were made in the context of a certain technological naiveté, bureaucratic carelessness, and/or corporate disinterest in the human rights of the end users. As this case study reveals, technology is not neutral to social injustice. People who are vulnerable by virtue of poverty, lack of education, or social discrimination are also uniquely vulnerable to the harmful application of new and potentially problematic technologies.

Second, this chapter helps to develop the still sparse legal framework of the right to science and proposes a violations approach to understand state duties regarding the right to science. Specifically, I argue that the right *to protection from being subjected to technology in ways detrimental to human rights and dignity* is immediately justiciable. The chapter uses the Phiri fact pattern to illustrate how human rights advocates can frame a claim in these terms. I also provide guidance as to how courts can evaluate the merits of such claims and fashion appropriate remedies. Even in jurisdictions where the right to science is not legally recognized, it remains possible and productive to rhetorically frame a problematic application of technology as an illegal experiment. This approach can complement invocation of the rights to health, life, and privacy, or any other right impacted by the particular technology at issue.

Third, and most ambitiously, it is my hope that engineers and others responsible for designing and implementing technical interventions can take lessons to avoid similar errors in the future. The framework of human rights safeguards should inform internal processes and policies around other new and potentially problematic technologies. The principles set forth and illustrated here can be applied by designers, technologists, in-house counsel, corporate social responsibility teams, and user communities to evaluate any proposed technological application or intervention. Building a pragmatic approach to conducting human rights impact assessments as a matter of routine is likely to be more impactful, in the end, than any campaign of human rights litigation. This is crucial to advancing the ultimate goal: to protect people from harm by anticipating and avoiding potential human rights missteps.

I "DEMAND MANAGEMENT DEVICES"

From 1948 to 1991, South Africa practiced an extreme form of racial discrimination known as apartheid. Between 1960 and 1983, millions of nonwhite South Africans were forcibly removed from their homes and resettled in racial enclaves under the Group Areas Act. Black South Africans could not vote and were denied basic civil rights. At one time, it was forbidden to publish newspapers in black languages or to teach black students in any language other than the language of the white Afrikaner minority that controlled the country. Decades of internal resistance and international pressure finally propelled the country to end apartheid in the 1990s.

Emerging from its first free election in 1994, the new democracy articulated bold goals for racial and economic inclusion. The new constitution was widely hailed as the most progressive in the world, committing to a long list of human rights. At the same time, South African policy-makers fully embraced capitalist development and modernization. South Africa needed to borrow money and attract foreign investment. This led to strong pressure to privatize public utilities as part of the Washington Consensus list of economic policies then viewed as essential for developing countries. An internal political struggle resulted in a compromise: the country would partially, but not fully, privatize key public services.

The political context of the introduction of prepaid water meters in South Africa has been thoroughly analyzed by Patrick Bond and Jackie Dugard.[4] In the wake of partial privatization, Johannesburg Water continued to be fully owned by the City of Johannesburg. Its operation, however, was largely contracted out to a French company, Suez. From that point forward, Johannesburg Water increasingly emphasized cost recovery as a fundamental principle, notwithstanding the constitutional commitment to water provision as a basic human right. An important component of this strategy was "water demand management," which focused on reducing the amount of water consumed by households too poor to pay standard fees. One component of demand management involved shifting very poor neighborhoods that had enjoyed piped water on credit to more limited water service.

One way this was accomplished was through the installation of prepaid meters in black areas. These meters did not simply measure the amount of water delivered to

---

[4] P. Bond and J. Dugard, "The Case of Johannesburg Water: What Really Happened at the Prepayment 'Parish Pump'" (2008) 12 *Law, Democracy, and Development* 1–28; J. Dugard, "Choice from No Choice; Rights for the Left? The State, Law and the Struggle against Prepayment Water Meters in South Africa," in S. Molta and A. G. Nilsen (eds.), *Social Movements in the Global South: Dispossession, Development and Resistance* (Basingstoke: Palgrave Macmillan, 2011), p. 59.

homes. Their more significant function was to *limit* the amount of water delivered. The devices were programmed to let through a set ration of free water each month, a nod to the national and constitutional commitment to water as a human right. To access additional water, however, residents had to purchase tokens in advance. The killer feature of the technology was its ability to ensure stricter water conservation and financial discipline. All this was accomplished while bypassing normal legal procedures for discontinuing water service to a property.

Like any technology, prepaid water meters should not be understood as inherently negative or positive from a human rights standpoint. In theory, interruptive water meters could be deployed in a manner supportive of human rights. The technology undoubtedly made it more cost-effective to provide free water to low-income households. The savings might have been reinvested in expanding access to improved water sources for families with no connection at all. Interruptive water devices could also have been used to create shared access points in water-scarce regions, with water tokens distributed like wartime ration coupons. This point is particularly important to appreciate: the practical impact of a technology depends almost entirely upon the myriad particular decisions made by individuals and organizations implementing the technology in a specific social context.

For example, the human rights impact of prepaid water meters depends significantly on the pricing of additional water supply in relation to the resources of each household. If the household has substantial disposable income and tokens are priced very cheaply, it would be merely inconvenient to maintain a supply of tokens to regularly unlock the water supply. However, if the tokens are quite costly compared to the household's limited resources, then the introduction of a prepaid meter imposes a new financial hardship. And if the household is truly desperately poor, then the monthly free water ration can become an oppressive ceiling rather than a supportive floor. From a human rights perspective, prepaid meters would have been least problematic in wealthier households. In practice, however, demand management devices were never introduced in areas where residents could easily afford to pay their water charges. The bureaucratic imperative that led to the adoption of this technology was to reduce unpaid water consumption by the extremely poor.

Another crucial design choice is what amount of free water to permit before restriction begins. With a very generous free water allowance, the devices might have no impact on existing water consumption patterns. With a moderate allowance, families might find that they had sufficient water for sanitation, bathing, drinking, cooking, and washing, but perhaps not enough to raise a kitchen garden, mop their floors, or allow children to play with water. With a meager allowance, residents might find that they faced frequent interruptions in water service and significant new pressure on household budgets already inadequate to meet basic needs.

In Phiri, the free water allowance proved to be meager. Johannesburg Water allotted each household just 6 kiloliters (kL) of free water per month. This was a steep 70 percent reduction from the previous estimated consumption of 20 kL per household. This should not be understood as an oversight or unintended consequence. Johannesburg Water deployed the devices specifically to dramatically reduce water consumption by nonpaying households. To achieve this goal, bureaucrats logically chose a ration well below current usage. Johannesburg Water defended the 6 kL figure as designed to provide a household of eight residents with 25 liters (6.6 gallons) of water per person per day. However, the average property in Phiri housed not eight, but sixteen residents. Thus, the free basic water ration actually worked out to a daily allowance of less than 15 liters (4 gallons) per person, on average. This level of water consumption is considered meager even within the setting of a desert refugee camp. Studies have found that limiting consumption to this level significantly increases the risk of disease.[5] Typical Americans use around 333 liters (88 gallons) of water per person per day for home uses.[6] Phiri residents were permitted only 5 percent of this amount.

Apart from the central question of whether water policy allocated *enough* free water to very poor households, the installation of prepaid meters impacted access to water in other ways. As might be expected with any new technology, the devices did not always work. Some households complained that their meters behaved erratically, accepting tokens but failing to release water. The devices also had not been well designed to allow residents to monitor their usage and estimate how long the remaining supply would last. Instead, the water supply would be cut off with very little warning. This could happen in the middle of cooking, or while caring for children, or at night, when it was not safe to leave the home to purchase additional tokens. Thus, water might not be available when urgently needed, even if the household was willing and able to pay. In one particularly tragic incident, two small children died when the monthly water supply ran out while residents were attempting to put out a shanty fire.[7]

Another policy decision surrounding the technology was the degree of information and autonomy provided to residents. Although the water company sought consent from each household prior to installing the prepaid meters, the consent was neither free nor informed. Residents were not given an accurate picture of the risks and benefits of prepaid water devices in advance, and they had no ability to withdraw their consent after experiencing the technology firsthand. Phiri residents who complained about problems after meters were installed were told they could not go back to the old system. Residents who resorted to "self-help" measures by

---

[5] A. Cronin et al., "A Review of Water and Sanitation Provision in Refugee Camps in Association with Selected Health and Nutrition Indicators – The Need for Integrated Service Provision" (2008) 6(1) *Journal of Water and Health* 1–18 at 2.

[6] US Environmental Protection Agency, "Statistics and Facts," www.epa.gov/watersense/statistics-and-facts.

[7] Bond and Dugard, "The Case of Johannesburg Water" 1–28.

attempting to disable or bypass the water-restriction devices were punished with total disconnection from water service.

Users thus did not get to choose the prepaid meter technology; they were subjected to it. Prepaid water meters were deployed in South Africa to serve the cost-saving and revenue-maximizing goals of the agency that had the power to impose the technology rather than the human needs of the persons subjected to it. The design was optimized to meet the needs of the organization that purchased and implemented the devices, but did not take into account the needs of the people who would be directly affected by the technology. Phiri residents were not in control of the technology and did not have input into its design or implementation. They were not consulted about its parameters or informed of its potential risks. And even after the harmful effects became clear, they did not have the freedom to reject it. Technology became yet another means by which poor, black, politically disempowered South Africans were abused.

II THE *MAZIBUKO* LITIGATION

Because Phiri residents lacked the power or autonomy to shape decisions around the prepaid meter technology, they were ultimately forced to seek redress in the courts. Human rights lawyers from the Center for Applied Legal Studies at the University of Witwatersrand represented them on a *pro bono* basis. The case was first heard in the High Court of South Africa, Witwatersrand Division.[8] The City of Johannesburg initially argued that it had no obligation to provide any amount of free water to the poor, but only a statutory obligation to offer water services for a fee. Justice Tsoka dispensed with that argument swiftly, confirming a constitutional obligation "to ensure that every person has both physical and economic access to water."[9] This interpretation was grounded upon the South African Constitution's explicit recognition of the right to water, read in light of international human rights law.[10] The High Court upheld the amount of 25 kL per person per day as a reasonable *minimum*, but emphasized that the city was "obliged to provide more than the minimum if its residents' needs so demand and they are able, within their available resources, to do so."[11] In light of conditions in the Phiri neighborhood, where

---

[8] *Mazibuko and Others v. City of Johannesburg and Others* (06/13865) [2008] ZAGPHC 491, April 30, 2008 (High Court).
[9] Ibid., ¶ 41. Here, Judge Tsoka referred to the International Covenant on Economic, Social, and Cultural Rights' guarantees of the right to an adequate standard of living and the right to health; General Comment No. 15 on the Right to Water, issued in 2002 by the UN Committee on Economic, Social, and Cultural Rights; the UN Convention on the Rights of the Child; the African Convention on the Rights of the Child; and the African Charter on Human and Peoples' Rights. For further background, see I. Winkler, *The Human Right to Water: Significance, Legal Status, and Implications for Water Allocation* (Oxford: Hart Publishing, 2012).
[10] Constitution of the Republic of South Africa, art. 27(1).
[11] *Mazibuko* (High Court), ¶ 126.

households are large, sanitation depends upon an adequate water supply, and the residents are "mainly poor, uneducated, elderly, sickly and ravaged by HIV/AIDS," the court concluded that the 6 kL free water allowance was insufficient.[12]

Judge Tsoka went further, however, characterizing the water restriction devices as *inherently* unconstitutional because of the automatic shutoff function. The opinion first reviewed judicial opinions regarding water disconnections in Brazil, Argentina, France, and the United Kingdom. The Brazilian and Argentinian precedents were characterized as holding that a water company may not interrupt the supply of water due to nonpayment, because this would be a violation of human rights.[13] In France, it was noted, water companies are required to make special arrangements to ensure access to water for poor households, and disconnection for failure to pay requires court authorization.[14] In the United Kingdom, prepayment meters had been declared illegal under national statutes because they offered no notice or opportunity for a hearing prior to the cutoff of water supply.[15] Similarly, Judge Tsoka held that the use of automatic shutoff devices violated the South African constitutional right to "lawful, reasonable and procedurally fair administrative action," at least for poor households.[16] Judge Tsoka further condemned the lower procedural protections given to Phiri's poor black residents as "unreasonable, unfair and inequitable ... [and] discriminatory solely on the basis of colour."[17] As a remedy, the High Court ordered that all Phiri residents should receive a free basic water supply of at least 50 liters per person per day and be able to opt out of prepaid meters.[18]

The City of Johannesburg and Johannesburg Water appealed the decision. The Appellate Court also found in favor of the Phiri residents. Its opinion upheld the lower court's determination that the restrictive meters were unauthorized by law, and set the standard for constitutionally adequate water provision only slightly lower, at 42 liters per person per day. At this point, the City of Johannesburg and Johannesburg Water accepted their loss and did not initiate a further appeal. But the applicants themselves decided to push forward. Despite having obtained most of what they sought, including recovery of attorney fees, they pushed forward in hopes that the Constitutional Court would reinstate the High Court's 50 liter per person per day benchmark and establish an even stronger precedent.

What happened next was a stunningly disappointing result for the *Mazibuko* plaintiffs. To everyone's surprise, the Constitutional Court reversed every aspect of the prior decisions. Instead, the Constitutional Court used the *Mazibuko* case as an

---

[12] Ibid., ¶¶ 169–79.
[13] Ibid., ¶¶ 86–87.
[14] Ibid., ¶¶ 88–89.
[15] Ibid., ¶ 90.
[16] Ibid., ¶¶ 91–93.
[17] Ibid., ¶ 94.
[18] The order left ambiguous whether this was to be multiplied by the actual number of residents in a household, the average number of sixteen residents, or some other multiplier.

opportunity advance a more conservative vision of "the role of courts in determining the content of social and economic rights."[19] While recognizing the constitutional right to water, the opinion refused to assign any real content to that right. It firmly rejected the notion of setting any quantitative threshold for a constitutionally adequate water supply.[20] Contrary to the lower courts' reasoning, the Constitutional Court sided with Johannesburg in insisting that the right to water "does not confer a right to claim 'sufficient water' from the state immediately."[21] Rather, it "requires the state to take reasonable legislative and other measures progressively to realise the achievement of the right of access to sufficient water, within available resources."[22] In judging the city's water policies as lawful and reasonable, the court went out of its way to give the state the benefit of every doubt. The opinion repeatedly emphasized that judicial review in the area of social and economic rights should be very deferential to the democratically accountable branches of government.[23] Rather than strictly scrutinizing the state's actions, as the High Court and appellate judges had done, the Constitutional Court looked only for a rational basis.

Indeed, the opinion strongly suggests that the sole function of courts with respect to social and economic rights should be to provide a venue where citizens can require government agencies to publicly explain the reasons behind their policies.[24] This spotlight effect can pressure the government to revise its policies, as happened in this case, with the water agency making several concessions as the case wound its way through the courts. The opinion concluded optimistically:

> This case illustrates how litigation concerning social and economic rights can exact a detailed accounting from government and, in doing so, impact beneficially on the policy-making process... Having to explain why the Free Water Policy was reasonable shone a bright, cold light on the policy that undoubtedly revealed flaws. The continual revision of the policy in the ensuing years has improved the policy in a manner entirely consistent with progressive realization.[25]

The opinion's frequent explicit references to "social and economic rights" make clear that this deferential approach to judicial review is reserved specifically for socioeconomic rights. Thus the secret to gaining stronger judicial protection against the use of technology to the detriment of human rights – at least in South Africa, and likely elsewhere as well – is to position it within more familiar frames of traditional civil rights protections against government abuses. The right to science can point the way to such an approach.

---

[19] *Mazibuko and Others v. City of Johannesburg and Others* [CCT 39–09 2009] ZACC 28, October 8, 2009, ¶¶ 45–46 (Constitutional Court).
[20] Ibid., ¶¶ 56–57.
[21] Ibid., ¶ 57.
[22] Ibid.
[23] Ibid., ¶¶ 59–168.
[24] Ibid., ¶¶ 160–65.
[25] Constitutional Court, *Mazibuko*, ¶ 153.

III THE RIGHT TO SCIENCE

When the *Mazibuko* case was being litigated in 2007–09, the conceptual foundation for the right to water had only recently been laid. At that time, "the right to science" was not yet even a term in human rights discourse. It was thus not possible for the *Mazibuko* lawyers to consider framing their case in terms of the right to science, in addition to the right to water. Since that time, however, the right to science has been the subject of significant normative development by scholars and within the United Nations system.[26]

The right to science offers two ways of framing the problems posed by prepaid meters. One is to focus on the aspect of the right to science that calls for expanding access to technology, particularly to what the Special Rapporteur has described as "technologies essential for a life with dignity."[27] Modern indoor plumbing, which permits washing, bathing, cooking, and sanitation, has a strong claim to be considered an essential technology. In this way, the right to science might provide a hook for constitutional litigation in the same way that the right to water did. This view, however, also lends itself to similar weaknesses. This aspect of the right to science is a positive rights claim, subject to the logic of progressive realization. Many courts are hesitant to impose minimum core standards for essential services. Yet without judicially enforced minimums, it often seems impossible to make out a clear violation.

There is a second and, in my view, more interesting way of approaching the right to science as it relates to prepaid meters: to focus on the aspect of the right to science that calls for states "to prevent and preclude the utilization of scientific and technological achievements to the detriment of human rights and fundamental freedoms and the dignity of the human person."[28] This approach is based on the notion that

---

[26] In July 2009, experts convened by UNESCO produced the Venice Statement on the Right to Enjoy the Benefits of Scientific Progress and Its Applications. In 2012, the UN Special Rapporteur in the field of cultural rights, Farida Shaheed, issued a report to the Human Rights Council that endorsed the view of recent scholarship conceiving of "the right to science" as a commitment to treating knowledge and technology as global public goods that must be made widely available. *Report of the Special Rapporteur in the Field of Cultural Rights, Farida Shaheed*, U.N. Doc. A/HRC/20/26 (May 14, 2012) ("*Shaheed Report*") (citing L. Shaver, "The Right to Science and Culture" (2010) *Wisconsin Law Review* 121–85). In 2013, the UN Office of the High Commissioner on Human Rights hosted a highly generative academic seminar on this subject, convened at the UN headquarters in Geneva. *Report on the Seminar on the Right to Enjoy the Benefits of Scientific Progress and Its Applications*, U.N. Doc. A/HRC/26/19 (April 1, 2014). Another seminar was organized by Prof. Samantha Besson at the University of Freibourg, resulting in a special edition of the *European Journal of Human Rights* ([2015] *European Journal of Human Rights*, Special Issue, 403–580). The UN Committee on Economic, Social, and Cultural Rights has also taken early steps to develop a General Comment on the right to science.

[27] *Shaheed Report*, ¶¶ 29, 74.

[28] See, e.g., *The Declaration on the Use of Scientific and Technological Progress in the Interests of Peace and for the Benefit of Mankind*, G.A. Res. 3384 (XXX) (November 10, 1975) ("All States

technologies are not inherently beneficial or harmful. Problems arise only when technologies are deployed in ways that are detrimental to human rights. Certain categories of technology are more likely to be problematic, however, an idea I will return to later in this chapter.

The right to science offers a framework for anticipating and responding to potential problems through human rights safeguards. This way of framing the legal challenge to demand management devices will refocus attention away from the question of *how much* water is sufficient – with all the attendant complexity, difficulty, and resistance to defining a quantitative minimum core to be enforced by courts – and instead toward recognition of the coercive application of the restrictive technology as the fundamental harm to be prevented. This way of looking at the right to science presents a classic negative rights claim, a right to *freedom from* harmful state action. Viewed from this perspective, Phiri residents suffered from a violation of their *right not to be subjected to technology in a way that was harmful to their human rights and human dignity*.

One way to understand the rollout of prepaid water meters in Phiri is as a massive technological experiment, conducted upon thousands of vulnerable people. Indeed, officials at Johannesburg Water understood the Phiri effort as an experiment. The stated intent was to pilot an unproven technical solution, in order to decide whether to roll it out more broadly. The Phiri prepaid water experiment was problematic for a number of reasons. It involved an unproven technology designed to restrict a substance vital to human life. It was conducted on a very large group of people who were particularly vulnerable by reason of their extreme poverty, including children, pregnant women, elderly people, and people suffering from HIV/AIDS, tuberculosis, and malaria. Subjects were coerced into participating and had no freedom to withdraw from the experiment once it started – in violation of the human rights principle that groups nonconsensual scientific experimentation with torture.[29] Fewer than twenty-five years after the end of apartheid, only black South Africans

---

shall take appropriate measures to prevent the use of scientific and technological developments ... to limit or interfere with the enjoyment of human rights and fundamental freedoms... All States shall take effective measures, including legislative measures, to prevent and preclude the utilization of scientific and technological achievements to the detriment of human rights and fundamental freedoms and the dignity of the human person."). That Declaration also mentioned, while giving lesser emphasis to, positive aspects of the right to science, e.g., "all states shall take measures to ensure that scientific and technological achievements satisfy the material and spiritual needs for all sectors of the population"; "[a]ll States shall co-operate in the establishment, strengthening and development of the scientific and technological capacity of developing countries"; and "the state shall take measures to extend the benefits of science and technology to all strata of the population." Ibid., ¶¶ 3, 5, 6.

[29] International Covenant on Civil and Political Rights, G.A. Res. 2200A (XXI), Mar. 23, 1976, art. 7 ("No one shall be subjected to torture or to cruel, inhuman or degrading treatment or punishment. In particular, no one shall be subjected without his free consent to medical or scientific experimentation.")

were selected for participation. Had university researchers proposed to run such an experiment, they would never have received ethical approval.

IV HUMAN RIGHTS SAFEGUARDS

After World War II, Nazi scientists were tried for war crimes for failing to follow internationally accepted ethical standards. The Nuremberg Code, the first international attempt to articulate these standards, remains a foundational document in scientific ethics. The World Medical Association's 1964 Declaration of Helsinki similarly articulates basic standards for human subjects research and is regularly updated. While both of these documents are specifically directed to medical research, the 1978 Belmont Report is framed more broadly.[30] Its intent was to elaborate "broader ethical principles [to] provide a basis on which specific rules may be formulated, criticized, and interpreted."[31] The Belmont principles continue to serve as touchstones for research ethics internationally, most significantly in the Universal Declaration on Bioethics and Human Rights.[32] The high level of generality of these principles makes them particularly useful for thinking about human rights obligations related to the application of new technologies.

A *The Threshold Question: When Is Human Rights Scrutiny Required?*

This first step in the Belmont Report was to establish threshold criteria for when special ethical review is required. Similarly, companies, administrative agencies, and courts need criteria to determine when the implementation of a new technology should require special human rights scrutiny.

The Belmont Report distinguished, on the one hand, traditional medical or behavioral "interventions that are designed solely to enhance the well-being of an individual patient or client and that have a reasonable expectation of success."[33] Similarly, a proven technology that is designed primarily to enhance the well-being of the end user should trigger no special human rights scrutiny. Examples include extending traditional water service to new homes, upgrading old or leaking pipes with newer ones, or providing users with a text-messaging hotline to report water problems.

---

[30] National Commission for the Protection of Human Subjects of Biomedical and Behavioral Research, *The Belmont Report: Ethical Principles and Guidelines for the Protection of Human Subjects of Research* (Washington, DC: US Government Printing Office, 1978).
[31] Ibid., p. 1.
[32] UNESCO, *Universal Declaration on Bioethics and Human Rights*, October 19, 2005; see also M. Kruger et al., *Research Ethics in Africa: A Resource for Research Ethics Committees* (Stellenbosch: Sun Press, 2014) (discussing the Belmont principles as "foundational" and citing the Belmont Report repeatedly).
[33] National Commission, *The Belmont Report*, p. 2.

In contrast, the Belmont Report cautions that "radically new procedures ... should ... be made the object of formal research at an early stage in order to determine whether they are safe and effective."[34] Prepaid water meters offer an ideal example of a radically new technology that requires safety and efficacy study. The technology had not previously been used in South Africa and had very limited application anywhere in the world. The prepaid meters also operated in a radically new way. Globally, the standard approach to water delivery has always been to provide a secure and consistent source of water controlled by the end user. Traditional water meters measure usage for billing purposes, but do not interrupt the supply against the user's wishes. The prepaid meters, in contrast, dispensed a limited water supply determined by someone other than the user, and required payment in advance for additional amounts, through a brand-new token system.

Moreover, the interruptive meter technology was not designed and selected "solely to enhance the well-being" of the user.[35] Instead, this technology was selected and designed to advance larger demand-management and cost-recovery goals of Johannesburg Water. This criterion is important, because it helps to determine whether a conflict of interest exists between the person recommending the intervention and the person who will bear the risks. Where such a conflict of interest or tension exists, external review and accountability are recognized as being particularly important. Here, a strong tension existed between the goals of the technology planners and the well-being of individuals who were subjected to the technology. This conflict of interest heightens the risk of harm and therefore the need for human rights safeguards.

A third consideration should be the degree of freedom that individuals have to opt out of using the technology. When people are free to adopt or reject a particular technology, the option to "exit" serves as an important safeguard. When a technology is revealed to be harmful, individuals can often protect themselves by discontinuing its use. If an experimental drug makes a person sick, he or she can stop taking it. If a restrictive water meter is cutting off the water needed to cook, an individual should be able to bypass it. The exit option incentivizes companies to offer well-designed technologies with adequate support, so that consumers do not abandon their products. In a context where individuals are not free to reject a technology, however, these important safeguards are destroyed. Where a government mandates a technology that individuals experience as harmful to their health, liberty, privacy, or other interests, the need for human rights safeguards is paramount.

Finally, technologies that are intentionally designed, in the normal situation of use, to limit human rights call for special scrutiny. Most technologies will not trigger this concern. Seatbelts, credit card security chips, vaccinations, and Internet

---

[34] Ibid., p. 3.
[35] Ibid., p. 2.

protocols are all examples of technology that is designed to enhance the user's welfare. Government mandates to use these technologies would not typically be concerning in the ordinary case. In contrast, technologies that are *designed* to limit privacy, freedom of expression, or access to basic services deserve greater scrutiny. Such technologies would include surveillance technology and water-restriction devices. Dual-use technologies, or technologies that are designed in ways that enable both rights-restricting and rights-enhancing activity, also warrant enhanced scrutiny. Similarly, technologies that automate processes previously subject to the due process of law also deserve special scrutiny.

Prepaid water meters raised all four of these red flags. They were radically new and designed for a purpose other than the user's well-being. People had little to no freedom to opt out of using them. They restricted access to a substance essential for human life. They bypassed existing legal processes by interrupting water access without the required court order. Following the logic of the Belmont principles, these characteristics should have triggered formal research to evaluate their safety and effectiveness. Arguably, the pilot application of the prepaid meter technology in Phiri *was* this required research. The problem, therefore, was not the complete lack of research, but rather the haphazard and unethical way in which the pilot was designed and carried out.

This can be seen by proceeding to examine the three principles that the Belmont Report lays out to guide ethical research: respect for persons, beneficence, and justice. First, *"respect for persons* demands that subjects enter into the research voluntarily and with adequate information."[36] Second, *beneficence* requires researchers to put the best interests of the research subjects front and center in order to do no harm, or at least to minimize possible harms while maximizing benefits.[37] Finally, *justice* requires that persons submitting to the risks of scientific research should benefit from the fruits of that research, and that vulnerable people should not be inappropriately targeted as experimental subjects.

In other words, the Belmont principles establish requirements related to "informed consent, risk/benefit assessment, and the selection of subjects for research."[38] The Phiri pilot had significant flaws with respect to each of these criteria. Consent was neither informed nor free, appropriate efforts were not taken to minimize risks, and the selection of subjects was discriminatory.

B *Respect for Persons: Ensuring Free and Informed Consent*

The first essential human rights safeguard for experimental technologies is insistence upon free and informed consent. The Belmont Report states, for example, that

---

[36] Ibid., p. 6.
[37] Ibid.
[38] Ibid., p. 10.

"[r]espect for persons requires that subjects, to the degree that they are capable, be given the opportunity to choose what shall or shall not happen to them. This opportunity is provided when adequate standards for informed consent are satisfied."[39] The Universal Declaration on Bioethics provides: "The autonomy of persons to make decisions, while taking responsibility for those decisions and respecting the autonomy of others, is to be respected. For persons who are not capable of exercising autonomy, special measures are to be taken to protect their rights and interests."[40] According to the Belmont Report, informed consent requires that "the subjects should understand clearly the range of risk and the voluntary nature of participation."[41] Individuals must be adequately informed of the risks presented by the technology before making a free choice whether to participate or not. "Consent may be withdrawn by the person concerned at any time and for any reason without any disadvantage or prejudice."[42] Experimentation in a context where consent may be unduly influenced by power dynamics, such as research on prisoners, is highly suspect.[43] Financial compensation may be offered for research participation, but care must be taken that such incentives do not corrupt freedom of choice, particularly with poor populations.

The consent process followed by Johannesburg Water was egregiously deficient by these standards. Johannesburg Water sent facilitators through the Phiri neighborhood with instructions to obtain signed consent forms from every property owner. Most of the property owners lacked the education to read the consent form, creating the distinct risk that facilitators would mislead them about the nature of the intervention in order to obtain their signatures. Respect for autonomy was so low, water meters were installed even despite some residents' explicit refusal. For example, the lead plaintiff in *Mazibuko* testified that the facilitator seeking her signature deceived her, stating that the repairs were necessary to replace old and rusty pipes and making no mention of a prepaid meter. Having heard about prepaid meter installations elsewhere, however, she cautiously refused to sign the paper and verbally refused a meter. Notwithstanding her refusal, Johannesburg Water installed a prepaid meter at her property the following day.[44]

Indeed, the vast majority of persons subjected to the Phiri prepaid meter experiment were never even asked for consent. A property owner's signature was deemed sufficient for all residents of the property, whether adults or children, and even for unrelated members of separate households renting backyard shacks. Compounding this problem, the property owner was offered a significant financial benefit in

---

[39] Ibid.
[40] Universal Declaration on Bioethics, art. 5.
[41] National Commission, *Belmont Report*, p. 11; *see also* Universal Declaration on Bioethics, art. 6.
[42] Universal Declaration on Bioethics, art. 6.
[43] National Commission, *The Belmont Report*, p. 6.
[44] Founding Affidavit, *Mazibuko*, ¶¶ 79–92.

exchange for subjecting other residents to the study. Johannesburg Water promised to erase years of past water debts in exchange for formally consenting to the installation of a prepaid meter.

To the extent that consent was sought, it was obtained under duress. Residents were advised that if they did not consent to the installation, their water supply would be completely disconnected. This threat was actually carried out when property owners did not sign the consent form. The lead plaintiff testified that after Johannesburg Water installed a prepaid meter on her property against her wishes, it disconnected her water supply. For several months, she walked 12 kilometers per day to transport water in a wheelbarrow for her household. Eventually, she broke down and "consented" to use a prepaid meter.[45]

An additional fundamental requirement of free and informed consent is that participants have the opportunity to exit at any time. This was not assured in Phiri. Property owners were asked to consent to the installation of an unfamiliar technology. Once they had the opportunity to become familiar with the technology and its limitations and risks, however, there was no procedure by which they could revoke consent and have the device removed.

Finally, no effort was made to inform residents of the risks entailed by water restriction devices, despite their significance. Water authorities knew that the system would force households to dramatically reduce their water consumption. They also knew that the free allowance would be inadequate for even the most basic needs of larger households. Yet this information was not shared with residents, who had no ability to predict how far the free water supply would go in their household or how much they might have to spend on tokens. Some households lost renters who were contributing desperately needed income because the water supply had been downgraded. Other households watched their gardens, which were supplying much-needed fresh foods, wither and die. Most households had to divert funds from other essential spending in order to meet new water expenses. Households that could not do this would be entirely without water for days or weeks, until the next month's supply began.

In sum, the Phiri prepaid water experiment is a case study in how *not* to practice informed consent. Gestures were made toward the need for consent, but only to the limited extent that was convenient for Johannesburg Water. There was never any actual respect for the individual's right to refuse. These deficiencies reflect the extremely lopsided power dynamic at play in this particular technological experiment. Johannesburg Water understood itself as entitled to install its technology whether the individuals who were subjected to it wished it or not. Residents' preferences were quite literally deemed to be irrelevant by actors who presumed to make these decisions in their best interest. Judge Tsoka, who had grown up as a black man under apartheid, reacted strongly to the patronizing racism he perceived

---

[45] Ibid., ¶ 94.

in this wholesale denial of options: "That patronization sustained apartheid: its foundational basis was discrimination based on colour and decisions taken on behalf of the majority of the people of the country as *'big brother felt it was good for them'*."[46]

C Beneficence: Maximizing Benefit and Minimizing Harm

The second principle outlined in the Belmont Report is beneficence. The ordinary meaning of this term is the quality of doing something for the benefit of others. In the context of research ethics, it means that designers of experiments should concern themselves with the welfare of research participants by taking steps to minimize the risks and maximize the benefits to them. The Hippocratic principle "do no harm" is cited as an ideal to be approached.[47] "In applying and advancing scientific knowledge, medical practice and associated technologies, direct and indirect benefits to patients, research participants and other affected individuals should be maximized and any possible harm to such individuals should be minimized."[48]

It is entirely possible that Johannesburg Water officials believed they *were* acting in the best interests of the Phiri residents. It is always easy for planners to overestimate the reliability and user-friendliness of a new and untried technology. It is also clear that Johannesburg Water officials believed that substantial water was being wasted due to residents' failure to value what they did not pay for. From their comfortable middle-class perspective, it may have been difficult to imagine that Phiri residents could not afford to pay such tiny charges.

What is clear from the Belmont Report, however, is that good intentions are not enough. Careful and independent scrutiny adds needed accountability to the risk-benefit calculation. "[T]here should first be a determination of the validity of the presuppositions of the research; then the nature, probability and magnitude of risk should be distinguished with as much clarity as possible."[49] A key goal of this process is not merely to prevent unethical experiments from occurring, but also to inform the design of the research in order to minimize the potential for harm and maximize the benefits for those involved.

In this case, a more gradual, careful rollout should have preceded a neighborhood-wide implementation. Water officials should have identified, at the outset, that the technology they sought to implement needed to be field-tested to

---

[46] *Mazibuko* (High Court), ¶ 153. The white author of the Constitutional Court opinion likely reminded Judge Tsoka of precisely this patronization in concluding that the policy was not "*unfairly* discriminatory" because the discriminatory impact upon black households could not be judged as harmful, given that Phiri's indigent consumers were, if anything, better off avoiding the "worrying measures" of debt accumulation and collection. *Mazibuko* (Constitutional Court), ¶¶ 148–57.
[47] National Commission, *Belmont Report*, p. 7.
[48] Universal Declaration on Bioethics, art. 4.
[49] National Commission, *Belmont Report*, p. 17.

ensure its safety in the context of acute poverty. They might have first tried the technology in their own offices or homes, to become familiar with its limitations and potential risks (such as device malfunction). The experimental design should also have been informed by research studying current water use patterns in the community. The trial should have started with a restriction at 30 kL and worked gradually downward toward the target of 6 kL, watching for problems along the way. Each home should have been provided with an emergency water supply in case of device failure. Study participants might have been given cell phones so that they could easily report any urgent problems they experienced. Early participants should have been provided with a supply of tokens, so they would not have to use scarce resources in order to participate in the study.

Another critical safeguard is the participants' ability to end their participation in the experiment, which is relevant to both freedom of consent and minimization of harm. Johannesburg Water should have designed pipes to permit either prepaid or conventional service, at the option of the consumer. This would have secured the exit option and the ability to quickly respond to problems. Had this level of care been put into the Phiri experiment, well-meaning persons at Johannesburg Water and the City of Johannesburg would have been better able to anticipate and manage the risks created by prepaid meters.

D *Justice: Selection of Experimental Subjects*

The third and final Belmont principle is justice. "The principle of justice gives rise to moral requirements that there be fair procedures and outcomes in the selection of research subjects."[50] The risks of experimentation should not be placed disproportionately upon stigmatized or disadvantaged persons, nor upon persons who already bear great burdens. Where an experiment is conducted on a population of vulnerable persons – such as children, racial or linguistic minorities, or the very poor – careful scrutiny is required to ensure that they are not involved solely because they are easier to manipulate.[51] The potential benefits to society cannot be used to justify the risks to individuals directly involved. "In applying and advancing scientific knowledge, medical practice and associated technologies, human vulnerability should be taken into account. Individuals and groups of special vulnerability should be protected and the personal integrity of such individuals respected."[52]

Phiri was a community of families already facing massive burdens of poverty, unemployment, and extremely poor health. They should not have been also asked to face the burden of piloting a new and risky technology. The principle of justice would have been advanced by initially testing the technology in middle-class

---

[50] Ibid., p. 18.
[51] Ibid.
[52] Universal Declaration on Bioethics, art. 8.

neighborhoods with high levels of education and English fluency. Those users would have been in a better position to cope with and report problems. Eventually, the technology would also need to be tested for safety in the specific social context of poverty, where it was intended to be used. At this stage, the study should not have been designed to involve more people than was necessary. Families with pregnant women, young children, and significant health problems should not have been included in the first wave of experimentation with the devices.

More broadly, we should also ask whether the principle of justice was honored not just in how the technology was field-tested, but also in the plan for its normal use. In litigation, the City of Johannesburg emphasized the public necessity of water conservation. Was it just for the burden of water conservation to be placed upon the extremely poor? If coercive measures were truly necessary to limit water usage, surely they were all the more necessary in affluent homes with much higher water usage per capita. Perhaps policy-makers chose to place this burden upon the poor and socially marginalized residents of Phiri because they were less likely to effectively complain and resist. Or perhaps the need for water conservation was a pretense all along, and the real need was simply to increase monetary recovery. The principle of justice would urge this to be accomplished by charging higher rates in affluent neighborhoods rather than squeezing additional payments from the very poor.

V CONCLUSION

The *Mazibuko* water dispute highlights the harm that even innocent-seeming technologies can do in the context of a power disparity between designers and users. From beginning to end, this problematic technology was controlled by a powerful institution that designed and implemented it to serve its own purposes, with little regard for the wishes of the vulnerable people subjected to it. The result was a series of decisions about implementation that resulted in serious harm to human dignity, health, and life.

Despite the country's having one of the most rights-protective constitutions in the world, South Africa's Constitutional Court felt unable to remedy this harm within the framework of the right to water. Reframing the facts of this case through the lens of the right to science, however, puts things in a very different perspective. The state is no longer engaged in a difficult process of progressive realization of socioeconomic rights, which courts are arguably ill-suited to second-guess. Instead, the state is more clearly seen to be actively violating the human rights of its citizens through its own reckless actions.

The implementation of a new, unproven, and potentially harmful technology calls for special ethical and legal safeguards to prevent abuse and harm. Even in jurisdictions where the right to science is not explicitly recognized, the framework of human rights safeguards as protection against problematic technologies still works. South African lawyers, for instance, might assert the right to freedom and security of persons, which is constitutionally defined to include the right "not to be subjected to medical or scientific experiments without their informed consent."[53] This could then be interpreted by South African courts with reference to international practice and norms related to science, such as the Belmont Report and the Universal Declaration on Bioethics and Human Rights.

Courts of law are well positioned to adjudicate human rights violations within this framework. This approach does not require courts to set minimum standards for technical performance or to reach the conclusion that a particular technology does or does not violate human rights. Instead, it asks courts to do something that is uniquely within their institutional competence: to evaluate the sufficiency of procedural safeguards. The Belmont principles offer a framework for defining the procedural rights that must be upheld when people are subjected to unproven and potentially harmful technologies, in order to prevent abuse and minimize harm. Where these procedural rights are not respected, a court might either order a technological experiment to be suspended or require specific new safeguards, such as improved consent procedures, special protections for vulnerable populations, or the freedom to opt out. A court might also order financial compensation, a formal apology, or other remedies to rectify past harm. Where the evidence demonstrates a propensity for repeated violations, a court could mandate human rights training or the creation of an ethical review board. These measures might also be stipulated by means of a settlement negotiated between the parties.

A central function of ethical scrutiny of human subjects research is to provide external oversight over potentially harmful action before it takes place. This can help to guard against conflicts of interest, where a particular researcher or agency may be insufficiently concerned with the impact on individuals involved because of their own interest in seeing the project through. Yet an equally important function of ethical review is to encourage researchers to be self-policing, in order to sensitize them to their professional duty to protect their subjects' interests and encourage them to design research protocols to minimize the potential for harm. Had the Johannesburg authorities subjected their plans for the Phiri trial of prepaid meters to an ethical evaluation of this sort, they would likely have chosen to redesign the technology, policy, and procedures, avoiding significant harm in the process.

The introduction of unproven and potentially dangerous technologies, whether formally conceptualized as research or not, should be subjected to similar safeguards. Individuals subjected to such technologies without their consent should be able to seek redress in a court of law. That court should consider whether the technology's introduction was undertaken with appropriate consideration for human rights, in the form of procedural protections such as informed consent, risk-benefit analysis, and concern for vulnerable populations. Legal accountability along these lines would focus the attention of public interest advocates, lawmakers, bureaucrats,

---

[53] Constitution of South Africa, art. 12(2)(c).

and corporations on the adoption of appropriate safeguards to minimize risks arising from the introduction of unproven and potentially harmful technologies. The possibility of legal liability would incentivize desirable internal caution. The same principles provide the framework for exercising that self-scrutiny within corporations and administrative agencies, as part of policy and technological design.

3

Climate Change, Human Rights, and Technology Transfer

Normative Challenges and Technical Opportunities

Dalindyebo Shabalala

I INTRODUCTION

This chapter will review the broad strategy to link human rights and climate change, focusing specifically on how well the strategy works to strengthen obligations on developed countries to transfer technology that can reduce or mitigate the effects of increased carbon emissions. Elements of this strategy include regulatory action such as exceptions to patent protection, performance and technology transfer requirements for investments, compulsory licensing, price controls, and other measures[1] to make climate technologies available and affordable for the populations that need them most. The chapter posits that the state-centered "development" approach that has dominated both economic development and climate discourse to date has failed to provide a sufficient foundation for realistically addressing the issue of technology transfer.[2]

This chapter argues that the human rights approach solves two key problems that the development framework does not. First, it enables differentiation to take place, not between states, but between more vulnerable and less vulnerable populations within countries. It thus enables a focus on the most vulnerable populations, and in doing so also provides a basis for limiting the scope and nature of the demand for technologies to address climate change. Second, by limiting the scope of needed

---

[1] This includes such tools that were historically used by countries to promote their development and move up the technology value chain: patent working requirements, patent revocations, antitrust measures, reduction of patent terms, and joint venture requirements for foreign direct investment. For a full listing and discussion of these measures, see D. Shabalala, *Climate Change, Technology Transfer and Intellectual Property: Options for Action at the UNFCCC* (Maastricht: Maastricht University, 2014), pp. 263–97.

[2] Transfer happens when technology is first transferred from one country to another and is then adopted by public or private firms, being built into either their means of producing products and services or into the products and services themselves. It involves transfer of goods, craft knowledge, and scientific and technical information.

technologies, a human rights approach makes it more likely that such technologies will be made available to populations in need. If they are not, and lower-resource governments must act to secure climate change mitigation technologies for their citizens, the human rights approach will limit the grounds upon which actors in developed countries can challenge these decisions.

The United Nations Framework Convention on Climate Change (UNFCCC), ratified in 1992, is the international treaty framework that governs global efforts to combat climate change. This treaty uses a state-centered development framework for its legal and political framework of duties and rights,[3] with the unit of analysis being the state for both climate mitigation[4] and adaptation. The 1997 Kyoto Protocol governs the UNFCCC States Parties' legal obligation to limit greenhouse gas GHG emissions as measured at the economy level.[5] The UNFCCC also creates a state-to-state obligation to provide technological and financial support for climate change mitigation and adaptation.[6]

The development approach embodied in the UNFCCC aggregates needs and achievements at the national level rather than focusing on various strata of society that often do not evenly embody the risks and benefits of technological, social, and environmental change. Cost-benefit calculations can end up obfuscating the losses suffered by some portion of the population by counting the benefits to other portions of the population against them. This is particularly problematic from a human rights perspective when the portion of the population that suffers the loss is already disadvantaged while the portion that benefits is already privileged.

A human rights approach ensures that the individual rather than the nation as a whole is the primary beneficiary of actions that include technology transfer. It strengthens the fairness claim from both sides: developing countries can make a strong claim that any action they take is in the interests of specific vulnerable populations, and that they are not engaging in pure protectionist mercantilism or industrial policy favoring well-connected industrial actors. Developed countries providing financial and technological support can insist that this support be targeted primarily at vulnerable populations, making it easier to justify the spending to their citizens.

The effectiveness of this fairness claim is directly linked to the instrumental effect of a human rights approach. The human rights approach shifts away from the question of state-to-state obligations and instead asks the more direct question about

---

[3] *See* United Nations Framework Convention on Climate Change ("UNFCCC"), New York, May 9, 1992, in force March 21, 1994, 1771 UNTS 107, arts. 4.1(c), 4.3, 4.5, and 4.7.

[4] Climate mitigation refers to efforts to reduce or prevent emissions of greenhouse gases through development, deployment, and use of technologies or changes in economic behavior. United Nations Environment Programme, "Climate Change Mitigation," www.unep.org/climate change/mitigation/.

[5] Kyoto Protocol to the United Nations Framework Convention on Climate Change ("Kyoto Protocol"), Kyoto, Japan, December 10, 1997, in force February 16, 2005, 2303 UNTS 148.

[6] *See* UNFCCC, art. 4.1(c).

who within each country should be the recipient of technology transfer aimed at fulfilling rights. This has several effects. Most importantly, it removes the danger of cost-benefit calculations that try to aggregate gains and allow particular individuals or groups to be sacrificed for the sake of the general welfare. It also removes from the set of options those that would more negatively impact vulnerable populations. A classic example of this is hydroelectric dam construction that involves the forced resettlement and land dispossession of marginalized communities. While the electricity generated may be cleaner and result in lower emissions, thus meeting the country's obligations in the UNFCCC, the displacement and dispossession inflicts irreversible harms on those affected by it.

The human rights approach solves another core problem in climate technology transfer: how to limit the scope of technologies for which action is needed and justified. At the aggregate level, the scope of technologies necessary to achieve climate mitigation and adaptation within the necessary time frames is enormous. There is no global or UNFCCC coordinating principle or mechanism for identifying and prioritizing technology needs in developing countries. This leaves decisions about which technologies to pursue at the discretion of *demandeur* developing countries, and the decision about whether to provide technology at the discretion of developed countries. This chapter argues that by focusing on impacts as felt by individuals on the ground, a human rights approach provides such a mechanism for prioritizing technologies needed to address climate impacts.[7]

The next section of this chapter describes the limits of the development framework, tracing it from its roots in postcolonial demands for a New International Economic Order (NIEO) to the bargains embodied in multilateral environmental treaties. The subsequent section describes the two key problems posed by the approach within the UNFCCC: the fairness issue and the scope of technologies. The final section discusses the ways in which a human rights approach could solve some of the problems inherent in the development approach.

II THE LIMITS OF THE DEVELOPMENT APPROACH

A *Modernization and Development*

The increase in GHG emissions since the advent of the industrial revolution can largely be attributed to the economic activity of what we now call developed countries,[8] which are somewhat coextensive with political groupings such as the

---

[7] There is, of course, a direct link between mitigation and adaptation. The faster and more extensively greenhouse gas mitigation action takes place, the lower the likely cost of action to address adaptation will be. As I note below, some impacts will be unavoidable no matter how much mitigation takes place.

[8] The historical responsibility of developed countries amounts to almost 79 percent of all greenhouse gas emissions in the period from 1850 to 2011. *See* Centre for Global Development,

G7 and the Organization for Economic Cooperation and Development (OECD).⁹ The definition of this group of early industrializers significantly overlaps with the colonial powers that dominated Africa, Asia, and South America through the middle of the twentieth century.

This chapter makes significant claims around differential treatment of populations in developing countries. There is a long history of similarly categorizing nations, whatever the limitations of such schema may be. At the height of the Cold War, for instance, public discourse in journalism and political science divided states into the First, Second, and Third Worlds.¹⁰ The Third World referred to those countries that remained colonized, or had recently been decolonized and remained on the periphery of world economic affairs, consisting of all of Africa, all of Asia (sometimes including Japan), and all of South America.¹¹ Second World referred to the specific group of Eastern European and Soviet-orbit countries, and First World referred primarily to the industrialized countries of the Western Hemisphere, encompassing Western Europe and the United States, but also Australia and New Zealand. In economics, the use of terms such as "industrialized," "industrializing," and "nonindustrialized" became more prevalent in the late 1970s and early 1980s, given the need for more objective descriptors of the economic status (as opposed to the geopolitical status) of different countries.¹²

This chapter traces the trajectory of the shift toward differentiation within and between developing countries. The term "developing country" fully came to the fore in 1974 with the publication of the United Nations Declaration for a NIEO.¹³ The Declaration did not define the term "developing country" except to frame it with regard to those countries that had received their freedom from

---

"Developed Countries Are Responsible for 79 Percent of Historical Carbon Emissions," August 8, 2015, www.cgdev.org/media/who-caused-climate-change-historically (citing CO_2 emissions excluding LUCF [CAIT v2.0]).

⁹ The OECD consists of thirty-five industrialized countries, primarily in Europe and North America, but also encompassing Australia, Israel, Japan, Korea, Mexico, New Zealand, and Turkey. See OECD, "Members," www.oecd.org/about/membersandpartners/. The group is largely self-selected, not simply as an outcome of the discourse of the postcolonial international political order but also as a creation of membership requirements. The members are defined as "developed economies," with three members defined as "emerging economies": Chile, Mexico, and Turkey.

¹⁰ See, e.g., L. Wolf-Phillips, "Why Third World?: Origin, Definition and Usage" (1987) 9(4) Third World Quarterly 1311–27 at 1313.

¹¹ C. S. Clapham, Third World Politics: An Introduction (Madison: University of Wisconsin Press, 1985), p. 1.

¹² See, e.g., E. Ferrill, "Clearing the Swamp for Intellectual Property Harmonization: Understanding and Appreciating the Barriers to Full TRIPS Compliance for Industrializing and Non-Industrialized Countries" (2006) 15 University of Baltimore Intellectual Property Law Journal 137–70 at 141.

¹³ UN General Assembly Resolution 3201 (S-VI), Declaration on the Establishment of a New International Economic Order, U.N. Doc. A/RES/S-6/3201 (May 1, 1974) ("NIEO"). For a fuller discussion, see also M. Bedjaoui, Towards a New International Economic Order (New York: Holmes and Meier, 1979).

colonialism.[14] The term "developing country" has now come to be used in conjunction with two other terms: "developed country" and "economy in transition." Usually, within the category of developing economies there is also a subset of least-developed countries (LDCs). LDCs are the only set of countries for which there is a legal, international definition within the UN system, under the responsibility of the UN General Assembly Committee for Development Policy.[15] LDC status does not legally require states to provide special measures to support these countries either bilaterally or multilaterally, but the clear demarcation and process makes it possible to do so where countries and international organizations wish.[16]

The broader category of "developing countries," in contrast, has no legal basis as such, but nonetheless is commonly used in discourse about international political economy in the United Nations and other international organizations. The Bretton Woods institutions (the World Bank and the International Monetary Fund) recognize the existence of the category of developing countries, primarily through differentiated status for concessionary lending for what it defines as lower-income countries, blended lending for lower-middle-income countries, and nonconcessionary lending for upper-middle-income countries.[17] The World Trade Organization (WTO) also recognizes the distinction between developed and developing countries, categories into which countries self-select.[18]

In the UN General Assembly, developing countries self-define as the Group of 77 plus China. The G77 was established at the first meeting of the UN Conference on Trade and Development in 1964 and originally consisted of seventy-seven developing countries, defined as such by a joint declaration.[19] The membership is

---

[14] "The greatest and most significant achievement during the last decades has been the independence from colonial and alien domination of a large number of peoples and nations which has enabled them to become members of the community of free peoples. Technological progress has also been made in all spheres of economic activities in the last three decades, thus providing a solid potential for improving the well-being of all peoples. However, the remaining vestiges of alien and colonial domination, foreign occupation, racial discrimination, apartheid, and neo-colonialism in all its forms continue to be among the greatest obstacles to the full emancipation and progress of the developing countries and all the peoples involved." NIEO, art. 1.

[15] *See* UN Development Policy and Analysis Division, "LDC Identification Criteria & Indicators," www.un.org/development/desa/dpad/least-developed-country-category/ldc-criteria.html.

[16] *See* Committee for Development Policy and the UN DESA, *Handbook on the Least Developed Country Category: Inclusion, Graduation and Special Support Measures*, 2nd ed. (Geneva: United Nations, 2015), pp. 11–12.

[17] *See* World Bank, "Country and Lending Groups," https://datahelpdesk.worldbank.org/knowledgebase/articles/906519.

[18] World Trade Organization, "Who are the developing countries in the WTO?" www.wto.org/english/tratop_e/devel_e/d1who_e.htm.

[19] *Joint Declaration of the Seventy-Seven Developing Countries Made at the Conclusion of the United Nations Conference on Trade and Development*, U.N. Doc. E/CONF. 46/138 (1965).

now 133 self-selected countries, primarily former colonies.[20] Membership in the G77 provides no specific legal status in the United Nations or other organizations, but provides a mechanism for this group of countries to make demands on "developed" countries on the basis of justice, fairness, and other principles, and to use their common weight to try to influence outcomes in international negotiations and decision-making bodies.

The G77 negotiates as a group to represent common interests in several organizations, such as the UNFCCC,[21] but does not serve as the sole organizational mechanism for developing countries. Within the broader UN system, countries are also categorized using the Human Development Index (HDI), labeled as having a low, medium, high, or very high level of human development.[22] The various methods of categorizing countries have significant overlap. The nations that rank lowest on the HDI tend to be LDCs, while those that rank in the middle of the HDI tend to be the same ones that lie between the least and most economically developed. A high or very high human development score tends to be associated with a high level of economic development.[23]

The existence of intermediate categories of developing countries, such as "emerging economies," "economies in transition," "middle-income countries," and BRICS,[24] is more than just an attempt to do a better descriptive job. Rather, it poses a challenge to the very concept of a cohesive set of developing countries with common interests largely defined against the interests of developed countries. The differentiation plays out in unique ways in various international fora. In this chapter, when discussing a specific forum, I will specify the ways in which it makes the distinction, but will generally refer to the broad categories of developed and developing countries.

In the WTO regime, specific differential treatment for developing countries is generally called special and differential (S&D) treatment.[25] S&D policy usually provides developing countries (mostly former colonies) more time and financial assistance to comply with international economic obligations. S&D treatment is

---

[20] See *The Group of 77 at the United Nations*, www.g77.org/doc/members.html.
[21] "Party Groupings," *United Nations Framework Convention on Climate Change*, http://unfccc.int/parties_and_observers/parties/negotiating_groups/items/2714.php.
[22] See United Nations Development Program, "Internal Human Development Indicators," http://hdr.undp.org/en/countries.
[23] For example, looking at BRICS, India and South Africa have a medium HDI, whereas Brazil and China have a high HDI. Russia has a very high HDI.
[24] BRICS is a group of countries – Brazil, Russia, India, China, and, more recently, South Africa – identified by Jim O'Neill of Goldman Sachs as having significant economic growth rates and increasing influence in the global economy. J. O'Neill, "Building Better Global Economic BRICS," *Goldman Sachs Economic Research Group Global Economics Paper no. 66* (London: Goldman Sachs Economic Research Group, 2001), p. 4.
[25] See, e.g., World Trade Organization, "Special and differential treatment provisions," www.wto.org/english/tratop_e/devel_e/dev_special_differential_provisions_e.htm.

accorded to developing countries under several of the WTO-covered agreements.[26] The WTO officially recognizes LDCs as a group within it based on the UN Committee on Development Policy List,[27] and these countries are entitled to additional concessions.[28] The justification for S&D treatment is that these countries are saddled by the ongoing and pernicious aftereffects of colonialism, which continue to retard their development. This policy is based in part on fairness claims; i.e., that differently situated and less capable countries should not be subject to the same obligations in international arenas as developed countries. It is also based on justice; i.e., that developing countries are owed some form of restitution and aid because developed countries dominate the current economic landscape based on their past success exploiting the natural resources and populations of their colonial holdings.

The demand for policies giving special treatment to developing countries was most strongly expressed in the NIEO.[29] The NIEO drew on traditional modernization theory,[30] which viewed industrialization as a stepwise process by which countries moved up the value chain from primary commodity producers to value-added manufactured product producers, resulting in a "modern" industrialized society. A key to modernization was integrating technology into production processes and relinquishing traditional extractive modes of economic activity. At the same time, dependency theory also gained traction as a critique of the international economic system, arguing that developed countries had deliberately deindustrialized their former colonies in order to ensure access to raw materials and ready markets for their own finished industrial products.[31]

These twin economic theories/ideologies framed the NIEO. On the one hand, modernization theory was an argument for access to similar opportunities to develop along the same paths and using the same policies historically available to developed economies, without having to adhere to newer rules and having limitations imposed. On the other hand, dependency theory was an argument for restitution for the deliberate damage caused by developed economies during their periods of colonial and postcolonial exploitation.

These twin strains can be seen in the text of the NIEO, which, among other things, demands sovereignty over all natural resources and the right to nationalize,

---

[26] Ibid.
[27] *See* Committee for Development Policy and the UN DESA, *Handbook on the Least Developed Country Category*.
[28] World Trade Organization, "Least-Developed Countries," www.wto.org/english/thewto_e/whatis_e/tif_e/org7_e.htm.
[29] NIEO.
[30] *See, e.g.*, D. C. Tipps, "Modernization theory and the comparative study of national societies: A critical perspective" (1973) 15(2) *Comparative Studies in Society and History* 199–226 at 208–10.
[31] *See, e.g.*, T. Smith, "The Underdevelopment of Development Literature: The Case of Dependency Theory" (1979) 31(2) *World Politics* 247–88 at 249.

free from external coercion by other states;[32] the right to restitution and compensation for colonialism;[33] fair and equitable pricing of raw minerals and resources in international markets;[34] special and differential, or preferential, treatment for developing countries in all international economic institutions; and, most importantly: "Giving to the developing countries access to the achievements of modern science and technology, and promoting the transfer of technology and the creation of indigenous technology for the benefit of the developing countries in forms and in accordance with procedures which are suited to their economies."[35]

This demand for transfer of technology as a means of achieving development was central to the vision of the NIEO, yet there is still no universally recognized or legally enforceable definition for what technology transfer is or what form it must take. In this chapter, the term "technology transfer" refers to the flow of technological goods and knowledge across borders.[36] Transfer happens when technology and associated know-how are first transported from one country to another and then adopted by public or private firms. The technology can either be built into the means of production (i.e., industrial or economic processes), or built into the products and services themselves. The clearest and most well-articulated provisions for technology transfer in the environmental arena can be found in Chapter 34 of Agenda 21 of the 1992 Rio Declaration on Environment and Development.[37] The language in Agenda 21 has been a guide for the kinds of actions expected of developed countries in multilateral environmental treaties, wherein developed countries agree to provide technology transfer in exchange for developing country participation in efforts to mitigate climate change. The NIEO vision lines up with the broader approach of Agenda 21, focusing on free or low-cost access to technology and relaxation of intellectual property restrictions.

The NIEO vision has been consistently resisted by developed countries as an imposition on the legitimate intellectual property rights of their corporations, especially in international fora such as the United Nations Conference on Trade and Development (UNCTAD)[38] and the World Intellectual Property Organization (WIPO). That resistance led to the failure to adopt the Draft Code of Conduct of

---

[32] NIEO, art. 4(e).
[33] Ibid. art. 4(f).
[34] Ibid. art. 4(j).
[35] Ibid. art. 4(p).
[36] This discussion on definitions is drawn from D. Shabalala et al., "Climate Change, Technology Transfer and Human Rights," Working Paper, CIEL/ICHRP (2010), www.ichrp.org/files/papers/181/138_technology_transfer_UNFCCC.pdf.
[37] United Nations Division of Sustainable Development, *Agenda 21: The Rio Declaration on Environment and Development, Adopted at United Nations Conference on Environment and Development (UNCED)*, Rio de Janeiro, June 3–14, 1992 (Geneva: United Nations 1992), http://sustainabledevelopment.un.org/index.php?page=view&nr=23&type=400&menu=35.
[38] For more on this history, see P. G. Sampath and P. Roffe, "Unpacking the International Technology Transfer Debate: Fifty Years and Beyond," Working Paper (International Centre for Trade and Sustainable Development, 2012).

International Transfer of Technology[39] in 1985, despite formal negotiations that had been ongoing since 1976.[40] That resistance also resulted in the failure of WIPO to adopt true S&D treatment provisions after the minimal achievements of the 1967 revisions of the Berne Convention[41] and the Paris Convention[42] in Stockholm.

By the time the 1971 revisions to the Berne and Paris Conventions were being considered, developing countries had become the majority of members in WIPO and could block any further adoption of stronger intellectual property protections that did not take their concerns into account. This majority of developing countries[43] influenced internal processes in most UN-affiliated international institutions, which operated on a one-country, one-vote structure. Developing countries could block further decision-making, but they could not impose their will on developed countries, because those countries would simply not negotiate or refuse to join treaties on economic matters proposed by developing countries. This developing country veto blocked any further norm-setting in WIPO. It also meant that developed countries had few venues for further development of norms on intellectual property other than bilateral mechanisms.

Developed countries retained primary decision-making power in the Bretton Woods institutions (the World Bank and the IMF) and the General Agreement on Trade and Tariffs (GATT), however, where clout was determined by levels of financial contribution. As these institutions gained prominence in the 1980s, the desire for access to the low trade tariffs enjoyed by developed country GATT members was a key impetus for developing countries to participate in the Uruguay Round of negotiations that led to the creation of the World Trade Organization[44] and resulted in the Agreement on Trade-Related Aspects of Intellectual Property Rights (TRIPS).[45]

---

[39] UNCTAD, "Compendium of International Arrangements on Transfer of Technology: Selected Instruments – Relevant Provisions in Selected International Arrangements Pertaining to Transfer of Technology," UNCTAD/ITE/IPC/Misc.5 (Geneva: UNCTAD, 2001), www.unctad.org/en/docs//psiteipcm5.en.pdf.

[40] See Sampath and Roffe, "Unpacking the International Technology Transfer Debate" at 26.

[41] Berne Convention for the Protection of Literary and Artistic Works, Berne, Switzerland, September 9, 1886; revised at Berlin, November 13, 1908; completed at Berne, March 20, 1914; revised at Rome, June 2, 1928; at Brussels, June 26, 1948; at Stockholm, July 14, 1967; at Paris, July 24, 1971; and amended September 28, 1979, 1161 UNTS 3.

[42] Paris Convention for the Protection of Industrial Property, Paris, March 20, 1883; as revised at Brussels, December 14, 1900; at Washington, DC, June 2, 1911; at The Hague, November 6, 1925; at London, June 2, 1934; at Lisbon, October 31, 1958; and at Stockholm, July 14, 1967; and as amended September 28, 1979, 828 UNTS 107.

[43] As was established at the 1964 UNCTAD Conference. See *Joint Declaration of the Seventy-Seven Developing Countries*, U.N. Doc. E/CONF. 46/138 (1965).

[44] Marrakesh Agreement establishing the World Trade Organization, Marrakesh, April 15, 1994, in force January 1, 1995, 1867 UNTS 4 ("WTO Agreement").

[45] Agreement on Trade-Related Aspects of Intellectual Property ("TRIPS"), Annex 1C to the Marrakesh Agreement establishing the World Trade Organization ("WTO Agreement"), Marrakesh, April 15, 1994, in force January 1, 1995, 1867 UNTS 4.

S&D treatment was enshrined in TRIPS, which laid out minimum standards for intellectual property protections in all countries but was limited to additional transition time from old rules to new ones, primarily for the subgroup of LDCs.[46] TRIPS provided for longer transition periods, but only a five-year period was provided for most developing countries.[47] In order to extend patent protection to products in addition to processes, countries such as India had only a ten-year transition. Only the LDCs received a full transition period of ten years for the entire agreement; this period has since been extended twice.[48]

TRIPS represents an almost complete transition in international economic policy from a system that provided some, albeit limited, recognition of fairness and justice claims by developing countries to one that now largely elides such differences in favor of harmonization. The dispute settlement mechanism at the WTO, which provides for financial sanctions for noncompliance, poses some of the strongest restrictions on policy discretion that developing countries have experienced since their independence. There remains some room for taking action around compulsory licensing, but the use of historically available measures to enable technology transfer – e.g., exclusions from patentability, local working requirements, exceptions to patent rights – was severely curtailed by TRIPS.[49]

Nevertheless, the demand for and discourse around S&D treatment have permeated other regimes in international law and remain a potent force in the environmental arena. The next section discusses how this development framework has also led to an impasse in international environmental negotiations.

B Environment, Development, and Climate Change

1 Differentiation, Fairness, and Justice Claims in the UNFCCC Framework

Differential categorization of developing countries – whether they are "middle income," "emerging," or "in transition" – is the backdrop for current debates regarding which countries are obligated to immediately take the first steps to address climate change and which should be given a grace period to start this process and be entitled to financial assistance from developed countries to meet their obligations. In that debate, "middle income," "emerging," "industrializing," or "newly industrialized" economies like India, China, Mexico, Argentina, Brazil, South Africa,

---

[46] Identification as a developing country is a matter of self-selection. *See* World Trade Organization, "Least Developed Countries," www.wto.org/english/thewto_e/whatis_e/tif_e/org7_e.htm.
[47] TRIPS, art. 65.
[48] "Extension of the Transition Period under Article 66.1 of the TRIPS Agreement for Least Developed Country members for Certain Obligations with respect to Pharmaceutical Products: Decision of the Council for TRIPS of November 6, 2015," WTO, IP/C/73 (November 6, 2015).
[49] *See* Shabalala, *Climate Change, Technology Transfer and Intellectual Property*, p. 363.

Taiwan, and Singapore are now seen by many economically dominant countries as viable competitors in the international economy rather than worthy recipients of aid and support. While this erosion of concern is almost complete in international economic law and institutions, a similar situation in environmental law, and climate change law in particular, is still in its early stages.

The UNFCCC was built on the concept of common but differentiated responsibilities (CBDR), under which developed countries have had the burden to act first and most, as the acknowledged contributors to the problem, for all but the last two decades of the twentieth century.[50] That division is specifically established in the UNFCCC classification of countries between Annex I and non-Annex I countries. Annex I countries includes the OECD countries plus twelve economies in transition.[51]

However, unlike the pure fairness and justice claims to differential treatment in the WTO setting that were based on colonial exploitation, CBDR also had another component – a transactional one. Threats to the international commons, such as climate change (or chlorofluorocarbon emissions damaging the ozone layer), require common action. The threats affect developed and developing economies alike, although with some greater impacts in developing countries due to lower resilience and adaptive capacity. There are, however, considerable opportunity costs associated with climate change mitigation. Differentiated treatment was needed to incentivize developing countries to do their part.

If a country avoids the use of a specific technology or product or reduces its GHG emissions, that imposes a cost on the country's development by forcing it to develop or adopt different products, technologies, and economic behaviors. Developed countries that wished to address climate change had to convince developing countries to participate in a process under which they would commit to reduce their emissions even when it cost them in economic terms or slowed their rate of development. This required a carrot: financial assistance to deal with adjustment costs and technology, at low or zero cost, to enable them to reduce their own emissions and adapt to climate impacts. This bargain is reflected in Article 4, which establishes a series of commitments that distinguish between Annex I and non-Annex I countries as well as between developed and developing countries more generally.[52]

---

[50] There is a basic philosophical and political objection to this concept that is an issue of attributability and knowledge-intent that I believe elides the broader international law presumption that successor states take on the obligations of previous states in terms of financial obligation. This was the basis on which developing countries took on international intellectual property obligations that have prevented them from accessing technology and knowledge held in developed economies. A concise and clear discourse on this can be found in D. Bell, "Global Climate Justice, Historic Emissions, and Excusable Ignorance" (2011) 94(3) *The Monist* 391–411.

[51] UNFCCC, Annex I.

[52] Ibid.

On technology transfer in particular, Articles 4.1(c), 4.3, 4.5, and 4.7 establish S&D treatment for developing countries. Article 4.2 establishes commitments that only apply to "developed countries and others in Annex I." The UNFCC also establishes a group of Annex II countries that have financial and technological support obligations on top of mitigation commitments under Article 4 generally. These countries are the traditional OECD group of early industrializers.[53]

The CBDR framework, however, has proven to be unworkable within the current international political climate, with the emergence of a category of "intermediate" developing countries not identified in the original UNFCCC. In particular, while historically emissions have been largely due to developed countries, developing countries like China, India, Brazil, and Mexico have begun to catch up, and China has become the largest source of emissions at present.[54] These are also countries that have become significant trading competitors for developed countries.[55] Despite large domestic populations that remain in poverty, the aggregate economic strength of these countries is such that they are no longer viewed by developed countries as having any right to financial and technological support.[56] Their emissions also erode the argument that they should delay taking mitigation action, as they continue to increase their share of emissions. This was already the case at the time the Kyoto Protocol was signed in 1997. It is worth noting that the US Senate refused to ratify it precisely because countries like China were excluded from quantified emissions-reduction obligations as a developing country in the UNFCCC.[57]

The transaction frame has not been successful, however, for a different reason: because developing countries believe that developed countries have not provided sufficient levels of finance and technology transfer.[58] This has created a chicken-or-egg problem for mitigation action by those intermediate developing countries that may be in a position to take on more obligations. These countries are driven by several concerns. The first is the perception that, until relatively recently, developed

---

[53] UNFCCC, Annex II. The countries are: Australia, Austria, Belgium, Canada, Denmark, European Economic Community, Finland, France, Germany, Greece, Iceland, Ireland, Italy, Japan, Luxembourg, Netherlands, New Zealand, Norway, Portugal, Spain, Sweden, Switzerland, Turkey, United Kingdom of Great Britain and Northern Ireland, and United States of America.

[54] M. Ge, J. Friedrich, and T. Damassa, "6 Graphs Explain the World's Top 10 Emitters," World Resources Institute, November 25, 2014, https://wri.org/blog/2014/11/6-graphs-explain-world%E2%80%99s-top-10-emitters.

[55] See, e.g., European Commission, *The EU's New Generalised System of Preferences (GSP)* (Brussels: European Commission, 2012), http://trade.ec.europa.eu/doclib/docs/2012/december/tradoc_150164.pdf.

[56] Ibid., p. 3. The EU GSP excludes upper-middle-income countries from its coverage.

[57] See C. R. Sunstein, "Of Montreal and Kyoto: A Tale of Two Protocols" (2007) 31 *Harvard Environmental Law Review* 1–65 at 2–3. The Kyoto Protocol only applied to implementation of Annex I country obligations to quantified economy-wide emissions reductions.

[58] Expert Group on Technology Transfer, *Report on the review and assessment of the effectiveness of the implementation of Article 4, paragraphs 1(c) and 5, of the Convention*, FCCC/SBI/2010/INF.4, May 10, 2010.

countries had failed to meet their obligations under the Kyoto Protocol to reduce their emissions.[59] The second is the perception that developed countries have failed to meet their obligations to provide financial and technological support to prepare them to take on obligations.[60] These perceived failures make intermediate developing countries less willing to take on obligations, while hardening other developing countries against any measures to weaken the differentiation between developed and developing countries in the UNFCCC framework. This conflict came to a head during the Bali Conference of the Parties in December 2007. It led developing countries to refuse to agree to negotiations for any new commitment period in the Kyoto Protocol or any new agreement that included developing country emissions-reduction commitments.[61]

The issue of differentiation and the push to have developing countries take on measurable, reportable, and verifiable emissions-reduction obligations has been at the center of climate negotiations since the post-Kyoto discussions. The end result of that process was the 2015 Paris Agreement,[62] which significantly erodes the distinction between developed and developing countries compared to the UNFCCC and the Kyoto Protocol. Article 1 notes that all parties are required to provide voluntary nationally determined contributions (NDCs),[63] a major departure from past agreements. Each country determines for itself the extent of action it will take, instead of the obligation being tied to country classifications. This solution was viewed as necessary to bring countries such as the United States into the agreement by addressing the major complaints about the Kyoto Protocol: the exclusion of major developing country economies from emissions-reduction obligations and the

---

[59] M. Le Page, "Was Kyoto climate deal a success? Figures reveal mixed results," *New Scientist: Daily News*, June 14, 2016, www.newscientist.com/article/2093579-was-kyoto-climate-deal-a-success-figures-reveal-mixed-results/. While the study examined notes that all Kyoto parties (excluding signatories US and Canada) had met their commitments, the article points out that much of that was attributable to already existing reduction in the economies in transition, the financial crisis, and purchase of carbon credits on trading markets in developing countries. Nevertheless, it does seem clear that while the overall aim of the Kyoto Protocol was met, there had long been arguments that the countries had failed.

[60] See M. Khor, *Climate Change, Technology and Intellectual Property: Context and Recent Negotiations* (Geneva: South Centre, 2012), p. 1.

[61] Summary of the Thirteenth Conference of Parties to the UN Framework Convention on Climate Change and Third Meeting of Parties to the Kyoto Protocol, December 3–15, 2007, *Earth Negotiations Bulletin* 12(354), December 18, 2007, p. 15, www.iisd.ca/climate/cop13/.

[62] The Paris Agreement was signed in December 2015 as an agreement to succeed the Kyoto Protocol and the Copenhagen Accord, under which countries agreed to reduce their greenhouse gas emissions. The Paris Agreement was adopted as a decision of the UNFCCC parties rather than a protocol or a treaty, so as to avoid the domestic ratification obligations of some countries regarding formal international agreements. See *Report of the Conference of the Parties on Its Twenty-First session, Held in Paris from November 30 to December 13, 2015: Decision 1/CP.21*, U.N. Docs. FCCC (FCCC/CP/2015/10/Add.1) (January 29, 2016), http://unfccc.int/resource/docs/2015/cop21/eng/10a01.pdf.

[63] Nationally determined contributions are the voluntary pledges made by countries on the extent to which they will reduce their emissions.

imposition of stronger measurable, reportable, and verifiable emissions reductions on UNFCCC Annex I countries.

The agreement does recognize in several instances that developing countries may make slower progress toward emissions reductions[64] and that some support will be needed to ensure their implementation of commitments,[65] but the broader obligation to commit to emissions reductions applies *to all* parties. Nevertheless, Article 4 (4) notes that developed countries are to express their NDCs as quantified economy-wide emissions reductions, whereas developing countries are only required to provide enhanced action. Additionally, Article 4(5) reiterates that developing countries should be supported. In terms of financial and technological support, Articles 9 and 10 state that developed countries remain obligated to provide support to developing countries.[66]

Nothing in the Paris Agreement necessarily redefines what constitutes a developing or developed country, but it does abandon the framework of Annex I, Annex II, and non-Annex I countries that previously framed obligations. It is difficult to escape the conclusion that this erosion of differentiation may force more capable developing countries to take on more concrete obligations even while they contain significant vulnerable populations. The majority of people in developing countries, even intermediate countries such as China and India, live in climate-vulnerable environments and ecosystems.[67] Coastal and island states risk sea-level rises leading to flooding, salinization of arable land, and increased vulnerability to extreme weather events.[68] For North African and West African states, the risks are associated with encroachment of the Sahara Desert and increased frequency and severity of drought events.[69] For central and eastern Africa, shifts in disease bands, as well as increased floods and droughts, may lead to more hunger and illness. For countries on the Asian subcontinent, such as India and Bangladesh, the melting of glaciers and the unpredictability of monsoons are likely to lead to increased coastal flooding and shortages of drinking water.[70]

Climate vulnerabilities are complicated by a lack of resources needed to prepare for and adapt to changes in climate. For developing countries, the costs of paying for

---

[64] Paris Agreement, art. 4.
[65] Ibid., art. 3.
[66] Ibid., art. 9(1). ("Developed country Parties shall provide financial resources to assist developing country Parties with respect to both mitigation and adaptation in continuation of their existing obligations under the Convention.")
[67] See UNEP, *Towards a Green Economy: Pathways to Sustainable Development and Poverty Eradication* (Nairobi: UNEP, 2011), p. 19.
[68] G. McGranahan et al., "The Rising Tide: Assessing the Risks of Climate Change and Human Settlements in Low Elevation Coastal Zones" (2007) 19(1) *Environment and Urbanization* 17–37 at 18.
[69] Intergovernmental Panel on Climate Change, *Climate Change 2014: Synthesis Report. Contribution of Working Groups I, II and III to the Fifth Assessment Report of the Intergovernmental Panel on Climate Change* (Geneva: IPCC, 2014), p. 50.
[70] Ibid.

activities needed to address climate change are astronomical and likely to be crippling to their development.[71] The current development framework represented in the UNFCCC approach provides little or no means by which differentiation can be made between newly industrialized countries and other developing countries. This erosion of differentiation means that newly industrialized countries will be increasingly asked to make greater sacrifices despite their lack of financial capacity and large populations living at the bottom of the HDI. Absent a fundamental realignment of interests, the CBDR framework has left countries mired in this no-man's land, where both the fairness and justice claims of developing countries to financial and technological support and the claims on them to take action equal to that of developed countries remain weak.

In the last section of this chapter, I discuss how a human rights approach may enable a way out by refocusing the justice and fairness claims within the countries rather than state-to-state. In the next section, however, I discuss the second major problem posed by the CBDR/development approach: that it is overly broad in the scope of technologies that are identified for technology transfer.

2 The CBDR/Development Approach and the Scope of Technologies

The Earth continues to experience record-breaking temperatures caused by increased atmospheric concentrations of carbon dioxide and other GHGs.[72] To keep warming well below 2 degrees, and to retain the possibility of stabilizing at the safe level of 1.5 degrees above preindustrial levels, it may be necessary for global emissions to peak by 2020.[73] Projections suggest that past emissions mean the Earth is already locked into a baseline increase in temperature that makes some impacts unavoidable by 2100.[74] None of the associated costs of climate change between now and 2050 are likely to be avoided because of this lock-in.[75]

The timing of impacts is crucial to determine who, both between and within countries, is likely to be impacted first, as well as which technologies are going to be most effective at addressing the mitigation and adaptation needs of those populations. That calculus of which technologies to transfer and who they should benefit differs depending on whether harm is viewed at the aggregate level or is based on the needs of the most vulnerable.

---

[71] For example, looking just at mitigation scenarios, the IEA projected that from 2010–2020, more than $2.3 trillion (US trillion) annually would need to be invested, the majority of that private flows. The share of developing countries was $1.3 trillion annually, of which China represented $500 billion. In contrast to the scale of projected need, total investment flows in 2010 and 2011 were $247 billion and $260 billion, respectively. See IEA, Table 4.3, *Energy Technology Perspectives 2012* (Paris: OECD/IEA, 2012), p. 139.
[72] IPCC, *Climate Change*, p. 40.
[73] Ibid., p. 82.
[74] Ibid., pp. 78–79.
[75] Ibid.

A CBDR/development framework begins with the question of what the state, at the aggregate level, must do to meet its adaptation and mitigation needs. For mitigation, this is determined by when the country will need to begin to lower emissions (generally at some point between 2015 and 2020), implying a need for almost *all* currently available technologies aimed at reducing emissions most efficiently and quickly. A cursory glance at the details of the implied technologies shows a large, unwieldy list of specific technologies.[76]

In energy, for example, the speed and scale of action required would seem to select for energy projects that are energy-efficient and easily and quickly deployable, and that target those countries currently consuming high rates of GHG-intensive energy sources. While these will not always be the largest-scale projects, this approach will tend to select for the kinds of projects that generate large amounts of energy per site, that can feed into the current grid, and that supply heavy industrial and urban users. This tendency can be seen in the discussion and projections of the International Energy Agency, which focuses on large-scale carbon capture and storage and nuclear generation.[77]

In adaptation, the challenge is quite clear, from sea-level rise to changes in the hydrological cycle.[78] When focusing on the state level, as the UNFCCC does, adaptation requires actions like "strengthening institutional capacities and enabling environments for adaptation, including for climate-resilient development and vulnerability reduction"[79] or "[b]uilding resilience of socio-economic and ecological systems, including through economic diversification and sustainable management of natural resources."[80]

There are few, if any, references to differentiation of populations within individual countries, and little focus on those most vulnerable to climate change and thus most in need of adaptation. The UNFCCC provides funding at the level of the aggregate need of each country and does not require states to prioritize the needs of the most vulnerable in the disbursement of this money. Thus, in the Green Climate Fund, the primary financing entity within the UNFCCC, for example, funding is generally focused on high-impact, systemic-oriented programs and projects.[81]

---

[76] For a full list, see the tables in Shabalala, *Climate Change, Technology Transfer and Intellectual Property*, p. 59.

[77] IEA, *Energy Technology Perspectives 2012*, p. 11.

[78] Intergovernmental Panel on Climate Change, *Climate Change*, p. 51.

[79] *The Cancun Agreements: Outcome of the work of the Ad Hoc Working Group on Long-Term Cooperative Action under the Convention*, Decision 1/CP.16, FCCC/CP/2010/7/Add.1 (2010), ¶ 14(c), http://unfccc.int/resource/docs/2010/cop16/eng/07a01.pdf#page=4.

[80] Ibid., ¶ 14(d).

[81] *See* Green Climate Fund, *Investment Framework*, GCF/B.07/06 (May 9, 2014), p. 9. www.greenclimate.fund/documents/20182/24943/GCF_B.07_06_-_Investment_Framework.pdf/dfc2ffe0-abd2-43e0-ac34-74f3b69764c0.

Vulnerability is primarily discussed at the level of the state.[82] In the illustrative indicators for the impact of its adaptation programs, there are only two criteria that look to the extent of impact on the most vulnerable populations, one defining reduced beneficiaries and the other defining it in terms of "enhancing adaptive capacity and resilience ... focusing particularly on the most vulnerable populations or groups."[83]

While nothing requires countries to focus on the most vulnerable, the formulation of national adaptation plans may involve such prioritization or focus on the most vulnerable, as noted in the UNFCC's *Technical Guidelines for the National Adaptation Plan Process*.[84]

The Adaptation Fund, an independent financing entity of the UNFCCC established under the Kyoto Protocol, is a "direct access" fund, which means that few or no conditions are placed on countries' priorities and none are required as long as these are carried out by an accredited national implementing entity.[85] Nothing in the Adaptation Fund policies requires a focus on the most vulnerable populations within countries in order to receive funding.[86] As such, no distinction is made between the hypothetical need to ensure that beach erosion does not destroy the second holiday homes of the wealthy (including a significant percentage of foreigners) in Cape Town, South Africa, versus the need to set up flood defenses for riverside villages in Kwazulu-Natal. The Paris Agreement refers only to "vulnerable parties."[87]

Adaptation presents a complex challenge that involves a network of existing capacity and vulnerability, with impacts and adaptations to impacts taking place within a network of cofactors such as poverty, population shifts, migration patterns, and changing land use.[88] The Stern Review, commissioned by the UK government in 2006 to examine the economics of climate change in order to inform the government's policy positions within the EU and international negotiations,[89]

---

[82] Ibid., p. 9.
[83] Green Climate Fund, *Decisions of the Board – Ninth Meeting of the Board, 24–26 March 2015. Annex III: Initial Investment Framework: Activity-Specific Sub-Criteria and Indicative Assessment Factors*, GCF/B.09/23 (April 16, 2015), www.greenclimate.fund/documents/20182/239759/Investment_Criteria.pdf/771ca88e-6cf2-469d-98e8-78be2b980940.
[84] *See* Least Developed Countries Expert Group, *National Adaptation Plans: Technical Guidelines for the National Adaptation Plan Process* (Bonn: UNFCCC, 2012) pp. 43, 94, www.unfccc.int/files/adaptation/cancun_adaptation_framework/application/pdf/naptechguidelines_eng_high__res.pdf
[85] Adaptation Fund, *Climate Adaptation Finance: Direct Access* (November 2016), www.adaptation-fund.org/wp-content/uploads/2016/11/Direct-Access-English-Nov2016-WEB.pdf.
[86] *See* Adaptation Fund, *Guidance on Accreditation Standards*, www.adaptation-fund.org/wp-content/uploads/2016/10/Guidance-on-Accreditation-Standards.pdf.
[87] Paris Agreement, arts. 7.6, 7.9(c), 11.1.
[88] Shardul Agrawala (ed.), *Bridge Over Troubled Waters: Linking Climate Change and Development* (Paris: OECD Environment Directorate, 2005), pp. 16–17.
[89] The Stern Review (sometimes called the Stern Report) was led by Sir Nicholas Stern, former chief economist at the World Bank. It proved highly influential in generating consensus

suggested that the key mechanisms for addressing adaptation were the same as those for generating economic wealth more generally. The report argued for prioritizing infrastructure, technology, information, knowledge, and skills,[90] especially in domains like agriculture, which makes up the majority of economic activity in most developing countries (up to 64 percent participation in South Asia and sub-Saharan Africa) and is very sensitive to climate variability.[91] A stable and sustainably growing framework for agricultural production and distribution is necessary to reduce vulnerability and enable adaptive capacity in developing countries.[92] Health interventions to deal with chronic diseases (both communicable and noncommunicable) in developing countries are also necessary to reduce vulnerability and adaptive capacity.[93] This implicates general health infrastructure and health management systems, but also the opportunity costs associated with the prices of medical products, devices, and services.

The breadth of the development approach points to a crucial weakness: under a development framework, technology transfer for adaptation covers an extremely broad range of technologies, aimed at generating economic growth and increasing the fungible wealth that enables resilience and increases adaptive capacity. Viewed in this way, the adaptation need is indistinguishable from the development need framed under the NIEO; i.e., developing countries need technology in order to engage in modernization and industrialization. From this view, the adaptation challenge is essentially a development challenge[94] and covers *all* sectors of technology relevant to ensuring rapid, non–fossil-fuel-dependent economic development. This makes the demand for access to all these technologies equally broad and limitless, and, within the context of the CBDR framework, imposes such a cost burden on developed countries that it fails in the current international framework.

As a result, industrialized countries are increasingly reluctant to fund adaptation, because they fear they are simply adding to their overseas development assistance (ODA) obligation without really addressing climate change-related issues. Further, the lack of a clear dividing line between adaptation and development makes it difficult to hold industrialized countries accountable for their climate change obligations, since they can simply point to existing ODA as fulfilling their UNFCCC obligation to adapt.

---

around the idea that taking action on climate would not have a negative impact on economic growth. N. Stern et al., *The Economics of Climate Change: The Stern Review* (Cambridge: Cambridge University Press, 2007).

[90] Ibid., p. 94.
[91] UNEP, *Towards a Green Economy*, p. 38; see also Stern, *The Stern Review*, p. 95.
[92] Stern, *The Stern Review*, pp. 38–40.
[93] Ibid., pp. 208–09.
[94] Ibid., p. 430.

III HOW DOES A HUMAN RIGHTS APPROACH SOLVE THE PROBLEMS POSED BY THE CBDR/DEVELOPMENT APPROACH?

Having discussed the weakness of the common but differentiated responsibilities approach for fairness and justice claims and for delimiting the scope of technologies, this section discusses how taking a human rights approach can be used to address these weaknesses. As an initial step, I focus on a key element of the human rights approach: the requirement that states prioritize action on behalf of citizens whose rights are least fulfilled or most under threat.[95]

Locating the right to development within the individual rather than the state has been central to a human rights-based approach to development.[96] The CBDR/development approach clearly does not focus on the fulfillment of individual human development, and primarily only acknowledges state interests in climate mitigation and adaptation.

This deficiency points us to the first step necessary for a human rights-based approach to climate change: the unit of climate impact analysis must be the individual and the community rather than the state. A rights-based approach is also helpful in the context of climate change, because it identifies both the developing country itself and other states in the international community as potential duty bearers with the obligation to provide for technology transfer.[97] In this chapter, I focus on those impacts that trigger a human rights claim against both the government of the developing country where the harmed individual resides and the developed country that fails to provide technological or financial support to prevent or address climate impacts.

A human rights approach also provides a basis for prioritizing policy choices that benefit vulnerable individuals even over other choices that might be more beneficial in the aggregate.[98] Aggregate data does not reveal how the benefits of development are distributed; a human rights approach, in contrast, attends to these distributional questions.[99] Finally, a human rights approach aims not just to identify rights that are threatened by climate change or to specify duty holders, but to further argue that the need to protect rights requires access to technologies to address the harm.

---

[95] United Nations Development Program, *Human Development Report 2000* (New York: Oxford University Press, 2000), p. 2, http://hdr.undp.org/sites/default/files/reports/261/hdr_2000_en.pdf; see also B. Hamm, "A Human Rights Approach to Development" (2001) 23(4) *Human Rights Quarterly* 1005–31 at 1011; A. Cornwall and C. Nyamu-Musembi, "Putting the 'Rights-Based Approach' to Development into Perspective" (2004) 25(8) *Third World Quarterly* 1415–37 at 1417.

[96] *See, e.g.*, *Human Development Report 2000*, p. 2; Hamm, "A Human Rights Approach to Development" at 1016.

[97] Ibid., pp. 22–23.

[98] Ibid., p. 23.

[99] Ibid., pp. 178, 186.

A human rights framework for responding to climate change begins with an analysis of the rights that are negatively impacted by climate change. Thus, the first limiting principle of a human rights approach is that it is defined by the very specific suite of internationally recognized, legally cognizable rights. These rights are enshrined in international human rights treaties, including the International Covenant on Civil and Political Rights (ICCPR) and the International Covenant on Economic, Social, and Cultural Rights (ICESCR), as they have been elaborated by human rights institutions, and, where appropriate, judicial and quasi-judicial mechanisms. In this, I take a conservative approach that does not rely on newer or derivative "rights," such as the right to development or the right to energy. My concern here is to evaluate and elaborate on how a relatively uncontroversial reading of traditional[100] human rights obligations can serve to: 1) strengthen fairness and justice claims to technology transfer; and 2) enable limitations on the scope and prioritization of technologies for technology transfer.

The human rights framework provides a very circumscribed justification for claims of access to mitigation technologies.[101] By focusing on impacts, we inevitably focus on adaptation. This, however, is a feature, not a "bug" of the approach. Without a focus on impacts, the causal link to states' obligations is missing, or at least attenuated.

A Which Rights Are Impacted by Climate Change?

The extensive academic literature has linked climate impacts to the following key rights established in the Universal Declaration on Human Rights, the ICESCR, and the ICCPR[102]: the right to life, the right to health (and to a healthy environment), the right to water, and the right to food.

---

[100] I recognize that even this categorization of traditional and newer rights is not uncontested, but it is the fact of contestation that I am focused on; i.e., those rights about which there exists a minimal amount of contestation as to their existence and for which the scope has been elaborated by treaty bodies and other relevant institutions. As such, I do not address the right to development or the right to energy, for example, both of which would provide much stronger purchase for justifying unilateral action to access mitigation technologies.

[101] Some arguments for mitigation arise from a claim that there is a broad "right to energy" that requires direct access to the means of generating it. In that sense, energy-generating technologies that do not create GHG emissions then become a necessity for fulfilling the right, a direct claim on renewable technologies. As noted above, this is a less developed right within traditional human rights discourse and remains contested, which is why I exclude it from the analysis in this chapter. See International Council on Human Rights Policy ("ICHRP"), *Beyond Technology Transfer: Protecting Human Rights in a Climate-Constrained World* (Geneva: International Council on Human Rights Policy, 2011), p. 25, www.ichrp.org/files/reports/65/138_ichrp_climate_tech_transfer_report.pdf.

[102] See M. Limon, "Human Rights and Climate Change: Constructing a Case for Political Action" (2009) 33 *Harvard Environmental Law Review* 439–76 at 444; see also ICHRP, *Beyond Technology Transfer*; E. Caesans et al., *Climate Change and the Right to Food: A Comprehensive Study* (Berlin: Heinrich Boll Stiftung, 2009); M. Darrow et al., *Human*

Rights to health and food are economic rights that are governed by the framework of "progressive realization." Although a state has an obligation to fulfill these rights, progressive realization recognizes the reality of limited resources. States are not required to fulfill these rights immediately, but rather must make progress toward realizing them.[103] The framework of progressive realization has two components: the establishment of baselines or thresholds below which rights may not fall, and a general movement toward fulfillment of rights. The first requires an immediate focus on the most vulnerable populations to ensure that no absolute harm occurs, and a focus on other groups as resources permit.

The right to life, on the other hand, requires states to actively refrain from action that would cause death and to prevent relatively foreseeable deaths from actions by others or by nature. This implies a measure of prevention to mitigate the effects of disasters (human or natural) and an obligation to take action when such disasters strike, in order to reduce or prevent deaths. In the context of climate change, the majority of disaster events are water-based and relate to coastal risks, although major rain events occur inland as well. Additionally, drought events are related to famine conditions that would trigger obligations related to the right to life.

Article 12 of the ICESCR affords the right to the "highest attainable standard of physical and mental health." In its General Comment 14,[104] the Committee on Economic, Social, and Cultural Rights (CESCR), the body charged with receiving reports from states on their compliance with the treaty, defines health broadly. The key link to climate change impacts, as well as climate change vulnerability and adaptive capacity, is the idea that health includes the right to a healthy environment. Thus there are two levels at which links can be made: at the level of direct health effects, such as disease burdens, but also at the level of the underlying determinants of health.

At the level of direct health impacts, climate change will result in changes in precipitation patterns, the length of rainy seasons, and the length of warm seasons.[105] The Intergovernmental Panel on Climate Change (IPCC), for example, points to significant uncertainty regarding the increased frequency and intensity of diseases,

---

Rights and Climate Change: A Review of the International Legal Dimensions (Washington, DC: World Bank, 2011); J. H. Knox, "Climate Change and Human Rights Law" (2009) 50(1) *Virginia Journal of International Law* 163–218; J. H. Knox, "Linking Human Rights and Climate Change at the United Nations" (2009) 33 *Harvard Environmental Law Review* 477–98; S. Humphreys (ed.), *Human Rights and Climate Change* (Cambridge: Cambridge University Press, 2009).

[103] Frequently Asked Questions on Economic, Social and Cultural Rights, *OHCHR Fact Sheet 33* (Geneva: Office of the United National High Commissioner for Human Rights, 2008), www.ohchr.org/Documents/Publications/FactSheet33en.pdf.

[104] Economic and Social Council, United Nations Committee on Economic, Social, and Cultural Rights (22nd Session, Geneva, April 25, 2000), *The Right to the Highest Attainable Standard of Health*, U.N. Doc. E/C, December 4, 2000 ("General Comment 14").

[105] IPCC, *Climate Change*, p. 69.

due in large part to a lack of long-term epidemiological data.[106] The IPCC notes that disease incidence may be due to social changes resulting from climate change, such as migration and subsequent changes in population density. Nevertheless, the IPCC points to four major categories of health impacts[107]: direct effects of heat or cold, vector-borne diseases, food- and water-borne diseases, and pollen- and dust-related diseases.

The right to water is not explicitly mentioned in the ICESCR. However, the CESCR has concluded that the right to water is implied in Article 11 as an aspect of the right to an adequate standard of living.[108] Climate is linked to the right to water in two ways. The first is that extreme weather events associated with climate change are likely to result in temporary but severe disruptions of water supply that deprive portions of the population of access to water. During an extreme weather or sea event, water supply can be cut off due to the malfunction of desalination plants, damage to rainwater collectors, or contamination of wells. The second linkage is a reduction in available freshwater and increased incidence of drought, as temperatures increase and surface moisture evaporates more quickly. The reduction in access to water due to climate change can be traced to increased glacial melt and a general reduction in the amount of water held in ice each winter season.[109] Such ice systems provide freshwater for much of the Indian subcontinent, for example. We can also expect increased intensity of droughts as well as expansion of dry areas.[110] These are impacts that are already being felt and are likely to increase in intensity through to 2050.[111]

The right to food is addressed in a number of international human rights documents.[112] Food production, both plant and animal, is the primary source of GHG emissions in many nonindustrialized developing countries. Thus agricultural practices that involve fertilizers, soil-tilling methods, and bovine farming contribute to GHG emissions through nitrous oxide and methane release.[113] Deforestation to

---

[106] Ibid.
[107] Ibid.
[108] Economic and Social Council, Committee on Economic, Social, and Cultural Rights (29th session, Geneva, November 11–29, 2002), *Substantive Issues Arising in the Implementation of the International Covenant on Economic, Social and Cultural Rights; General Comment No. 15* (2002), *The Right to Water (arts. 11 and 12 of the International Covenant on Economic, Social and Cultural Rights)*, U.N. Doc. E/C.12/2002/11.
[109] IPCC, *Climate Change*, p. 51.
[110] Ibid.
[111] Ibid.
[112] Universal Declaration of Human Rights, December 10, 1948, U.N. G.A. Res. 217 A (III), art. 25 (right to adequate standard of living, including food); International Covenant on Economic, Social, and Cultural Rights, December 16, 1966, 993 UNTS 3 (1976), arts. 11.1 and 11.2; Convention on the Elimination of All Forms of Discrimination against Women, December 18, 1979, 1249 UNTS 13 (1981), art. 12 (adequate nutrition during pregnancy and lactation).
[113] K. A. Baumert et al., *Navigating the Numbers: Greenhouse Gas Data and International Climate Policy* (Washington, DC: World Resources Institute, 2005) p. 85, www.wri.org/publication/navigating-numbers.

create more agricultural land also removes carbon sinks. On the other hand, food production is also one of the areas affected by increased dry areas and drought, as well as flooding. The loss of productive land may result in lower food production. For coastal lands, increased sea-related extreme weather events such as storm surges can also lead to the loss of cultivable land due to salination of the soil. In addition, changing weather patterns are affecting the lengths of growing seasons as well as humidity levels, soil acidity, and other factors. This can render existing plant varieties less productive.[114] Finally, the IPCC report also points to increased vulnerability to extreme drought events in the near term.[115]

B *Instrumental Approaches: Limiting the Scope of Technologies*

Simply by describing human rights impacts on the most vulnerable populations within a country, we see a manageable set of technologies that might be needed to mitigate the effects of climate change. While the development approach to addressing adaptation is focused primarily on economic growth as a means of ensuring adaptive capacity, a human rights approach focuses the set of choices, narrowed down by: 1) identifying a specific set of beneficiaries within each country; and 2) selecting for specific types and categories of technology.

1 The Beneficiaries of Action

With respect to the beneficiaries of action, a human rights approach requires a focus primarily on the needs of the most vulnerable rather than those already well-off or capable of adapting.[116] This means that while middle-class residents in each country are clearly targets of mitigation and adaptation action, from a human rights perspective they are not the top priority.

In addressing, for example, the right to food, who, then, are the "most vulnerable" that the CESCR is concerned about and how does focusing on them limit the scope of technologies? A development approach has no specific mechanism for distributional concerns. At the aggregate level, a general focus on increasing food production encompasses technologies within the entire agricultural production and distribution chain. A human rights approach focuses on vulnerable populations in both rural and urban areas and can focus on specific technologies within the agricultural value chain. Looking first at rural populations of small-holder and subsistence farmers most vulnerable to hunger, the action would focus primarily on ensuring that subsistence farming remains viable and productive where drought

---

[114] IPCC, *Climate Change*, p. 51.
[115] Ibid.
[116] *See* Hamm, "A Human Rights Approach to Development" at 1011; Cornwall and Nyamu-Musembi, "Putting the 'Rights-Based Approach' to Development into Perspective" at 1417.

or flood events increase. This means prioritizing small-hold farming methods and practices over large-scale industrial agriculture, which is already supported by the powerful multinational agribusiness industry.

However, given the increased urbanization that is likely to occur due to the displacement of rural populations by increased land stress (i.e., reductions in quality and availability) as a result of climate change, it will also be necessary for small-holder farmers to generate sufficient surpluses to feed those urbanized populations. This means focusing on transportation, distribution, and food storage systems. To the extent that those urban populations would need food, some form of more intensive agriculture may be needed, but a human rights approach would allow that only to the extent that it did not involve dispossession and displacement of small-holder and subsistence farmers.

2 The Selection of Human Rights-Appropriate Technologies

A human rights approach also directs the selection of technologies. Although human rights would not necessarily require any particular kind of technology, it would create a bias toward those technologies that were inexpensive, easily deployable by the government or market providers, and easily maintained by the target beneficiaries. This is a result of the need to address near-term impacts on vulnerable populations without significant purchasing power or adaptive capacity of their own.

However, relevant technologies would not be limited to basic or low-cost ones, but could involve sophisticated micro-grid applications or big data programs for weather monitoring linked to mobile telephony and Internet access. The key is accessibility and sustainability of the technologies for the most vulnerable, as compared to something like energy management software for household appliances, which is low-hanging fruit as far as mitigation measures are concerned.[117] The type of technology could shift depending on the nature of the right and the needs of the population. It may make sense for the state to prioritize sensor and data technology to keep track of rainfall in water-stressed areas in order to deliver on its obligation to fulfill the right to water. It also makes sense for the state to enable access by individuals and communities to specific water-purification technologies, such as water-purification tablets, that are easily deployable and usable in a sustained manner. Adopting this approach does not necessarily mean that states should neglect national-level infrastructure, but it does mean at least a partial prioritization of technologies that address harms to the most vulnerable populations, or at least ensuring that these needs are addressed concurrently with broader investments in infrastructure.

---

[117] Appliances make up a significant portion of global electricity end use. Increased efficiency and use of best available technologies have large GHG emissions-reduction potential and are considered low-hanging fruit. See UNEP, *Towards a Green Economy*, p. 343.

IV CONCLUSION

By refocusing the discourse away from the obligations of states to one another and toward the obligations of all states to vulnerable populations, a human rights approach can bypass the continuing impasse at the UNFCC on how developed and developing countries should be categorized and assigned obligations. This chapter describes the means by which developing countries can frame their adaptation action (and, secondarily, their mitigation action) to make a stronger claim on developed countries to assist them in meeting their human rights obligations to their vulnerable populations. This kind of framing has the potential to transform the discourse from a "donor" framework, in which developed countries provide technology at their discretion, if at all, to a "*demandeur*" framework, in which vulnerable citizens of developing countries can make direct claims to technologies in ways that developed countries may find more difficult to challenge on legal, political, or economic grounds.

4

Judging Bioethics and Human Rights

Thérèse Murphy

I INTRODUCTION

Bioethics is, without doubt, the premier mode of governing the biomedical and human life sciences and their technologies: sciences that "have ethics"[1] are widely lauded as good sciences – the epitome of responsible research and innovation. But international human rights law and practice also has something to say about science and technology, and about bioethics. There is, for instance, a European Convention on Human Rights and Biomedicine, which has a preamble invoking "the need for international cooperation so that all humanity may enjoy the benefits of biology and medicine," as well as a range of protocols on matters such as biomedical research, human cloning, and genetic testing.[2] There is also a Universal Declaration on Bioethics and Human Rights,[3] which is part of a series of UNESCO initiatives on science, technology, and rights.[4] And recently, the UN Special Rapporteur on cultural rights expressed the view that it is "essential" for the ethics codes of professional scientific organizations to be "explicitly informed by human rights."[5]

---

[1] C. Thompson, *Good Science: The Ethical Choreography of Stem Cell Research* (Cambridge, MA: MIT Press, 2013), p. 25.
[2] Convention for the Protection of Human Rights and Dignity of the Human Being with regard to the Application of Biology and Medicine (Convention on Human Rights and Biomedicine, as amended) (Oviedo Convention), opened for signature April 4, 1997, entered in force January 12, 1999, ETS 164, Art. 28.
[3] UNESCO Universal Declaration on Bioethics and Human Rights (adopted by acclamation on October 19, 2005, by the General Conference). For commentary, see, e.g., H. A. M. J. ten Have and M. S. Jean (eds.), *The UNESCO Universal Declaration on Bioethics and Human Rights* (Paris: UNESCO, 2009).
[4] See, e.g., UNESCO International Declaration on Human Genetic Data (adopted by acclamation on October 16, 2003, by the General Conference); Universal Declaration on the Human Genome and Human Rights (adopted by acclamation on November 11, 1997, by the General Conference, and endorsed by the UN General Assembly in 1998).
[5] *Report of the Special Rapporteur in the Field of Cultural Rights, Farida Shaheed*, ¶ 53, U.N. Doc. A/HRC/20/26 (May 14, 2012).

For me, the coexistence of bioethics and international human rights law and practice in governing the biomedical and life sciences and their technologies gives rise to two questions. First, how do bioethics and bioethicists see human rights in general, and international human rights law in particular? Second, how do human rights, and international human rights lawyers in particular, see bioethics? These questions, to be clear, are not about turf war or about acclaiming interdisciplinarity. They are about starting over, about engaging in fresh conversation between fields that have a history of shared interests but little in the way of mutual understanding.

Fresh conversation is made easier by fresh content. In this chapter, I suggest international human rights case law as a source of such content. I see its appeal as twofold. First, human rights case law focuses us on what "is," obliging us to look at how the legal and the ethical are figured in judicial practice. Second, it is less likely to alienate than either philosophy, on the one hand, or stronger forms of legalism, on the other.

To develop these claims, I focus on an institution that has been described as "the conscience of Europe,"[6] the European Court of Human Rights. This court hears allegations of violations of the European Convention on Human Rights (ECHR), the "jewel in the crown" of international human rights law.[7] Its decisions bind contracting states, and its views tend to be cited both by other international human rights courts and quasi-courts and by national courts within and beyond Europe. It operates with a doctrine of deference to state decision-making in certain circumstances, which can irritate some who are purist about human rights, but appeals to others as an appropriate way to recognize that universal values are instantiated in specific local contexts.

Taking my lead from bioethics' longstanding interest in both the start of life and biotechnology, I examine the court's case law on assisted reproductive technologies (ARTs)[8] – a total of six cases, four of which include a judgment from the Grand Chamber of the court. The cases range across the technologies of assisted insemination (AI), *in vitro* fertilization (IVF), and pre-implantation genetic diagnosis (PGD),[9] and they focus primarily on limits and prohibitions on access to ARTs.

---

[6] Council of Europe, *The Conscience of Europe: 50 Years of the European Court of Human Rights* (London: Third Millennium Publishing, 2010).

[7] Convention for Protection of Human Rights and Fundamental Freedoms (European Convention on Human Rights, as amended), adopted November 4, 1950, entered into force September 3, 1953, ETS 5.

[8] For a science and technology studies reading of several of these cases, *see* T. Murphy and G. Ó Cuinn, "Taking Technology Seriously: STS as a Human Rights Method," in M. Flear et al. (eds.), *European Law and New Health Technologies* (Oxford: Oxford University Press, 2013), pp. 285–308.

[9] AI is the process whereby sperm from a partner or a donor is placed inside a woman's uterus; IVF is the process of fertilizing an egg, or eggs, outside the human body; PGD is a process used alongside IVF, whereby diagnostic testing is performed on an embryo to determine if it has inherited a serious genetic condition.

In *Evans* v. *United Kingdom*, the first of the cases, the applicant sought access to embryos that were her only chance of having a child to whom she would be genetically related.[10] The embryos had been created using her eggs and the sperm of her then-fiancé before she underwent treatment for ovarian cancer. When the couple separated, he withdrew permission for use or continued storage of the embryos, and under the law in the United Kingdom, this meant the embryos would have to be destroyed. The next case, *Dickson* v. *United Kingdom*, involved a married male prisoner serving a life term and his female partner, who challenged a refusal to allow them to access AI, which was their only option if they were to try for a genetically related child.[11] Next, in *S.H. and Others* v. *Austria*, two married heterosexual couples challenged a legislative ban on IVF with donor gametes.[12] In *Costa and Pavan* v. *Italy*, a married heterosexual couple who carried a serious inheritable genetic condition challenged a blanket ban on PGD.[13] In *Knecht* v. *Romania*, a woman complained *inter alia* of the refusal by medical authorities to allow her to transfer her embryos from the location where they were being stored to a clinic of her choice.[14] Finally, in the most recent case, *Parrillo* v. *Italy*, a woman who no longer wanted to use IVF embryos for reproductive purposes following the unexpected death of her male partner brought a challenge to a legislative ban on donating embryos to scientific research.[15]

So what am I aiming to do by looking at these six cases? Overall, the chapter is more exploratory and exhortatory than normative, which makes it unusual as a piece of legal scholarship. I am stepping back from what the law "ought to be" in order to encourage the fresh conversation that I want to see take place between bioethics and international human rights law and practice. To be openly and actively normative could encourage hubris – a sense that international human rights law and practice "does it better" than bioethics. It would almost certainly encourage allegations of hubris – a sense that international human rights lawyers think they and their field are the best. As we shall see, the relationships between bioethicists and international human rights lawyers are complicated enough; to focus on what a human rights-based approach to ART "ought to be" would only add to the trouble and hinder conversation.

I use the ART cases to address two questions. First, does the European Court of Human Rights have a view on how bioethics relates to international human rights law? Second, do the cases offer food for thought, not just for human rights lawyers,

---

[10] *Evans* v. *United Kingdom*, Eur. Ct. H.R., App. No. 6339/05 (Grand Chamber, April 10, 2007).
[11] *Dickson* v. *United Kingdom*, Eur. Ct. H.R., App. No. 44362/04 (Grand Chamber, December 4, 2007).
[12] *S.H. and Others* v. *Austria*, Eur. Ct. H.R., App. No. 57813/00 (Grand Chamber, November 3, 2011).
[13] *Costa and Pavan* v. *Italy*, Eur. Ct. H.R., App. No. 54270/10 (August 28, 2012).
[14] *Knecht* v. *Romania*, Eur. Ct. H.R., App. No. 10048/10 (October 2, 2012).
[15] *Parrillo* v. *Italy*, Eur. Ct. H.R., App. No. 46470/11 (Grand Chamber, August 27, 2015).

but for bioethicists too? Above all, do they incite us to think about the place of international human rights law in the space of bioethics, and the place of bioethics in the space of international human rights law?[16] These latter questions can, of course, be addressed in other ways as well. For instance, as Lea Shaver demonstrates in Chapter 2, we could concentrate on the right to science, framing any problematic application of technology as an "experiment" and foregrounding ways in which particular bioethics principles could be incorporated into human rights law and practice, and technological design and implementation, in order to prevent or mitigate harm.

This chapter, by contrast, begins by offering a short account of interactions to date between bioethics and international human rights law and practice. It describes the lack of interest displayed by the latter, and the harsh, persistent complaints concerning human rights that come from the former. The next sections offer an introduction to the ART cases. They explore the court's descriptions of the technologies involved and the issues to be addressed. They also examine the court's account of good lawmaking in this field (including how it sees its own role). I also consider the extent to which the court has broached the relationship between the legal and the bioethical. The last section provides a conclusion.

II ENTWINING AND ESTRANGING: A SHORT HISTORY OF THE RELATIONSHIP BETWEEN BIOETHICS AND INTERNATIONAL HUMAN RIGHTS LAW AND PRACTICE

In this section, I sketch how bioethics and international human rights law and practice view each other, and how they are viewed by critics. I want, in particular, to look at how the two domains have both entwined and estranged. To do this, I begin with bioethics, the word itself coming from two Greek words, *bios* (life) and *ethos* (values or morality), and used today in multiple registers. It is difficult to parse these registers, so my plan is to focus on the principal ones – namely, bioethics as an intellectual field and as a governance practice.[17]

I know that immediately some will insist that bioethics is not a field, but a discipline or set of disciplines, an expert domain, or a topic. I am no expert, but I suspect that diverse national histories and diverse priorities for the future of bioethics make it important to accommodate a range of terms. However, because I need to move forward, my hope is that even if the idea of bioethics as a field is controversial, there will be common ground on the following: first, bioethics has an interdisciplinary character; second, philosophy, law, and medicine have been key contributors to that

---

[16] Adapted from S. S. Silbey and P. Ewick, "The Architecture of Authority: The Place of Law in the Space of Science," in A. Sarat, L. Douglas, and M. Umphrey (eds.), *The Place of Law* (Ann Arbor: University of Michigan Press, 2003), pp. 75–108.

[17] See the incisive account in J. Montgomery, "Bioethics as a Governance Practice" (2016) 24 *Health Care Analysis* 3–23.

interdisciplinary character; third, bioethics has had a longstanding focus on patients and research participants; and fourth, more recently, it has been focusing on technology, in particular technologies of human reproduction and enhancement.[18]

As to the second register, which I have labeled "bioethics as a governance practice," here too it might be best to acknowledge that contemporary bioethics hosts a range of such practices,[19] including the ethics committees of hospitals and research institutions. But my particular interest is "public ethics,"[20] by which I mean the practices of political or advisory bodies on bioethics. These bodies may be appointed by one state acting alone or by a group of states; they can also be independent of the state. The United Kingdom's Nuffield Council on Bioethics is one example of the sort of body I have in mind; the Council of Europe's Committee on Bioethics is another. Other examples include the Deutscher Ethikrat (Germany's National Ethics Council), the US Presidential Commission for the Study of Bioethical Issues, the Comitato Nazionale per la Bioetica (Italy's National Bioethics Committee), UNESCO's International Bioethics Committee, and the European Commission's European Group on Ethics in Science and New Technologies.[21]

Interestingly, what I have just said about bioethics could also be said of international human rights law and practice. The latter, in other words, is both an intellectual field and a range of governance practices, with mechanisms and institutions ranging from courts and quasi-courts to the UN Special Procedures.[22] There are other parallels, too. There is, for instance, a range of resonant preoccupations, including "the vulnerable human subject," dignity, autonomy, and relatedly the requirement for informed consent prior to treatment. There are also stories of shared origins. Typically, this includes the Doctors' Trial in Nuremberg in the aftermath of World War II, but there are also country-specific accounts that emphasize particular braidings of bioethics and rights; in the United States, for instance, some commentators point to the significance of the civil rights era for both fields.[23]

---

[18] For a sense of the field, see, e.g., H. Kuhse and P. Singer (eds.), *Bioethics: An Anthology* (Oxford: Blackwell, 1999); B. Steinbock (ed.), *The Oxford Handbook of Bioethics* (Oxford: Oxford University Press, 2007). Both public and global health have become prominent in recent years, and climate change is the focus of at least one recent book: C. C. Macpherson (ed.), *Bioethical Insights into Values and Policy: Climate Change and Health* (Basel: Springer, 2016).

[19] See Montgomery, "Bioethics as a Governance Practice."

[20] See J. Montgomery, "Reflections on the Nature of Public Ethics" (2013) 22 *Cambridge Quarterly of Healthcare Ethics* 9–21 (contrasting ethical deliberation in the academy with its counterpart in the public domain).

[21] For a more complete list, it would be important to add the ad hoc groupings that tend to be convened in the wake of scandal or of growing public concern about a specific aspect of medicine, science, or technology.

[22] See generally A. Nolan, R. Freedman, and T. Murphy (eds.), *The UN Special Procedures System* (Antwerp: Brill, 2017).

[23] See, e.g., D. J. Rothman, "The Origins and Consequences of Patient Autonomy: A 25 Year Retrospective" (2001) 9 *Health Care Analysis* 255–64.

In a further parallel, both bioethics and international human rights law and practice seem rife with insider and outsider criticism. The critics of international human rights law and practice typically denounce it as both hegemonic and replete with out-of-date and inappropriate ideas and techniques. Meanwhile, the critics of bioethics accuse it of becoming "institutionalized ethics," both too focused on conforming to standardized rules and regulations (and too keen to wrap diverse settings and states in a timeless, placeless, one-size-fits-all "global bioethics") and too quick to endorse and promote technology's promised utopias.[24]

What's more, bioethics and international human rights law and practice have faced some similar criticisms. In particular, each has been accused of neglecting lived experience and structural injustice, and of overemphasizing freedom, autonomy, and consent and, more broadly, principles-based reasoning.[25] Typically, the most trenchant critics of each field are blunt about what needs to happen: bioethics (and, equally, international human rights law and practice) should "get out of the way." For some of these critics, "progress" is the preferred guide; for others, it is "justice." To be fair, though, whether we are talking about bioethics or international human rights law and practice, a majority of critics argue for reform and reorientation rather than abandonment.

Parallels are, however, just part of the story; there is considerable contrast too. Thus, bioethicists have spent far more time criticizing and engaging with international human rights law and practice than vice versa.[26] And bioethicists, not surprisingly, are far more skeptical of law than their international human rights counterparts. There are plenty of bioethicists who are also highly skeptical of human rights. More broadly, as the American Association for the Advancement of Science has observed: "Human rights *per se* are often viewed as irrelevant to the practice of ethics."[27] Why is that?

The lack of a convincing theory of human rights is the biggest stumbling block. It has become a particular irritant in the context of the international right to health, where bioethicists and philosophers line up to insist that neither law nor legal scholarship offers a way to make sense of this right in the context of limited resources.[28] Other stumbling blocks to bioethical engagement with rights include

---

[24] See, e.g., C. A. Heimer and J. Petty, "Bureaucratic Ethics: IRBs and the Legal Regulation of Human Subjects Research" (2010) 6 *Annual Review of Law and Social Science* 601–26.

[25] See, e.g., P. Farmer and N. G. Campos, "New Malaise: Bioethics and Human Rights in the Global Era" (2004) 32 *Journal of Law, Medicine and Ethics* 243–51.

[26] For a concise, engaging account of responses to the UNESCO Declaration on Bioethics and Human Rights, see R. Ashcroft, "The Troubled Relationship between Bioethics and Human Rights," in M. D. A. Freeman (ed.), *Law and Bioethics* (Oxford: Oxford University Press, 2008), pp. 31–51.

[27] American Association for the Advancement of Science, Science and Human Rights Coalition, *Intersections of Science, Ethics and Human Rights: The Question of Human Subjects Protection: Report of the Science Ethics and Human Rights Working Group* (February 2012) p. 2.

[28] See, e.g., J. P. Ruger, *Health and Social Justice* (Oxford: Oxford University Press, 2010).

the advocacy and activist orientation of international human rights law and practice, and its substantial focus on the state. There are also some bioethicists who would take the "human" out of human rights, either because they prefer to focus on "persons," or because they see human rights as an obstacle to enhancement technology's promise of "post-humans."[29] Others, focused on the doctor–patient relationship, believe that human rights law risks denuding medical ethics of its capacity to be the "soul of medicine."[30]

Thus, despite the fact that bioethics has engaged with rights-based approaches in the process of broadening its own interests from the doctor–patient relationship and questions concerning new technologies, there is an overall sense that bioethicists do not see human rights as a legitimate form of analysis and reasoning. There is a sense, too, that for those who play a formal part in public ethics, consensus might be hard to achieve if all parties were to come to the table claiming "their rights."

Another clear contrast is that philosophy is central to bioethics, whereas it is less common and less valued in international human rights law and practice. Equally, by contrast with bioethics, international human rights law and practice has had low levels of interest in ART.[31] As far as technology is concerned, advocates of human rights have tended to engage with information and communication technology (including how it can be used to expose human rights violations), and with the uptake of technology in the criminal justice sphere. And as regards reproduction, the central foci of international human rights advocacy have been safe motherhood, forced sterilization, and the vast unmet need for access to modern contraception and associated information and services.[32]

"Safe motherhood" requires engagement with abortion, but as the term itself indicates, the emphasis has been on unsafe abortion as a major public health concern, rather than competing rights claims. States have been encouraged to decriminalize abortion and to guarantee access to quality post-abortion care in order to reduce the high levels of maternal mortality and morbidity that stem from unsafe abortion. Rights arguments have been put aside in order to avoid conflict and dissent

---

[29] See, e.g., J. Harris, "Taking the Human out of Human Rights" (2011) 20 *Cambridge Quarterly of Healthcare Ethics* 9–20.

[30] See, e.g., J. Montgomery, "Law and the Demoralisation of Medicine" (2006) 26 *Legal Studies* 185–210.

[31] If we take case law as a measure of interest, apart from the six ECHR cases, there is just one further case: *Artavia Murillo et al. ("In Vitro Fertilization") v. Costa Rica*, Judgement, Inter-Am. Ct. H.R. (ser. C) No. 257 (November 28, 2012), a decision of the Inter-American Court in which Costa Rica's complete ban on IVF was held to be contrary to the American Convention on Human Rights.

[32] A shift may be under way. See UN Committee on Economic, Social and Cultural Rights, *General Comment no. 22 on the Right to Sexual and Reproductive Health (art. 12)*, U.N. Doc. E/C.12/GC/22, ¶ 21 (May 2, 2016) ("The failure or refusal to incorporate technological advances and innovations in the provision of sexual and reproductive health services, such as medication for abortion, assisted reproductive technologies and advances in the treatment of HIV and AIDS, jeopardizes the quality of care.").

that could stall state action and leave pregnant women at risk of injury and death. Relatedly, when asked to adjudicate upon state practice with respect to abortion, international courts (including the European Court of Human Rights) have avoided definitive statements on the status of the fetus. Their principal focus has been whether states, where abortion is lawful in at least some circumstances, have taken steps to ensure that abortion on these grounds is available in practice.[33]

III A POOR PERCH? INTRODUCING THE ART CASES FROM THE EUROPEAN COURT OF HUMAN RIGHTS

The aim of this section, and the two that follow, is to explore the extent to which the six ART cases from the European Court of Human Rights can open up conversation between bioethicists and international human rights lawyers. The conversation could simply be about the regulation of ART, or it could be about the broader question of the optimum relationship between the legal and the ethical in regulation of the life sciences and their technologies – in particular, the relationship between human rights-based approaches to science and sciences that "have ethics."

On first reading, the cases lack promise in that none of them makes explicit reference to bioethics.[34] There is also little by way of general comment on ART and human rights, and the technological dimensions of assisted reproduction seem to be under the radar too. Thus, IVF is described as a "fast-moving" technology, "subject to particularly dynamic development in science." And its use is said to raise "sensitive moral and ethical issues."[35] Overall, however, the cases lack the level of engagement one finds in cases from other fields of technology, notably *S and Marper* v. *United Kingdom*, in which the court issued a warning that any uptake of technologies in the criminal justice sphere must not "unacceptably weaken" the right to respect for private life, family, home, and correspondence in Article 8 of the ECHR.[36]

The judges also seem to be inconsistent in how they understand the technology involved, with some evidence that different labels align with different levels of scrutiny of state practice. In *S.H.*, which involved an unsuccessful challenge to

---

[33] *See, e.g., Karen Noelia Llantoy Huamán* v. *Peru*, U.N. Doc. CCPR/C/85/D/1153/2003 (Nov. 22, 2005); *Tysiąc* v. *Poland*, Eur. Ct. H.R., App. No. 5410/03 (March 20, 2007); *R.R.* v. *Poland*, Eur. Ct. H.R., App. No. 27617/04 (May 26, 2011).

[34] Interestingly, the court's research division produced a report that collects and classifies what it describes as the court's bioethics cases. *See* Research Division, "Bioethics and the case-law of the Court" (2009, updated 2012), www.coe.int/t/dg3/healthbioethic/texts_and_documents/Bioethics_and_caselaw_Court_EN.pdf.

[35] *See, e.g., Evans*, ¶ 81; *Parrillo*, ¶¶ 169, 174.

[36] *S and Marper* v. *United Kingdom*, Eur. Ct. H.R., App. Nos. 30562/04 and 30566/04 (Grand Chamber, December 4, 2008) ¶ 112; *see also* T. Murphy and G. Ó Cuinn, "Works in Progress: New Technologies and the European Court of Human Rights" (2010) 10 *Human Rights Law Review* 601–38.

Austria's ban on IVF using donor gametes, the joint dissenting opinion spoke exclusively of "medically assisted procreation,"[37] thereby framing ART as medical, not technological, and as an ordinary, understandable way of "assisting nature." The majority, by contrast, used the phrase "artificial procreation," taking its lead from the title of the law in the respondent state.

Overall, the cases display little depth or range. They indicate that the right to respect for private life encompasses the right to respect for the decision to become or not to become a parent, including in the genetic sense.[38] The right of a couple to conceive a child and to make use of medically assisted reproduction for that purpose is protected as an "expression of private and family life."[39] But this barely scratches the surface of the issues that arise. Relatedly, there is no sense of engagement across "different worlds of lived experience" of ART – engagement of the sort that is standard in fields like anthropology and sociology.[40] The court, of course, does not review legislation or practice in the abstract; it focuses on examining the issues raised by the case before it.[41] Nonetheless, there is a sense of missing context and problematic assumptions.

In S.H., for instance, the majority spent a good part of its judgment looking at claims made by the respondent state as to the risks of permitting IVF with donor eggs. But then, with no discussion of whether the context was the same or not, it took its reasoning on this matter and applied it to donor sperm.

Similarly, in *Evans*, in which the applicant unsuccessfully sought access to six embryos that were her only chance of having a child to whom she would be genetically related, the court made no reference to the differing levels of sophistication of egg and sperm freezing – even though the opportunity to freeze eggs, rather than embryos created with her then-fiancé's sperm, would have avoided the difficulties that followed when their relationship broke down and he withdrew his consent to the storage or use of the embryos. The *Evans* court also made no reference to the requirement in domestic law (since amended) that treatment providers take account of the child's "need for a father," a requirement that could well have blocked the applicant's access to treatment if she had asked to use donor sperm to fertilize her eggs, rather than the sperm of her then-fiancé.

---

[37] S.H. and Others, joint dissenting opinion of Judges Tulkens, Hirvelä, Lazarova Trajkovska, and Tsotsoria.
[38] See, e.g., Evans, ¶ 71.
[39] S.H. and Others, ¶ 82; Dickson, ¶ 66.
[40] See, e.g., C. Thompson, *Making Parents: The Ontological Choreography of Reproductive Technologies* (Cambridge, MA: MIT Press, 2005); S. Franklin and C. Roberts, *Born and Made: An Ethnography of Preimplantation Genetic Diagnosis* (Princeton, NJ: Princeton University Press, 2006); S. Franklin and M. C. Inhorn (eds.), "Symposium: IVF – Global Histories" (2016) 2 *Reproductive Biomedicine & Society Online* 1–136. From a legal perspective, see K. Lõhmus, *Caring Autonomy: European Human Rights Law and the Challenge of Individualism* (Cambridge: Cambridge University Press, 2015).
[41] See, e.g., S.H. and Others, ¶ 92; Knecht, ¶ 59.

The more recent decisions are just as frustrating. In *Costa and Pavan*, the applicants, who were carriers of cystic fibrosis, succeeded in their challenge to Italy's blanket ban on PGD. Crucial to the court's decision was the fact that Italian law was inconsistent: the law banned PGD (which allows the identification of genetic abnormalities in embryos created via IVF), but allowed abortion on grounds of fetal malformation. Under Italian law, in other words, the applicants could commence a pregnancy, take a prenatal genetic test, and request an abortion if the test suggested a malformation, but they could not have their embryos screened prior to implantation. For the court, it was entirely wrong that the applicants were put in such a position.

It is, I think, easy to empathize with the court on this, but where I have less empathy is the court's limited engagement with PGD. The court focused exclusively on the inconsistency of Italian law, making no real comment on PGD from a rights perspective. It did note that the applicants had not complained of a violation of a "right to have a healthy child." The applicants were, it said, relying on a more "*confined*" right, namely, "the possibility of using ART and subsequently PGD for the purposes of conceiving a child unaffected by cystic fibrosis, a genetic disease of which they are healthy carriers."[42] It also referred to the applicants' "*desire*" to conceive a child unaffected by cystic fibrosis, and to use ART and PGD to this end, as a "*choice*" that was "a form of expression of their private and family life" and thus within the scope of Article 8 of the ECHR.[43]

This leaves open many questions regarding both discrimination and choice. On different facts, would the court see PGD not as an ART but as a "selective reproductive technology"; i.e., a technology that is not only about overcoming infertility, but also "used to prevent or allow the birth of certain kinds of children"?[44] Evidence suggests that, in places, PGD is being used to select against sex and against disabilities, which compounds powerful prior practices of discrimination. The question that arises is: What would be a human rights-based approach to PGD?

PGD, and ART more broadly, also raise questions concerning choice and its companion, informed consent. How, for instance, would the European Court of Human Rights respond to ethnographic evidence that reproductive responsibility, rather than reproductive choice, can weigh heavily on would-be parents, leading some (who have given consent in the manner expected and promoted by international human rights law) to say they felt they had "no choice" but to use PGD?[45]

---

[42] *Costa and Pavan*, ¶ 53–54 (emphasis added).
[43] Ibid., ¶ 57 (emphasis added).
[44] T. M. Gammeltoft and A. Wahlberg, "Selective Reproductive Technologies" (2014) 43 *Annual Review of Anthropology* 201–16 at 201.
[45] Franklin and Roberts, *Born and Made*; see generally I. Karpin and K. Savell, *Perfecting Pregnancy: Law, Disability, and the Future of Reproduction* (New York: Cambridge University Press, 2012); R. Mykitiuk and I. Karpin, "Fit or Fitting In: Deciding against Normal When Reproducing the Future" (2017) 31 *Continuum* 341–51.

These would-be parents seem to be reporting obligations, not options. Evidence also suggests that prenatal diagnostic tests are often undertaken without full knowledge of their implications and, above all, in the hope of reassurance, which leaves would-be parents ill-equipped for the "choices" that follow. Reading between the lines, it is clear that the court in *Costa and Pavan* saw the applicants as responsible would-be parents and Italian law as deeply irresponsible. That analysis, however, offers no guidance as to what would count as a human rights-based approach to technologies that encourage would-be parents to keep trying for a child and also offer increasing options for identifying, and eliminating, unwanted traits in any future children.

There are two other points. The first concerns what counts as an ART case; the second concerns the court's position on the status of the human embryo. Those who follow the court's case law may ask why my list of ART cases does not include decisions such as *Mennesson v. France*,[46] *Foulon and Bouvet v. France*,[47] and *Paradiso and Campanelli v. Italy*,[48] each of which involved a surrogate pregnancy facilitated by IVF. The question, in other words, is: Where does technology-enabled surrogacy fit in the court's ART frame? My answer is that we need to wait and see, as the most recent decision, *Paradiso and Campanelli*, signals that several members of the court are deeply opposed to surrogacy and keen to limit the scope of Article 8 of the ECHR in this context.

The court's focus in the surrogacy cases is narrow: it is looking only at the post-birth context, in particular at the steps that can legitimately be taken when a child has been born following an overseas surrogacy arrangement and the contracting state believes there has been a contravention of its domestic law.[49] But this did not constrain the four concurring judges in *Paradiso and Campanelli*, who expressed regret that the court had not taken a clear stance against surrogacy. Surrogacy, according to these judges, treats people "not as ends in themselves, but as means to satisfy the desires of other persons." And, whether it is remunerated or not, "[it] is incompatible with human dignity. It constitutes degrading treatment, not only for the child but also for the surrogate mother." It is, the judges conclude, incompatible with the values underlying the Convention.[50] This is a very strong stance. It does not, to be clear, represent the decision of the court. The decision does not rule on the compatibility of surrogacy with the Convention. It simply says that surrogacy

---

[46] *Mennesson v. France*, Eur. Ct. H.R., App. No. 65192/11 (June 26, 2014); *see also Labassee v. France*, Eur. Ct. H.R., App. No. 65941/11 (June 26, 2014).

[47] *Foulon and Bouvet v. France*, Eur. Ct. H.R., App. Nos. 9063/14 and 10410/14 (July 21, 2016).

[48] *Paradiso and Campanelli v. Italy*, Eur. Ct. H.R., App. No. 25358/12 (Grand Chamber, January 24, 2017).

[49] For a broader international human rights perspective on surrogacy, see J. Tobin, "To Prohibit or Permit: What Is the (Human) Rights Response to the Practice of International Commercial Surrogacy?" (2014) 63 *International and Comparative Law Quarterly* 317–52.

[50] *Paradiso and Campanelli*, concurring opinion of Judges De Gaetano, Pinto de Albuquerque, Wojtyczek, and Dedov, ¶ 7; *see also* the separate concurring opinion of Judge Dedov.

raises "sensitive ethical questions" on which there is no consensus amongst contracting states.⁵¹ Hence my claim that we need to wait and see.

As to the status of the human embryo, by and large the court has sought to stay away from this issue, emphasizing that it is the state, not the court, that is the key decision-maker. But several of the ART cases suggest that this might not hold going forward. More importantly, the court could find itself in a tangle as it tries to steer clear of the question of personhood and when the right to life, protected in Article 2 of the ECHR, begins (which it sees as a question for states),⁵² while also protecting women's rights under the Convention and developing Article 8's qualified right to respect for private and family life. The developing idea of human dignity could play a part too, given that the court has said that "the potentiality of [the embryo/fetus] and its capacity to become a person ... require protection in the name of human dignity, without making it a 'person' with the 'right to life' for the purposes of Article 2."⁵³

In *Evans*, the court stated that "the embryos created by the applicant and [her ex-fiancé] [did] not have a right to life within the meaning of Article 2 of the Convention,"⁵⁴ a position described, in extreme terms, by a concurring judge in a later case as "the *Evans* anti-life principle."⁵⁵ Then, in *Costa and Pavan*, the court said that "the concept of 'child' cannot be put in the same category as that of 'embryo',"⁵⁶ while also seeming to indicate that the status of the embryo falls within Article 8(2) of the ECHR, under which infringement of the right to respect for private and family life may be justified if necessary in a democratic society to protect the "*rights and freedoms of others.*"⁵⁷

Most recently, in *Parrillo*, the majority made a range of similarly difficult-to-reconcile statements. On the one hand, and perhaps in line with *Costa and Pavan*, they framed the applicant's embryos as "others"; on the other hand, they went on to say that they were not determining whether "others" extends to embryos.⁵⁸ In another difficult-to-interpret position, the majority said that "the embryos contain the genetic material of [the applicant] and accordingly represent a constituent part

---

⁵¹ *Paradiso and Campanelli*, ¶ 203.
⁵² *See, e.g., Vo v. France*, Eur. Ct. H.R., App. No. 53924/00 (Grand Chamber, July 8, 2004) at ¶¶ 82, 85. Art. 2 provides, *inter alia*, that "Everyone's right to life shall be protected by law." In *X v. United Kingdom*, No. 8416/79, Commission decision of May 13, 1980, Decisions and Reports (DR) 19, p. 244, the former Commission ruled out recognition of an absolute right to life of the fetus under art. 2 (at p. 252, ¶ 19).
⁵³ *Vo v. France*, ¶ 84. Immediately prior to this, the court noted that although there is no European consensus on the status of the embryo and/or fetus, "they are beginning to receive some protection in the light of scientific progress and the potential consequences of research into genetic engineering, medically assisted procreation or embryo experimentation."
⁵⁴ *Evans*, ¶ 56.
⁵⁵ *Parrillo*, concurring opinion of Judge Pinto de Albuquerque, ¶ 31.
⁵⁶ *Costa and Pavan*, ¶ 62.
⁵⁷ Ibid., ¶ 59 (emphasis added).
⁵⁸ *Parrillo*, ¶ 167; see also *Costa and Pavan*, ¶ 59.

of [her] genetic material and biological identity," but also said that the applicant's relationship with her embryos did "not concern a particularly important aspect of [her] existence and identity."[59] It is hard to know why the majority made these clumsy statements. It seems the judges wanted to emphasize the importance of genetic material, while also ensuring that because the case was about the use of human embryos, the final decision would rest with the state, not the court.

Parrillo also featured a claim by the applicant to a property right over her embryos under Article 1 of the EHCR's Protocol 1. This was dismissed with little consideration. The majority concluded on the point by noting that because Article 1 of Protocol 1 has an "economic and pecuniary scope," human embryos cannot be reduced to "'possessions' within the meaning of that provision."[60] The majority also noted that Article 2 was not at issue in the case, which meant that "the sensitive and controversial question of when human life begins"[61] did not have to be examined.

IV THE ETHICS OF THE MARGIN: HOW HAS THE COURT USED THE DOCTRINE OF THE MARGIN OF APPRECIATION IN ITS ART CASES?

The court's particular identity is part of the reason it holds back from definitive statements on the status of the embryo and fetus: it is a supranational human rights institution, not a domestic one. To reflect this, the court has crafted what is called the doctrine of the margin of appreciation. The doctrine allows the court to control the amount of discretion it is willing to give to the contracting state against which a complaint of a rights violation has been levelled. Where the court declares that the margin should be wide, it takes a step back – in essence, positioning itself as the "international judge," less well-placed than national authorities to be the central decision-maker on the issue at stake.[62]

For some, the granting of a wide margin of appreciation is a sign that the court is too keen to constrain itself and too quick to shelter the Convention from the dynamic or evolutive style of interpretation that could, they say, secure its status as a living instrument. Others take a different view: for them, the legitimacy problem stems from cases in which the court limited the margin accorded to the respondent state. These cases, they complain, are evidence of an international court acting like a

---

[59] *Parrillo*, ¶¶ 158 and 174, respectively.
[60] Ibid. *See also* the concurring opinion of Judge Pinto de Albuquerque at n. 32, describing the applicant's property claim as inconsistent with her right-to-private-life claim.
[61] Ibid., ¶ 215.
[62] *See, e.g., Evans*, ¶ 78 (recalling the words of Lord Bingham in the UK case *Quintavalle*: "Where ... there is no consensus within the member States of the Council of Europe either as to the relative importance of the interest at stake or as to how best to protect it, the margin will be wider. This is particularly so where the case raises complex issues and choices of social strategy: the authorities' direct knowledge of their society and its needs means that they are in principle better placed than the international judge to appreciate what is in the public interest."); *see also Knecht*, ¶ 59.

constitutional court and involving itself in issues that are none of its business, issues that ought to be determined at the national level.[63]

For my purposes, the obvious question is: What would bioethicists think of the doctrine and its use-in-practice by the court? I anticipate strong views, given the debate about relativism versus universalism in bioethics. Four of the ART cases, *Evans, S.H. and Others, Knecht,* and *Parrillo,* involved the granting of a wide margin of appreciation to the states concerned. It appears there were two principal reasons for this: first, what the court saw as the absence of consensus within the member states of the Council of Europe on the issues raised ("either as to the relative importance of the interest at stake or as to the best means of protecting it"[64]), and second, and relatedly, the presence of "sensitive moral or ethical issues."[65] In *Evans,* the court provided a third reason for greater deference: "There will usually be a wide margin of appreciation accorded if the State is required to strike a balance between competing private and public interests or Convention rights."[66]

Dickson, by contrast, is a reminder that the court will restrict the margin accorded to a state "where a particularly important facet of an individual's existence or identity is at stake."[67] In *Dickson,* in which a prisoner and his wife successfully challenged the process by which they were refused access to AI facilities, the court indicated that "the choice to become a genetic parent" is one such facet of an individual's existence and identity.[68]

In *S.H. and Others,* however, this was not enough to assist the applicants; their challenge to Austria's ban on using donor gametes for IVF was rejected by the court. In rejecting their challenge, the court mentioned, *inter alia,* their option to seek assisted reproduction abroad and have parenthood recognized at home.[69] Relatedly, in *Parrillo,* having described the applicant's cryopreserved embryos as a "constituent part of [her] identity," a majority of the court later justified the granting of a wide margin of appreciation in part by drawing a distinction between the case before it and cases concerning "prospective parenthood." The majority's position was that

---

[63] For discussion, *see* M. K. Land, "Justice as Legitimacy in the European Court of Human Rights," in H. Cohen et al. (eds.), *Legitimacy and International Courts* (New York: Cambridge University Press, forthcoming); A. Legg, *The Margin of Appreciation in International Human Rights Law: Deference and Proportionality* (Oxford: Oxford University Press, 2012); G. Letsas, "The ECHR as a Living Instrument: Its Meaning and Legitimacy," in A. Føllesdal, B. Peters, and G. Ulfstein (eds.), *Constituting Europe: The European Court of Human Rights in a National, European and Global Context* (Cambridge: Cambridge University Press, 2013), pp. 106–41.

[64] *Parrillo,* ¶ 169.

[65] *Evans,* ¶ 81 (noting also that these issues arise "against a background of fast moving medical and scientific developments"); *see also S.H. and Others,* ¶ 94; *Knecht,* ¶ 59; *Parrillo,* ¶¶ 169, 174.

[66] *Evans,* ¶ 77.

[67] Ibid.

[68] *Dickson,* ¶ 78.

[69] See below text at nn. 90–93.

although the right to donate embryos to scientific research was "important," it was not "one of the core rights attracting the protection of Article 8 of the Convention as it does not concern a particularly important aspect of the applicant's existence and identity."[70]

When the court grants a wide margin of appreciation, it does not abdicate all power and responsibility to the respondent state. As the court explained in *Costa and Pavan*, "the solutions reached by the legislature are not beyond the scrutiny of the Court."[71] In conducting this scrutiny, the court tends to focus on the lawmaking process and whether the impugned law strikes a "fair balance" between all competing private and public interests.[72]

But the judges do not always agree on what is required, and there have been strongly worded dissenting opinions that accuse the majority of using the margin of appreciation in a formal or mechanical way – i.e., granting deference but without the companion scrutiny that would weigh the state's arguments, evidence, and expertise. In *Evans*, for instance, the joint dissenting opinion cautioned that the doctrine must not be used by the court "as a merely pragmatic substitute for a thought-out approach to the problem of proper scope of review."[73] Furthermore, none of the ART cases provides an explanation of the category of "sensitive moral and ethical issues" that will trigger a wide margin of appreciation. This leaves a range of questions. How does the court see the relationship between "moral" issues and "ethical" ones – are the terms synonymous? What is the significance of the adjective "sensitive" (or "delicate," another qualifier used by the court)? And in what circumstances is there "acute sensitivity" or "profound moral views" (both of which were used by the court to justify a wide margin of appreciation in *A, B and C v. Ireland*, which involved a challenge to a restrictive abortion law)?[74]

Those interested in debates concerning universalism versus relativism will want to pay particular attention to the "European consensus," another determining factor with respect to the degree of deference the court is willing to accord to contracting states. The "emerging European consensus" is more interesting still. It featured in *S.H. and Others*, in which the court made reference to "a clear trend in the legislation of the Contracting States towards allowing gamete donation for the purpose of *in vitro* fertilisation."[75] Ultimately, the trend was not used by the court to limit the wide margin of appreciation granted to the respondent state; it was not based on "settled and long-standing principles established in the law of the member

---

[70] *Parrillo*, ¶ 174; see also the concurring opinion of Judge Pinto de Albuquerque, ¶ 34.
[71] *Costa and Pavan*, ¶ 68; see also *S.H. and Others*, ¶ 97.
[72] See, e.g., *S.H. and Others*, ¶ 100.
[73] *Evans*, joint dissenting opinion, ¶ 12; see also *Paradiso and Campanelli*, concurring opinion of Judge Dedov: "For the first time when ruling in favour of the respondent State the Court has placed greater emphasis on values than on the formal margin of appreciation."
[74] *A, B and C v. Ireland*, Eur. Ct. H.R., App. No. 25579/05 (Grand Chamber, December 16, 2010).
[75] *S.H. and Others*, ¶ 96.

States," but rather reflected "a stage of development within a particularly dynamic field of law."[76] Nonetheless, the idea of the emerging consensus as a signaling tool – a shot across the (state) bow, a sign that the future could be different – is intriguing.

One final point about the emerging consensus and its counterpart, the European consensus: both concentrate the court's attention on collectives. Notice that this also can be said of the court's increasing reference to international instruments (which may not have been ratified by all contracting states) and to the evolution of societal thinking. So my question would be: How is this attention to collectives seen by those who criticize international human rights law and practice for being obsessed with the individual and the universal and with crude, decontextualized norms? Bioethicists feature heavily among those critics, so this would be one obvious opportunity for fresh content to stimulate fresh conversation.

V LAW'S CAPACITIES: DOES THE COURT HAVE FAITH IN LAW AND LAWMAKING ON ART?

I turn finally to the question of whether the court has expressed a view on what counts as good law and good lawmaking in the field of ART. As I see it, any such views would contribute to my hoped-for conversation between bioethicists and international human rights lawyers.[77] For instance, does the court share bioethics' enthusiasm for processes or procedures of decision-making? Equally, does it share the view of some bioethicists that law is a way of capping off bioethical consensus, a way of "putting [bioethics'] words and ideas into practice"?[78] Or does it, by contrast, prefer the negative view offered by others, that law generally gets in the way of both scientific progress and decent ethical argument?[79] Relatedly, does the court itself see law expansively, or is there a closing down of law's forms and capacities – a sense that in the end, law talk is about bans, moratoria, and limits of other sorts? Finally, to what extent does the court endorse the popular view that while law might be "marching alongside" medicine, science, and technology, it is "always in the rear and limping a little"?[80]

The ART cases do not provide much explicit detail on the court's views about what counts as good law and good lawmaking in the field of ART. In part the reason for this is that the wide margin of appreciation granted in four of the cases extends, in principle, "both to the State's decision whether or not to enact legislation

---

[76] Ibid.
[77] See also *Vo*, concurring opinion of Judge Costa, joined by Judge Traja.
[78] C. E. Schneider, "Bioethics in the Language of Law" (1994) 24 *The Hastings Center Report* 16–22 at 16–17.
[79] For discussion in the context of the cloning report produced by President George W. Bush's Council on Bioethics, see J. Lezaun, "Self-Contained Bioethics and the Politics of Legal Prohibition" (2008) 4 *Law, Culture and Humanities* 323–38.
[80] *Mount Isa Mines Ltd v. Pusey* (1970) 125 CLR 383, at 395.

governing the use of IVF treatment and, having intervened, to the detailed rules it lays down in order to achieve a balance between competing public and private interests."[81] Still, there are some pointers. In *Evans*, the court encouraged contracting states to take steps to recognize that embryo-freezing produces "an essential difference between IVF and fertilisation through sexual intercourse": "the possibility of allowing a lapse of time, which may be substantial, to intervene between the creation of the embryo and its implantation in the uterus." In light of this, the court advised states to devise legal schemes that take the possibility of delay into account. Such schemes were, it said, "legitimate – and indeed desirable."[82]

Second, in both *Evans* and *S.H. and Others*, the court made clear that even when important private life interests are engaged, it is open to legislators to adopt "rules of an absolute nature" – "[rules] which serve to promote legal certainty."[83] Put differently, legislators do not have to provide for "the weighing of competing interests in the circumstances of each individual case."[84] But this is not a free pass for legislators. For instance, in *S.H. and Others*, having noted that "concerns based on moral considerations or on social acceptability must be taken seriously in a sensitive domain like artificial procreation," the court went on to insist that such concerns "are not in themselves sufficient reasons for a complete ban on a specific artificial procreation technique such as ovum donation."[85] The court then reminded contracting states that, even where a wide margin of appreciation is granted, the impugned domestic legal framework "must be shaped in a coherent manner which allows the different legitimate interests involved to be adequately taken into account."[86]

But what is not clear from *S.H. and Others*, or from the ART cases as a whole, is what these different legitimate interests are. There are, at best, some pointers. Thus, in *Evans*, in determining whether the competing private and public interests had been weighed in a way that achieved fair balance, the majority emphasized two particular public interests: first, that the impugned provisions of the UK's ART law upheld the principle of the primacy of consent, and second, that their "bright-line," no-exceptions approach promoted legal clarity and certainty.[87] On the private interests side, the majority focused on the right to be, or not to be, a parent,

---

[81] See, e.g., *Evans*, ¶ 82; *S.H. and Others*, ¶ 100.
[82] *Evans*, ¶ 84.
[83] Ibid. ¶ 89; *S.H. and Others*, ¶ 110.
[84] *S.H. and Others*, ¶ 110.
[85] Ibid ¶ 100.
[86] Ibid. See also *Dickson*, in which the court determined that the applicants' interest in becoming genetic parents had not been given due weight; the relevant policy on prisoner access to assisted insemination imposed an undue "exceptionality" burden on the applicants; there was no evidence that the minister responsible weighed different legitimate interests, or assessed proportionality in fixing the policy; and because the policy was not part of primary legislation, the legislature also had not engaged in weighting or assessment.
[87] *Evans*, ¶ 89.

including in the genetic sense. In *Parrillo*, the majority noted that the drafting process of Italy's ban on donating human embryos to scientific research had taken account of both "the State's interest in protecting the embryo" and the private interest in exercising the right to individual self-determination "in the form of donating [one's] embryos to research."[88] In *S.H. and Others*, which upheld Austria's ban on donor gametes in IVF, the majority referred to the need to take account of "human dignity, the well-being of children thus conceived and the prevention of negative repercussions or potential misuse."[89] They also noted that there was "no prohibition under [the law] on going abroad to seek treatment of infertility"[90] and having the resulting maternity and paternity recognized upon returning to Austria.

Frustratingly, the majority made no attempt to engage with the particular challenges of cross-border reproduction. The joint dissenting opinion did engage to a degree,[91] and some of the potential challenges were raised by one of the interveners. The latter saw what it described as "procreative tourism" as a "negative side-effect of the ban" – one that left "couples seeking infertility treatment abroad ... exposed to the risk of low quality standards and of suffering from considerable financial and emotional stress."[92] No one, however, mentioned the impact on donors in the state visited by the couple (including the risk of exploitation), or the potential costs to the health service in the home state. Equally, no one mentioned how would-be parents viewed treatment abroad, or whether they cared about recognition of legal parentage.[93] And no one mentioned the difficulties that might be faced by a donor-conceived child seeking information about his or her genetic origin.

More positively, the ART cases as a whole are clear about the court's enthusiasm for process. The court encourages states to be proactive: In upholding Austria's ban on IVF with donor gametes in *S.H. and Others*, the court emphasized that "this area, in which the law appears to be continuously evolving and which is subject to a particularly dynamic development in science and law, needs to be kept under review by the Contracting States."[94] But the court also observed that it was "understandable that the States find it necessary to act with particular caution in the field of artificial procreation"[95] and applauded Austria for an approach that was "careful and cautious."[96] Similarly, in *Evans*, the United Kingdom's ART law was described as "the

---

[88] *Parrillo*, ¶ 188.
[89] *S.H. and Others*, ¶ 113.
[90] Ibid., ¶ 114.
[91] *S.H. and Others*, joint dissenting opinion of Judges Tulkens, Hirvelä, Lazarova Trajkovska, and Tsotsoria, ¶ 13.
[92] *S.H. and Others*, ¶ 74.
[93] For discussion, see E. Jackson et al., "Learning from Cross-Border Reproduction" (2017) 25 *Medical Law Review* 23–46.
[94] *S.H. and Others*, ¶ 118.
[95] Ibid., ¶ 103.
[96] Ibid., ¶ 114. For a bioethical view on the case, *see, e.g.*, W. V. Hoof and G. Pennings, "The Consequences of *S.H. and Others* v. *Austria* for Legislation on Gamete Donation in Europe:

culmination of an exceptionally detailed examination of the social, ethical and legal implications of developments in the field of human fertilisation and embryology, and the fruit of much reflection, consultation and debate."[97] Most recently, in *Parrillo*, the court highlighted the drafting process behind Italy's ban on donating to scientific research human embryos that had been created for the purpose of medically assisted reproduction.[98]

This procedural lens is not without difficulties. In *Parrillo*, the court turned to the drafting process of the impugned law, having said it was about to examine whether a fair balance had been struck between the competing public and private interests at stake. In what followed, however, it focused exclusively on the formalities of the legislative process; there was, in other words, no engagement with the substantive arguments that had emerged during that process. *Parrillo* also raises questions about why, precisely, the court is relying on legislative process. Is it to determine the breadth of the margin of appreciation? Is it an element of a broader proportionality analysis, or is it simply a poor substitute for that analysis?[99] The dissenting opinion of Judge Sajó in *Parrillo* takes a much more probing stance on both the legislative process and the impugned law itself, which suggests that the court's judges do not have an agreed stance on my questions.

There is a sense from some of the ART cases that, unlike some bioethicists and much of the media, the court has faith in law's capacity to handle the challenges of a rapidly evolving set of technologies like ART. We should, of course, expect a court to have faith in law, so what is interesting is the detail. In *S.H. and Others*, the court seemed to suggest that law could handle what Austria described as the risks of "harm to women" arising from egg donation (in particular, the risk of exploitation of donors). In this regard, it drew attention to the success of adoption law, though it accepted that "the splitting of motherhood between a genetic mother and the one carrying the child differs significantly from adoptive parent-child relations and has added a new aspect to this issue."[100]

There have also been indications of what irks the court in the context of law, law-making, and application of law – of what it sees as falling short of democratic legitimacy. In *Knecht v. Romania*, although the court dismissed the complaint, it took the time to describe the Romanian regulator's attitude as "obstructive and oscillatory."[101] In *Costa and Pavan*, it was the incoherence of Italian law that irked the court. The applicants, a married couple carrying cystic fibrosis, had been barred

---

An Ethical Analysis of the European Court of Human Rights Judgments" (2012) 25 *Reproductive BioMedicine Online* 665–69.

[97] *Evans*, ¶ 86.
[98] *Parrillo*, ¶¶ 184–88.
[99] *See generally* M. Saul, "The European Court of Human Rights' Margin of Appreciation and Processes of National Parliaments" (2015) 15 *Human Rights Law Review* 745–74.
[100] *S.H. and Others*, ¶ 105.
[101] *Knecht*, ¶ 60.

from using IVF and PGD to screen their embryos so that only healthy ones could be implanted. Were they to conceive, however, the law permitted them to terminate the pregnancy on the grounds of fetal genetic condition. This, the court said, "caused a state of anguish for the applicants,"[102] and, as noted earlier, it went on to hold that Italy's ban on IVF with PGD constituted a disproportionate interference with the applicants' right to respect for private and family life.

VI CONCLUSION

In drawing to a close, two points stand out from my review of the court's ART cases. First, the court invokes "the ethical" but largely leaves us guessing about what would constitute a human rights-based approach to the relationship between "the legal" and "the bioethical." Second, and relatedly, the court leaves us guessing as to what would constitute a human rights-based approach to ART.

These gaps are problematic. Rights holders and duty bearers would benefit from clarity on what is a legal responsibility and what is a moral or an ethical one. But I suggest we see the gaps as productive, too. They allow us to think about the scope and content of rights and responsibilities in the ART field, without stalling on what is legal and what is ethical. More broadly, they demonstrate that international human rights case law is a starting point for discussion, not a full stop – that it is much more contested than the stereotypes of freedoms, entitlements, and duties suggest, much more provisional than the word "law" indicates, and much more interwoven with the domestic than the word "international" implies.[103]

I am not saying that there is no need to grasp what is, or could be, unique to human rights law, just that it would be useful to approach that question in new ways. With that in mind, and focusing on the contribution that international human rights law and practice could make to "good science," this chapter has explored whether the ART case law of the European Court of Human Rights could generate fresh conversation between human rights lawyers and bioethicists. My conclusion is that it could.

The ART cases, as we have seen, do not set down what "ought to be." They flag the centrality of the right to respect for private life, which covers the decision to become or not to become a parent, including in the genetic sense, and to use medically assisted procreation to that end. In a less-developed manner, the cases flag the relevance of other Convention rights (notably the rights to life and to respect for family life), the Convention's guarantee of nondiscrimination,[104] and the state's

---

[102] *Costa and Pavan*, ¶ 66.
[103] *See* C. McCrudden, "The Pluralism of Human Rights Adjudication," in L. Lazarus, C. McCrudden, and N. Bowles (eds.), *Reasoning Rights: Comparative Judicial Engagement* (Oxford: Hart Publishing, 2014), pp. 3–27.
[104] Article 14's nondiscrimination guarantee is "parasitic," i.e., it only prohibits discrimination in the enjoyment of other Convention rights. Recently, however, the European Court of Human

interest in protecting "the rights and freedoms of others." Considerable ebb and flow should be expected as the court and states continue to grapple with the balance between these rights and interests.

The court's particular identity as a supranational human rights court is flagged by the cases, too. ART is presented as a field that raises "sensitive moral and ethical issues," which is one of the triggers for a wide margin of appreciation. But a wide margin is not a free pass, nor is it guaranteed. It can be limited by a European consensus (and potentially by an emerging one as well[105]). It can also be limited where a particularly important aspect of an individual's existence or identity is at stake, which includes the right to respect for the decision to be or not to be a parent in the genetic sense. And even where a wide margin is seen as apt, the court has the power to scrutinize the democratic legitimacy of the impugned domestic measure. Each of these elements offers scope for study and, of course, for different views.

There are other potentially interesting elements that I have not addressed, such as: Did the gender of the judge count in any of the court's ART cases?[106] In *S.H. and Others*, four women judges authored the biting dissent,[107] which suggests this question merits further study. Equally, how many of the court's ART cases have featured third-party interventions, and is there a pattern in terms of who is intervening, in what types of cases, and with what sorts of claims?

Case selection is relevant too. For instance, would this chapter have read differently if I had framed the court's ART cases as part of medical jurisprudence,[108] reproductive rights jurisprudence,[109] the jurisprudence of pregnancy,[110] or even a jurisprudence of kinship? And what would have emerged if I had looked at all of the court's bioethical cases, which address issues including consent to medical examination and treatment, abortion, and assisted suicide? Equally, what would have

---

Rights has been developing this right by reference to views on substantive equality from other courts and in scholarship; *see* S. Fredman, "Emerging from the Shadows: Substantive Equality and Article 14 of the European Convention on Human Rights" (2016) 16 *Human Rights Law Review* 273–301.

[105] *See Goodwin v. United Kingdom*, Eur. Ct. H.R., App. No. 28957/95 (Grand Chamber, July 11, 2002).

[106] *See* S. H. Vauchez, "More Women – But Which Women? The Rule and the Politics of Gender Balance at the European Court of Human Rights" (2015) 26 *European Journal of International Law* 195–221.

[107] *S.H. and Others*, joint dissenting opinion of Judges Tulkens, Hirvelä, Lazarova Trajkovska, and Tsotsoria. Women judges were part of the majority opinion, but the dissent was authored exclusively by women.

[108] *See* J. Harrington, *Towards a Rhetoric of Medical Law* (London: Routledge, 2016).

[109] *See* L. Oja and A. E. Yamin, "'Woman' in the European Human Rights System: How Is the Reproductive Rights Jurisprudence of the European Court of Human Rights Constructing Narratives of Women's Citizenship?" (2016) 32 *Columbia Journal of Gender and Law* 62–95.

[110] *See, e.g.*, M. Ford, "*Evans v. United Kingdom*: What Implications for the Jurisprudence of Pregnancy?" (2008) 8 *Human Rights Law Review* 171–84.

emerged if either a particular concept (say, dignity or autonomy[111]) or a particular practice (say, cross-border care or informed consent) had been the focal point? And, changing tack again, what would we have seen if we had looked at all of the court's technology cases? In particular, do we expect the question of ethics to feature in the court's cases on other technologies, and if not, why not?

These, I suggest, are just some of the questions that arise. Moreover, if the aim is to delineate the relationship between the legal and the bioethical, and relatedly to capture what is distinctive about the authority of international human rights law, it will be important to locate the European Court of Human Rights within a broader frame. That frame should include not just other human rights actors, both domestic and international, but also other "palaces of hope"[112] from the field of international law that play a part in the regulation of the life sciences and their technologies. And finally, because human rights practice is not exclusive to legal actors, we should also be exploring how national bioethics bodies (such as those mentioned earlier in this chapter) and professional scientific organizations view the bioethics–human rights relationship. These explorations are not standard fare. They could, however, engage bioethicists and international human rights lawyers in ways that are important and overdue. And over time, they could contribute to "good science" as well.

---

[111] *See, e.g.*, Lõhmus, *Caring Autonomy*; J.-P. Costa, "European Dignity and the European Court of Human Rights," in C. McCrudden (ed.), *Understanding Human Dignity* (Oxford: Oxford University Press, 2013), pp. 393–402; C. McCrudden, "Human Dignity and Judicial Interpretation" (2008) 19 *European Journal of International Law* 655–724.

[112] R. Niezen and M. Sapignoli (eds.), *Palaces of Hope: The Anthropology of Global Organizations* (New York: Cambridge University Press, 2017).

5

Drones, Automated Weapons, and Private Military Contractors

Challenges to Domestic and International Legal Regimes Governing Armed Conflict

*Laura A. Dickinson**

I INTRODUCTION

The development of unmanned aerial vehicles and increasingly autonomous weapons systems is radically reshaping the nature of modern warfare. The US military now depends on unmanned aerial systems, often referred to as drones, to provide intelligence and surveillance, while armed drones that can drop bombs on terrorists form a core part of the US arsenal in the war on terror. From the long-standing Phalanx system, deployed by the US Navy to automatically detect and neutralize missiles that breach a warship's protective envelope, to the Counter Rocket, Artillery, and Mortar (C-RAM) system, the US Army's comparable land-based defense, to the X-47B drone that can fly by itself during takeoff and landing, the United States has relied increasingly on weapons systems that are unmanned or have at least some autonomous capability. Other countries have developed such systems as well, from Israel's Harpies, which autonomously ferret out enemy radar and then dive-bomb to destroy them, to South Korea's robotic sentries protecting the border.[1]

In addition to altering the nature of warfare on the ground and in the air, the use of these new weapons systems has shaken up existing legal frameworks, both domestic and international, regulating the use of force abroad. Scholars and policy-makers have frequently grappled with the legal implications of deploying unmanned vehicles and increasingly autonomous weapons systems.[2] For example, some have discussed the legality of US efforts to target individual terrorists for

---

* Thanks to Bryan Cenko and Jiyoon Moon for useful research assistance.
[1] For a discussion of these systems, see text and notes below.
[2] J. Elsea, *Legal Issues Related to the Lethal Targeting of U.S. Citizens Suspected of Terrorist Activities* (CRS Legal Memorandum) (Washington, DC: Congressional Research Service, 2012), https://perma.cc/M26G-PR2W.

assassination, often through drone strikes.[3] Others have focused specifically on the question of whether such targeted killings can be justified anywhere outside the battlefields of Afghanistan and Iraq.[4] Still others debate whether drones are worse (or better) than conventional weapons at limiting civilian casualties.[5]

Drones and automated weapons also raise broader questions about the asymmetry inherent in this sort of warfare, where one side has few if any casualties and the other is subject to multiple strikes without an effective way to respond. Such asymmetry, some argue, violates fundamental principles of humanity undergirding the entire body of International Humanitarian Law (IHL).[6] Finally, debates under US domestic law focus

[3] See, e.g., J. Daskal, "The Geography of the Battlefield: A Framework for Detention and Targeting Outside the Hot Conflict Zone" (2013) 161(5) *University of Pennsylvania Law Review* 1165–234 (surveying approaches of legal scholars); A. Hudson, "Beyond the Drone Debate: Should US Military and CIA Be Judge, Jury, and Executioner?" *Truthout*, June 2, 2015, https://perma.cc/8EE4-4FEK (discussing reactions by the legal world, noting that Philip Alston, Christof Heyns, and Naz Modirzadeh have condemned drone strikes as outside the norms of international humanitarian law).

[4] See, e.g., M. O'Connell, "Unlawful Killing with Combat Drones: A Case Study of Pakistan 2004–2009," in S. Bonitt (ed.), *Shooting to Kill: The Law Governing Lethal Force in Context* (London: Hart Publishing, 2012) (arguing that drone strikes are not a lawful use of force, that there is no armed conflict in Pakistan, and that drones are ineffective at killing only the intended target); *see also* Daskal, "The Geography of the Battlefield."

[5] For example, some suggest that tallies undercount civilian casualties and that civilians are being unnecessarily sacrificed in failed attempts to reach terrorists. See, e.g., C. Friedersdorf, "Flawed Analysis of Drone Strike Data Is Misleading Americans," *The Atlantic* (July 18, 2012), https://perma.cc/PEY3-DTLB; "You Never Die Twice: Multiple Kills in the US Drone Program," *Reprieve*, https://perma.cc/KVY6-8UZ8. Others argue that the numbers of civilian casualties have dropped as targeting technologies have improved. See P. Bergen and J. Rowland, "Civilian casualties plummet in drone strikes," CNN, July 14, 2012, https://perma.cc/ZT4X-8M6G. Overall, the New America Foundation has calculated that Pakistani casualties ranged between 2,003 and 3,321 from 2004 to April 2013. See P. Bergen, *Drone Wars: The Constitutional and Counterterrorism Implications of Targeted Killing: Testimony presented before the U.S. Senate Committee on the Judiciary, Subcommittee on the Constitution, Civil Rights and Human Rights*, 113th Cong. (2013) (testimony of Peter Bergen, director of the National Security Studies Program, New America Foundation), https://perma.cc/64HV-B9AE. In Yemen, where the number of drone strikes has increased, estimates of civilian casualties in 2013 range from 467 to 674, the vast majority of which occurred under the Obama administration. See Bergen, *Drone Wars*. In late 2013, a drone strike mistakenly attacked a wedding party in Yemen, killing numerous unarmed civilians. See J. Serle, "American drone suspected in wedding-day massacre," *Salon*, December 16, 2013, https://perma.cc/FWF6-87FJ. And in April 2015, the United States government was forced to admit that two civilians had been killed as a result of a mistaken drone strike. See J. Diamond, "U.S. drone strike accidentally killed 2 hostages," CNN, April 23, 2015, https://perma.cc/S6KF-2E73.

[6] *See generally* R. Sparrow, "War Without Virtue?" in B. Strawser (ed.), *Killing by Remote Control* (Oxford: Oxford University Press, 2013) (questioning the effects drone warfare will have on the military); B. Strawser, "Moral Predators: The Duty to Employ Uninhabited Aerial Vehicles" (2010) 9(4) *Journal of Military Ethics* 342–68 (arguing that drones are justified because they allow for greater protection of the military, providing that the drones do not interfere with the military's ability to make just judgment calls); J. McMahan, "The Ethics of Killing in War" (2004) 114(3) *Ethics* 693–733 (arguing that combatants fighting with an unjust cause require no reciprocity morally, so the drone program violates no ethics); J. McMahan,

largely on whether the Constitution can be invoked extraterritorially to protect the rights of citizens (or even noncitizens) targeted in such strikes, given that so far individuals have only been targeted while outside the territorial borders of the United States.[7]

Concurrent with this trend toward the use of weapons that are more autonomous, the United States and many other countries have also privatized a broad array of military and security functions, using private contractors to an arguably unprecedented degree.[8] For example, at the high point of the conflict in Iraq and Afghanistan, the US government employed more than 260,000 contractors, which at times exceeded the total number of US military personnel deployed in those two countries.[9] Such contractors have performed a myriad of roles, from constructing military bases and refugee camps to cutting soldiers' hair, serving meals in mess halls, maintaining weapons on the battlefield, interrogating detainees, and guarding diplomats and military facilities.[10] Scholars and policy-makers have confronted the legal implications of this shift, and many have called for increasing oversight.[11]

Yet few, if any, studies have addressed the potentially incendiary mix of legal challenges arising from the *combined* use of drones and automated weapons, on the one hand, and private contractors, on the other. These trends have occurred side-by-side and even reinforce each other, because drones and automated weapons themselves often depend on contractors to function. Each twenty-four-hour combat air patrol of the US armed Predator and Reaper drones, for example, requires at least 350 people, many of whom are contractors. Contractors often invent and produce autonomous systems, and contractors may also be responsible for maintaining and operating these systems on the battlefield.

This chapter charts the rapid and intertwined growth of unmanned and increasingly autonomous weapons, on the one hand, and private military and security contractors, on the other. And it grapples with the particular challenges this combination of forces creates under both domestic and international law. The first part describes the increased use of drones and more fully autonomous weapons as well as

"On the Moral Equality of Combatants" (2006) 14(4) *Journal of Political Philosophy* 377–496 (same).
[7] *Memorandum for the Attorney General re: Applicability of Federal Criminal Laws and the Constitution to Contemplated Lethal Operations Against Shaykh Anwar al-Aulaqi* (Washington, DC: Office of Legal Counsel, 2010), https://perma.cc/AKQ5-GBVQ; *see also* Daskal, "The Geography of the Battlefield."
[8] *See, e.g.*, D. Avant, *The Market for Force: The Consequences of Privatizing Security* (Cambridge: Cambridge University Press, 2005); L. Dickinson, *Outsourcing War and Peace: Preserving Public Values in a World of Privatized Foreign Affairs* (New Haven, CT: Yale University Press, 2011); S. McFate, *The Modern Mercenary: The Rise of the Privatized Military Industry* (Oxford: Oxford University Press, 2014); P. Singer, *Corporate Warriors: The Rise of the Privatized Military Industry* (Ithaca, NY: Cornell University Press, 2007).
[9] *Transforming Wartime Contracting: Controlling Costs, Reducing Risk – Final Report to Congress* (Arlington, VA: Commission on Wartime Contracting, 2011).
[10] Ibid.
[11] Dickinson, *Outsourcing War and Peace*; *see also* discussion below.

the growing role of contractors in developing and operating these systems. The next part discusses the destabilizing impact that this trend is having on one of the foundations of the US constitutional framework itself: the allocation of power between the president and Congress in deciding whether or not to use force overseas. By reducing the political cost of war, the rise of these weapons systems and the growing use of contractors have together emboldened the president to deploy force unilaterally. At the same time, these twin trends have made possible legal arguments that the use of force is itself not significant enough to warrant congressional involvement. Such arguments are premised upon the idea that if there are few US military casualties there is less likely to be a war for constitutional purposes, or hostilities within the meaning of the War Powers Resolution. This reality is made possible in part because the use of autonomous weapons and contractors radically reduces the official casualty count.

The final part charts the similarly disruptive impact these trends have had on accountability and oversight under IHL. Together, the use of autonomous weapons and privatization have fragmented decision-making over the use of force, rendering accountability for violations of IHL principles much more difficult to achieve. Accordingly, from the perspective of both domestic and international law, the rise of autonomous weaponry combined with the increased use of contractors raises significant challenges to the rule of law during armed conflict and the protection of human rights in noncombat contexts.

II THE RISE OF DRONES AND AUTONOMOUS WEAPONS, AND THE INCREASING ROLE OF MILITARY CONTRACTORS

A Drones

Unmanned aerial vehicles (UAVs), popularly referred to as "drones," are now a seemingly permanent and pervasive fixture of the United States' strategy to combat terrorism. These remote-controlled flight technologies are probably best known for their ability to drop bombs on terrorism suspects without putting US troops in harm's way. Less discussed is their even bigger role in conducting surveillance. Such remote surveillance has radically reshaped and augmented military capabilities, both because UAVs can provide so-called enhanced situational awareness without endangering US troops on the ground and because they gather an enormous store of information that can then be used for multiple strategic purposes.[12] Given these important benefits, it is not surprising that drone use has surged

---

[12] *See* J. Levs, "CNN Explains: U.S. Drones," CNN, February 8, 2013, https://perma.cc/QUK5-UH6M; J. Gertler, *U.S. Unmanned Aerial Systems* (CRS Report No. R42136) (Washington, DC: Congressional Research Service, 2012), https://perma.cc/HS6H-UU9A.

dramatically in recent years.¹³ The 2016 fiscal budget included $2.9 billion (US billion) for known costs in drone research, development, and procurement, and the Pentagon has at least 7,000 drones under its control.¹⁴ Drone technology has become "fundamental" to military operations, and demand for the technology far outstrips the actual supply.¹⁵

While the US overseas drone program has garnered significant public scrutiny, the integral role of contractors in developing and operating drones has attracted much less attention. It should be no surprise that leading defense firms develop and manufacture the broad range of drones that the United States uses, including the enormous high-flying RQ-4 Global Hawk (Northrup Grumman), with its powerful intelligence, surveillance, and reconnaissance capabilities; the deadly MQ-9 Reaper and MQ1 Predator (General Atomics), tactical strike and imaging drones that form the core of the US fleet unleashing bombs on terrorists; and tiny, handheld, battery-powered drones like the Dragon Eye (AeroVironment) and FQM Pointer (AeroVironment), which soldiers release from their backpacks to gather information.

But contractors also play a central role in the operation of drones, which for the larger models requires enormous teams of people. For example, each twenty-four-hour surveillance mission of the Global Hawk requires roughly 300–500 people. Each twenty-four-hour combat air patrol of the slightly smaller Predator and Reaper could require up to 350 people. Even the tiny handheld Dragon Eye and FQM Pointer depend on background teams of maintenance staff and intelligence analysts.

Contractors participate in almost every step of the complex array of tasks that make up a drone mission. Once the government purchases a drone, the manufacturer often provides continuing maintenance and logistical support services.¹⁶ For example, in addition to assembling drones and providing maintenance for them, contractors have also reportedly fueled them and loaded ammunition onto them at military bases overseas.¹⁷ Contractors also help operate drones, including, at times,

---

13 *U.S. Unmanned Systems Integrated Roadmap (fiscal years 2009–2034)* (Washington, DC: US Department of Defense, 2009), p. 2, https://perma.cc/T88M-BYMA; *Statement of Commander Gerald Lloyd J. Austin III before the House Armed Services Committee*, 114th Cong. (2015), p. 41, https://perma.cc/M856-L7VL; see also Bergen, *Drone Wars*, p. 12 (chart outlining the differences between the number of air strikes and drone strikes comparing 2002 through 2013).
14 "Understanding Drones," Friends Committee on National Legislation, https://perma.cc/9AV8-VWTP (drawing numbers from the 2016 budget and military reports).
15 Statement of Commander Gerald Lloyd J. Austin III; C. Whitlock, "How crashing US drones are exposing secrets about US war operations," *The Washington Post*, March 25, 2015, https://perma.cc/SP5F-S88P.
16 D. Cloud, "Civilian contractors playing key roles in U.S. drone operations," *Los Angeles Times*, December 29, 2011, https://perma.cc/6VSZ-P3M3.
17 A. Sundby, "CIA Hired Blackwater to Arm Afghan Drones," CBS News, August 21, 2009, https://perma.cc/6PSU-GGQL; J. Risen and M. Mazzetti, "C.I.A. Said to Use Outsiders to Put Bombs on Drones," *The New York Times*, August 20, 2009, www.nytimes.com/2009/08/21/us/21intel.html; Cloud, "Civilian Contractors."

steering them.[18] The Air Force has stated that while contractors do not "pilot" drones during the targeting portion of their missions, contractors do assist with other aspects of flight, such as takeoff and landing. In addition, the ever-present demand for drone missions, particularly as the conflict against ISIS escalated in Iraq and Syria, has led to pilot shortages and a growing need for contractors.

Finally, contractors also collect and analyze the data that drones gather.[19] In many cases, they review the live footage captured by drones.[20] Indeed, contractors frequently become the subject matter experts and can more easily distinguish patterns on the ground than their military counterparts.[21] Thus, while the Air Force has said that contractors do not make targeting decisions, contractor analysis of data can often shade into decision-making. For example, a 2013 *Washington Post* report suggested that a contractor who reviewed live footage from a drone essentially directed a Navy Seal to a particular target.[22] A more recent report by the Bureau of Investigative Journalism indicates that contractors working for the Department of Defense not only analyze the data from drones, but also provide evaluative judgments that may affect targeting decisions.[23]

B *Autonomous Weapons*

The development of increasingly autonomous weapons systems is also reshaping modern warfare. Although there are multiple potential definitions of "autonomous," the International Committee for the Red Cross defines an autonomous weapon as one "that is able to function in a self-contained and independent manner although its employment may initially be deployed or directed by a human operator" and that can "independently verify or detect a particular type of target object and then fire or detonate."[24] An autonomous weapon system can also "learn or adapt its functioning

---

[18] K. D. Clanahan, "Drone-Sourcing? United States Air Force Unmanned Aircraft Systems, Inherently Governmental Functions, and the Role of Contractors" (2012) 22 *Federal Circuit Bar Journal*, https://perma.cc/C238-2JVA; A. Fielding-Smith et al., "Revealed: Private firms at heart of US drone warfare," *The Guardian*, February 12, 2015, https://perma.cc/XG4Q-XVKL; see also A. Fielding-Smith and C. Black, "Pentagon's 'Insatiable Demand' for Drone War Intelligence," The Bureau of Investigative Journalism, July 30, 2015, https://perma.cc/F95V-6SZT (describing the many layers of a drone mission, including the number of people reviewing the instructions, relaying information, and operating the chain of command).

[19] Cloud, "Civilian Contractors"; Fielding-Smith et al., "Revealed."

[20] See Fielding-Smith et al., "Revealed"; A. Fielding-Smith and C. Black, "Civilians who are drone pilots' extra eyes," *The Guardian*, August 2, 2015.

[21] Fielding-Smith et al., "Revealed."

[22] D. Priest, "NSA growth fueled by need to target terrorists," *The Washington Post*, July 21, 2013, https://perma.cc/4JYS-BJQ6.

[23] Fielding-Smith and Black, "Pentagon's 'Insatiable Demand'."

[24] "International Humanitarian Law and the Challenges of Contemporary Armed Conflicts," *Report of the 31st International Conference of the Red Cross and Red Crescent* (Geneva: International Committee of the Red Cross, 2011), p. 39, https://perma.cc/ML3F-RUZB. Other organizations have defined autonomous weapons differently. For example, The US

in response to changing circumstances in the environment in which it is deployed."²⁵ Most drones would not qualify as autonomous weapons, because human operators control them remotely. Yet many weapons systems, including a few types of drones, can now function autonomously to some degree. While fully autonomous weapons may still be decades away, current systems deployed on land, at sea, and in the air can now operate with an extraordinary degree of independence from individualized human decision-making.

In practice, militaries have deployed these systems primarily for defensive purposes, and they have retained a degree of human involvement even when the systems could operate without human participation. More than thirty countries are using such systems to defend military vehicles and bases.²⁶ They typically protect predetermined areas by detecting incoming munitions, such as mortars, rockets, or other projectiles, and then automatically responding by neutralizing these objects. Governments tend to deploy them in fixed, rather than mobile, positions in unpopulated and relatively simple and predictable environments, such as at sea or in remote areas, and they typically target weapons and objects rather than persons. In most cases, the reaction time required for engagement is so short that human interaction with the machines is minimal. Often, the human being is only given a brief opportunity to accept or reject the system's choice of action before it deploys, or to override a course of action that the machine will otherwise take automatically.²⁷

> Department of Defense (DoD) distinguishes between fully and semi-autonomous weapon systems based on the machine's ability to make an autonomous choice. It defines fully autonomous weapon system as a "weapon system that, once activated, can select and engage targets without further intervention by a human operator." See "DoD Directive on Autonomy in Weapon Systems," ICRAC, November 27, 2012, https://perma.cc/U46R-9WUS. Human Rights Watch, by contrast, focuses on human involvement, classifying autonomous weapons as either "human-in-the-loop weapons," "human-on-the-loop weapons," or "human-out-of-the-loop weapons." Human-in-the-loop weapons are "robots that can select targets and deliver force only with a human command," while human-on-the-loop weapons are "robots that can select targets and deliver force under the oversight of a human operator who can override the robots' actions." Finally, human-out-of-the-loop weapons are "capable of selecting targets and delivering force without any human input or interaction." B. Docherty, "Losing Humanity: The Case against Killer Robots," Human Rights Watch, November 19, 2012, p. 2, https://perma.cc/N77J-EASX. The United Nations defines autonomous weapons as systems that "once activated, can select and engage targets without further intervention by a human operator." *Report of the Special Rapporteur on Extrajudicial, Summary or Arbitrary Executions, Christof Heyns*, U.N. Doc. A/HRC/23/47 (April 9, 2013).

[25] Ibid., p. 39; *see also* "Autonomous weapons: States must address major humanitarian, ethical challenges," International Committee of the Red Cross, February 9, 2013, https://perma.cc/PD33-Z5M7.
[26] P. Scharre and M. Horwoit, "An Introduction to Autonomy in Weapon Systems," Center for a New American Security (February 13, 2015), https://perma.cc/3BDX-SGW6.
[27] These countries are: Australia, Bahrain, Belgium, Canada, Chile, China, Egypt, France, Germany, Greece, India, Israel, Japan, Kuwait, the Netherlands, New Zealand, Norway, Pakistan, Poland, Portugal, Qatar, Russia, Saudi Arabia, South Africa, South Korea, Spain, Taiwan, the United Arab Emirates, the United Kingdom, and the United States. Ibid.

A semi-autonomous offensive system, which mostly consists of projectiles, tends to require a human operator to launch,[28] but then autonomously guides the weapon to a preselected target, either by relying on passive sensors that respond to signals from the environment or by using active sensors that send out signals and look for a signal in return.[29] Known as guided munitions, these projectiles can be divided into two categories: "go-onto-target" projectiles, designed to hit a particular target, and "go-onto-location" projectiles, designed to hit a particular geographic location.[30] In some cases, humans can "control, abort, or retarget" such weapons in flight, but other times the projectiles cannot be adjusted once launched.[31]

As with drones, private military contractors have been major players in inventing and producing these autonomous systems.[32] But perhaps even more significant is the potential role contractors play in testing, maintaining, and operating them. The military typically hires private contractors to test autonomous weapons systems in order to determine whether they will be effective and whether any changes need to be made before deployment. Sometimes contactors actually test their products in the field. For example, integrated teams of Northrop Grumman contractors and Navy personnel have tested the unmanned X-47B to prove that it is capable of operating in tandem with manned aircraft.[33] Based on such tests, private companies often receive contracts to upgrade autonomous weapons systems.[34] Finally, in some cases private contractors provide maintenance on autonomous weapons being used in the field in active deployment situations. For example, the US Army contracted with Northrop Grumman to install and maintain the C-RAM interceptor system at forward operating bases during Operation Enduring Freedom and the ongoing US mission in Iraq.[35] And Raytheon is currently contracted to provide ongoing technical and support services to the US Army on this system.[36]

---

[28] Scharre and Horwoit, "An Introduction to Autonomy in Weapon Systems," pp. 8–10.
[29] Ibid.
[30] Ibid.
[31] Ibid.
[32] See, e.g., P. Jones, "Aegis Combat System's Newest Baseline Demonstrates Over the Horizon Capability in Series of Three Tests," Lockheed Martin, July 8, 2014, https://perma.cc/APE6-5YUL; "Patriot Missile Long-Range Air-Defence System, United States of America," army-technology.com, https://perma.cc/3JUS-FZE7; "Terminal High Altitude Area Defense, United States of America," army-technology.com, https://perma.cc/T88S-Y4NG; "Taranis is an unmanned combat aircraft system advanced technology demonstrator programme," BAE Systems, https://perma.cc/K3J8-46AJ.
[33] S. J. Freedberg Jr., "X-47B Drone & Manned F-18 Take Off & Land Together in Historic Test," Breaking Defense, August 17, 2014, https://perma.cc/5JZE-9NHR.
[34] For example, Raytheon received a $115.5 million contract in 2014 to "remanufacture, overhaul and provide upgrades" to the MK 15 Phalanx system. "Raytheon awarded $115.5 million Phalanx upgrade contract," Raytheon, June 11, 2014, https://perma.cc/DA5P-U8MS.
[35] "Northrop wins US Army's C-RAM contract," army-technology.com, January 31, 2012, https://perma.cc/WH37-44ZM.
[36] "Raytheon receives $109 million contract for Patriot Air and Missile Defense System: US and international Patriot partners strengthen defense against evolving threats," Raytheon, September 10, 2014, https://perma.cc/B462-8D4Q.

Contractors have even been hired to provide support to the autonomous systems being used in the field. Lockheed Martin has served as the AEGIS combat system engineering agent for the US Navy for the last forty years.[37] Currently, Northrop Grumman is fulfilling a $122 million contract with the US Army to "provide systems engineering, production, deployment and logistics support services for the C-RAM systems."[38] Similarly, the foreign firm Rheinmetall Air Defence has received a contract for €20 million to provide documentation and training services for the Mantis, a short-range air defense protection system developed to protect the German Army in Afghanistan.[39]

III THE UNITED STATES SEPARATION OF POWERS FRAMEWORK

This distinctive mix of autonomous devices and contractors upends traditional frameworks, both domestic and international, for regulating the conduct of war. This part examines how the rise of unmanned systems affects the implementation of the US constitutional and statutory regime governing the decision to use force abroad. The next part turns to the impact on the international humanitarian law regime.

The US Constitution explicitly and purposefully divides responsibility over the use of force abroad between the Congress and the president. Under the Constitution, Congress declares war[40] and holds the purse strings,[41] while the president is commander-in-chief of the armed forces.[42] Debates about the scope of the president's power in relation to Congress date back centuries[43] and have grown more fraught in the last fifty years as formal declarations of war have largely become a historical relic.[44] Some scholars and government lawyers have asserted broad theories of the president's inherent power to use force unilaterally,[45] while others have taken a more restrained approach, articulating a view that shared responsibility

---

[37] "Aegis Combat System's Newest Baseline."
[38] "Northrop Wins US Army's C-RAM Contract."
[39] "NBS MANTIS Air Defence Protection System, Germany" army-technology.com, https://perma.cc/Y49B-FHEC.
[40] US Const., art. I, § 8, cl. 11.
[41] US Const., art. I, § 8, cls. 12–13.
[42] US Const., art. I, § 2.
[43] Compare the Prize Cases, 67 U.S. (2 Black) 635 (1863) (upholding the constitutionality of President Lincoln's unilateral decision to impose naval blockade during the Civil War, when Congress was not in session) with *Little* v. *Barreme*, 6 U.S. (2 Cranch) 170 (1804) (concluding that a US naval captain's seizure of a ship sailing from a French port during the naval war with France was unjustified because Congress had only authorized seizures of ships sailing to French ports).
[44] The United States has not formally declared war since World War II. *See* Joint Resolution Declaring That a State of War Exists Between the Government of Rumania and the Government and the People of the United States and Making Provisions to Prosecute the Same, ch. 325, 56 Stat. 307 (June 5, 1942) (last declaration of war by the United States).
[45] *See, e.g.*, L. Meeker, "The Legality of United States Participation in the Defense of Viet-Nam" (1966) 54 *Department of State Bulletin* 474, reprinted in (1966) 75(7) *Yale Law Journal*

between the president and Congress should continue.[46] After the Vietnam War, Congress tried to protect its position by enacting the War Powers Resolution, which requires the president to halt "hostilities" if Congress has not approved them within sixty days.[47] Since then, while Congress has, at times, spoken loudly to authorize the use of force – as it did in 2001 to allow a muscular response to Al Qaeda and the Taliban[48] and in 2003 to authorize the war in Iraq[49] – at other times, Congress's voice has been muted at best. Years into the US operations against the Islamic State in Iraq and Syria (ISIS), no new authorization of force was forthcoming from Congress.

In the face of congressional silence, the rise of unmanned systems, combined with the growing use of contractors, has enabled the president to claim greater scope for unilateral action. New technology has not merely emboldened the president politically, but has actually made possible new legal arguments to justify an expanded conception of the president's role in relation to Congress. In order to see the development of these new legal arguments, this part focuses on three examples where the president used force overseas without receiving specific congressional approval: Kosovo in 1999, Libya in 2011, and the campaign against ISIS in Iraq and Syria beginning in 2014. In each case, we can see that the growing role of unmanned systems and contractors has opened space for legal arguments expanding the power of the executive branch.

A Kosovo

To understand the power shift that new technologies and the use of contractors have produced, and the impact of these trends on legal arguments about the use of force, it is useful to go back to 1999, when US President Bill Clinton sought to halt the ethnic cleansing campaign of the Yugoslav dictator Slobodan Milosevic in Kosovo. With the humiliating deaths of US troops in Mogadishu, Somalia, in 1993 still fresh in Americans' minds, President Clinton aimed to intervene in a way that would minimize the risk of US casualties by relying primarily on an air

---

1085–108; J. Yoo, "The Continuation of Politics by Other Means: The Original Understanding of War Powers" (1996) 84(2) *California Law Review* 167–305.

[46] *See, e.g.*, H. Koh, *The National Security Constitution* (New Haven, CT: Yale University Press, 1990); D. Barron and M. Lederman, "The Commander in Chief at the Lowest Ebb – Framing the Problem, Doctrine, and Original Understanding" (2008) 121(3) *Harvard Law Review* 689–804.

[47] War Powers Resolution, 50 U.S.C. §§ 1541–48 (1973); *see* P. Holt, *The War Powers Resolution* (Washington, DC: American Enterprise Institute for Public Policy Research, 1978).

[48] Joint Resolution to use the United States Armed Forces against those responsible for the recent attacks launched against the United States, Pub. L. No. 107–40, § 2(a), 115 Stat. 224 (2001).

[49] Authorization for Use of Military Force Against Iraq Resolution of 2002, Pub. L. No. 107–243, 116 Stat. 1498 (2002).

campaign carried out jointly with the North Atlantic Treaty Organization (NATO).[50] This strategy endeavored to stop the killing without spilling American blood.

The military campaign was swift. President Clinton ordered US air strikes to begin in the province of Serbia on March 24, 1999, without explicit congressional approval.[51] The effort, conducted along with other NATO countries, lasted seventy-eight days and consisted entirely of an air campaign.[52] Of 37,000 sorties, the United States flew 23,208 (62 percent).[53] When President Milosevic agreed to withdraw Yugoslav forces from the region and allow the presence of a United Nations peacekeeping force on June 10, 1999, NATO suspended the bombing.[54] There were no combat casualties, a point the president highlighted at the end of the campaign.[55] Civilian casualties among the Yugoslav population, however, were in the range of 480–530.[56]

Drones were also a part of the story. The US military had harnessed unmanned aircraft for surveillance as far back as the Vietnam War, when "Lightning Bugs" gathered information on the North Vietnamese. By 1999, drones had evolved sufficiently to give the military improved situational awareness and enable better targeting. During Operation Allied Force in Kosovo, the unmanned vehicles operated as "remote-controlled intelligence and surveillance platforms."[57] The US military employed three tactical systems: the Air Force Predator, the Army Hunter, and the Navy Pioneer.[58] The Defense Department after-action report to Congress highlighted the significance of the new technology, emphasizing that the drones were used at "unprecedented levels" and "played an important role in our overall success."[59] Specifically, the report noted, drones "enabled commanders to see the situation on the ground without putting aircrews at risk and provided continuous

---

[50] *See* President William Jefferson Clinton, "Statement on Kosovo," March 24, 1999; President William Jefferson Clinton, "Letter to Congressional Leaders Reporting on Airstrikes Against Serbian Targets in the Federal Republic of Yugoslavia (Serbia and Montenegro)" (1999) 35 *Weekly Compilation of Presidential Documents* 527, https://perma.cc/ZQ2Q-U8SK; *see also* "The Kosovo Air Campaign: Operation Allied Force," North Atlantic Treaty Organization, November 11, 2014, https://perma.cc/ESU5-4C27.

[51] Clinton, "Letter to Congressional Leaders."

[52] S. Bowman, *Kosovo and Macedonia: U.S. and Allied Response* (CRS Issue Brief) (Washington, DC: Congressional Research Service, 2003), https://perma.cc/VDP5-M25F.

[53] Ibid.

[54] Ibid.

[55] President William Jefferson Clinton, Address on the Kosovo Agreement, June 10, 1999, https://perma.cc/F99V-DCMJ.

[56] Human Rights Watch, *The Crisis in Kosovo* (2000), https://perma.cc/MBX4-7M3P.

[57] *Report to Congress: Kosovo Operation Allied Force After-Action Report* (Department of Defense, 2000), https://perma.cc/VF2L-NH99.

[58] Ibid.

[59] Ibid.; *see also* E. Becker, "Military Leaders Tell Congress of NATO Errors in Kosovo," *The New York Times*, October 15, 1999.

coverage of important areas."[60] Their ability to convey live video feeds for "real-time targeting" was key. At this point, the drones did not conduct strikes, but they could fly lower than military aircraft, scout out targets, and transmit information in real time to minimize civilian casualties and make the air strategy more effective.

Contractors also formed a crucial element of the military strategy. During the air campaign, contractors provided intelligence support, served as linguists, transported fuel from barges on the Adriatic to locations inland, and built three refugee camps for displaced Kosovars.[61] After the primary air campaign ended, military contractors poured into the region, augmenting the US contingent of peacekeepers. The contractors performed an unprecedented variety of roles, including logistics, construction, training, and policing.[62] Significantly, their presence allowed the US military to contribute a much smaller number of troops (7,000) than might otherwise have been needed.[63] Overall, in the Balkans, the number of contractors far exceeded the number of troops, at a ratio of approximately 1.5–1.[64] It would be fair to say that the combined power of drones and contractors delivered a one-two punch that enabled the US military to reduce its footprint and the risk of military casualties.

This strategy of waging war from a safe distance, made possible in part by new technologies and contractors, gave the president more political leeway to act without clear approval from Congress. With regard to Kosovo, Congress debated the conflict extensively but neither explicitly supported nor opposed it, because those in favor of military action never mustered enough votes to pass a resolution permitting the use of force. The US House of Representatives narrowly defeated a concurrent resolution that would have authorized the air strikes (213–213)[65] and voted down a declaration of war by an overwhelming majority (427–2).[66] At the same time, those who opposed the intervention also could not garner enough support to pass resolutions condemning the strikes. Indeed, a concurrent resolution introduced in the House by Rep. Tom Campbell (R-California) directing the president to remove armed forces from Serbia within thirty days lost by a large majority.[67] The House did

---

[60] Department of Defense, *Report to Congress* (describing statements of Pentagon officials that drones provided "indispensable" assistance in US air campaign).
[61] Ibid., p. 116.
[62] M. Schwartz and J. Church, *Department of Defense's Use of Contractors to Support Military Operations: Background, Analysis and Issues for Congress* (Congressional Research Service, 2013), https://perma.cc/JUT9-6JY8.
[63] Department of Defense, *Report to Congress*.
[64] *See* R. Fontaine and J. Nagl, "Contractors in American Conflicts: Adapting to a New Reality," Center for a New American Security, December 16, 2009, https://perma.cc/V4RP-HVSH; *see also* "Contingency Contracting throughout U.S. History," Office of Defense Procurement and Acquisition Policy, https://perma.cc/W588-NBWB.
[65] H.J. Res. 44, 106th Cong. (1999), 145 Cong. Rec. H2441 (daily ed. April 28, 1999); ibid., H2451-52 (recording vote).
[66] 145 Cong. Rec. H2474 (daily ed. April 28, 1999); ibid. at H2440-41 (recording vote).
[67] H. Con. Res. 82, 106th Cong. (1999), 145 Cong. Rec. H2414 (daily ed. April 28, 1999) (reprinting H. Con. Res. 82); ibid. at H2426-27 (recording vote).

vote (249–180) to block funding for ground troops,[68] but Congress subsequently passed an emergency supplemental appropriations bill allocating $12.9 billion (US billion) for the campaign.[69]

While Congress remained split, the aerial military strategy emboldened Clinton administration lawyers to justify the unilateral use of force as a legal matter. The Department of Justice Office of Legal Counsel (OLC) did not explicitly explain the constitutional basis of the president's authority to conduct the Kosovo campaign in the absence of Congress's authorization. Instead, an OLC memo articulated in general terms the purpose of the military action: "to demonstrate the seriousness of NATO's purpose so that the Serbian leaders understand the imperative of reversing course; to deter an even bloodier offensive against innocent civilians in Kosovo; and, if necessary, to seriously damage the Serbian military's capacity to harm the people of Kosovo."[70] However, the memo did contend that the president had not run afoul of the sixty-day clock imposed by Section 5(b) of the War Powers Resolution. Although that provision stops the president from using force in "hostilities" without congressional approval, OLC argued that Congress had in fact authorized the use of force by funding the campaign, even though Section 8(a) of the War Powers Resolution purports to disallow such an approach.

Despite the ambiguity at the time, when OLC lawyers sought to explain the legal basis of President Barack Obama's use of force in Libya in 2011 a decade and a half later, they drew a comparison to Kosovo. Specifically, they suggested that the low risk of US casualties in the Kosovo campaign justified, in part, exempting the president from needing to seek congressional approval either as a constitutional matter or within the framework of the War Powers Resolution. The memo reasoned that the Kosovo campaign "avoided the difficulties of withdrawal and risks of escalation that may attend commitment of ground forces – two factors that this Office has identified as 'arguably' indicating 'a greater need for approval [from Congress] at the outset'."[71] The memo also emphasized that, "as in prior operations conducted without a declaration of war or other specific authorizing legislation, the anticipated operations here served a 'limited mission'." Thus, OLC implied that Clinton never needed to get Congress's blessing, in part because of the minimal risk

---

[68] H.R. 1569, 106th Cong. (1999), 145 Cong. Rec. H2400 (reprinting measure); ibid. at H2413-14 (recording votes).

[69] H.R. 1664, 106th Cong. (1999), 145 Cong. Rec. H2634 (daily ed. May 4, 1999); see also A. Taylor, "Paying for the Kosovo Air War: How Much Is Too Much?," *CQ Weekly*, May 1, 1999, p. 1014. Following a floor debate on May 6, the House passed H.R. 1664 the same day by a vote of 311–105. 145 Cong. Rec. H2895 (daily ed. May 6, 1999).

[70] "Authorization for Continuing Hostilities in Kosovo" (2000) 24 *Opinions of the Office of Legal Counsel* 327, https://perma.cc/C52J-TSA7; see also "Proposed Deployment of United States Armed Forces into Bosnia," 19 *Opinions of the Office of Legal Counsel* 327 at 333, https://perma.cc/9V49-4FEF.

[71] "Authority to Use Military Force in Libya," 35 *Opinions of the Office of Legal Counsel* 1 at 13, https://perma.cc/9A2C-56XS.

of casualties. Of course, the casualty count was so low largely because harm to drones and contractors is not tallied in the official casualty figures.

B *Libya*

A decade after the Kosovo intervention, another US president ordered a military campaign to halt the atrocities of another dictator without first seeking congressional approval. When Libyan President Muammar Qadhafi attacked civilians protesting his iron-fisted rule, and continued to do so in flagrant defiance of UN Security Council condemnation,[72] President Obama joined a coalition of nations to initiate an air campaign to prevent further atrocities. In contrast to Kosovo, the Security Council had approved the multilateral intervention in Libya, imposing a no-fly zone and authorizing the use of force to protect civilians.[73] As Qadhafi's forces were preparing to retake the city of Benghazi, the dictator pledged that his troops would show "no mercy and no pity" against protesters and that "[w]e will come house by house, room by room."[74] Qadhafi, President Obama later noted, "compared [his people] to rats, and threatened to go door to door to inflict punishment... We knew that if we ... waited one more day, Benghazi, a city nearly the size of Charlotte, could suffer a massacre that would have reverberated across the region and stained the conscience of the world."[75]

The US military intervention, dubbed Operation Odyssey Dawn, commenced on March 19, 2011.[76] The same day, France and the United Kingdom, which had played leading roles in advocating for military action and had cosponsored the Security Council resolutions, commenced Operation Harmattan and Operation Ellamy, respectively, in coordination with the United States.[77] US and coalition

---

[72] *See, e.g.*, Security Council Res. 1970, U.N. Doc. S/RES/1970 (February 26, 2011), https://perma.cc/QAR9-6VLR ("condemning the violence and use of force against civilians" and "deploring the gross and systematic violation of human rights" in the country); "In Swift, Decisive Action, Security Council Imposes Tough Measures on Libyan Regime, Adopting Resolution 1970 in Wake of Crackdown on Protesters," UN Security Council Press Release SC/10187/Rev. 1 (February 26, 2011), https://perma.cc/BKQ7-99CH.

[73] *See* Security Council Res. 1973, U.N. Doc. S/RES/1973 (March 17, 2011), https://perma.cc/65UX-XLFC; "Security Council Approves 'No-Fly Zone' Over Libya, Authorizing 'All Necessary Measures' to Protect Civilians, by Vote of 10 in Favour with 5 Abstentions," UN Security Council press release SC/10200 (March 17, 2011), https://perma.cc/63RT-SJAX.

[74] *See* D. Bilefsky and M. Landler, "Military Action against Qaddafi Is Backed by U.N.," *The New York Times*, March 18, 2011, p. A1.

[75] The White House, "Remarks by the President in Address to the Nation on Libya," Press Release, March 28, 2011, https://perma.cc/2UU6-F6M6.

[76] J. Gertler, *Operation Odyssey Dawn (Libya): Background and Issues for Congress* (CRS Report No. R41725) (Washington, DC: Congressional Research Service, 2011), https://perma.cc/74WT-WJN4.

[77] Ibid.

forces quickly established command of the air over Libya's major cities, destroying portions of the Libyan air defense network and attacking pro-Qadhafi forces deemed to pose a threat to civilian populations.[78]

At the time, President Obama emphasized that US military involvement in the operation would be limited. In his report to Congress on March 21, 2011, he explained that the use of military force in Libya served important US interests by preventing instability in the Middle East and preserving the credibility and effectiveness of the United Nations Security Council.[79] He also stated that the goal of US operations was to "set the stage" for further action by coalition partners.[80] The president pledged that no US ground forces would be deployed, except possibly for search-and-rescue missions, and he asserted that the risk of substantial casualties for US forces would be low.[81]

Establishment of the no-fly zone over Libya proceeded relatively smoothly. One US aircraft was lost due to mechanical malfunction, but the crew was rescued. Estimates of the cost of the initial operation ranged between $400 million and $1 billion. NATO assumed control of military operations on March 30, 2011,[82] and international military operations concluded later that year, on October 31, after the death of Qadhafi, when the Security Council ended the mandate for military action.[83]

Drones assumed an even larger role in Libya than they had during the Kosovo campaign. The technology of unmanned systems had evolved considerably. Drones now enhanced the Air Force's capability to better gather information for targeting because of their greater ability to fly low for extended periods, hover in densely populated areas, and transmit high-quality information in real time.[84] This capacity increased the precision of conventional air power and helped reduce civilian casualties. Even more significantly, in addition to providing intelligence, drones could now themselves also drop bombs. The United States announced that it was sending its first armed Predator drones to the region in April 2011.[85] During the roughly seven-month air campaign, US unmanned systems launched 145 of the strikes on Libya, almost half of the overall total of US strikes, while also doing most

---

[78] Ibid.
[79] President Barack Obama, Letter from the President regarding the commencement of operations in Libya, March 21, 2011, https://perma.cc/2HAL-P93Q.
[80] Ibid.
[81] Ibid.; see also "Authority to Use Military Force in Libya."
[82] Gertler, *Operation Odyssey Dawn*.
[83] "Security Council votes to end Libya operations," BBC News, October 27, 2011, https://perma.cc/7WZZ-SCHX.
[84] See C. Pellerin, "Gates: Obama OKs Predator Strikes in Libya," *Department of Defense News*, April 21, 2011, https://perma.cc/56SW-YTJ9; see generally J. Walsh, *The Effectiveness of Drone Strikes in Counterinsurgency and Counterterrorism Campaigns* (Carlisle, PA: Strategic Studies Institute and US Army War College Press, 2013).
[85] See Pellerin, "Gates: Obama OKs Strikes in Libya."

of the spotting that enabled the manned strikes.[86] Moreover, even after NATO announced that the conflict was over in October 2011, the drones did not go home, but rather continued to fly combat air patrols, primarily for surveillance.[87]

As in Kosovo, contractors also played a key role after the international air campaign ended. Because the new Libyan government expressed concerns about an extensive contractor presence, however, the contractor footprint was smaller and quieter.[88] The US State Department did not contract with well-known security firms such as Blackwater, a company that had guarded US diplomats in Iraq but earned a poor reputation after multiple abuse-of-force incidents, such as the notorious September 2007 killing of civilians in Baghdad's Nisoor Square. Nevertheless, the US Embassy in Libya did employ contractors from lesser-known firms, some of which hired local nationals as employees.[89] The CIA also used contractors in Libya.[90] Indeed, in a brutal attack on September 11, 2012, militants killed not only the US ambassador, but also two American CIA contractors, Tyrone Woods and Glen Doherty.[91] Thus, while the contractor role in Libya may have been more muted, the combination of drones and contractors again enabled the military to shrink its visible image, just as it had in Kosovo.

This seemingly smaller footprint both emboldened the president politically to take action without congressional approval and lent credence to legal arguments that he was neither constitutionally nor statutorily required to obtain such approval. As in the Kosovo context, Congress did not authorize the military action, although debate within the legislature resulted in a mix of actions on several resolutions pointing in multiple directions. In the House of Representatives, two resolutions came to the floor on June 3. The first, introduced by Rep. Dennis Kucinich (D-Ohio) with Republican cosponsors, directed the removal of US armed forces from Libya within fifteen days, but it failed 148–265. The second, introduced by House Speaker John Boehner (R-Ohio), succeeded by a vote of 268–145. It noted

---

[86] See S. Ackerman, "Libya: The *Real* U.S. Drone War," *Wired*, October 20, 2011, https://perma.cc/24XG-JYRB; C. Woods and A. Ross, "Revealed: US and Britain Launched 1,200 Drone Strikes in Recent Wars," The British Bureau of Investigative Journalism, December 4, 2012, https://perma.cc/T787-QL6U.

[87] See S. Ackerman, "U.S. Drones Never Left Libya; Will Hunt Benghazi Thugs," *Wired*, September 12, 2012, https://perma.cc/XR56-LG9K.

[88] J. Risen, "After Benghazi Attack, Private Security Hovers as an Issue," *The New York Times*, October 12, 2012, www.nytimes.com/2012/10/13/world/africa/private-security-hovers-as-issue-after-embassy-attack-in-benghazi-libya.html?_r=1; see also A. Mehra, "Time to Put Security Contractors Under the Gun," *Huffington Post*, February 28, 2013, https://perma.cc/K7HN-F8ER.

[89] Ibid.

[90] Associated Press, "New Benghazi Investigation Finds No Fault in Response," *The New York Times*, November 21, 2014, www.nytimes.com/2014/11/22/us/new-benghazi-investigation-finds-no-fault-in-response.html.

[91] Ibid.; *see also* D. Kirkpatrick, "A Deadly Mix in Benghazi," *The New York Times*, December 28, 2013, www.nytimes.com/projects/2013/benghazi/#/?chapt=0.

that President Obama had "failed to provide Congress with a compelling rationale" for military activities in Libya and directed him to describe in detail within fourteen days US interests and objectives in Libya, as well as the reason why he did not seek authorization from Congress to use military force. Although not legally bound by the resolution, the president did submit a report within the requested time frame. On June 24, the House voted down (123–295) a resolution that would have authorized the military action, and rejected by a smaller margin (180–230) a measure that would have provided limited funding for the war. A Senate resolution put forward by Senator John Kerry (D-Massachusetts) and Senator John McCain (R-Arizona) on June 24 would have authorized the use of force (while prohibiting boots on the ground) and agreed with the president that the military intervention did not require congressional authorization under the Constitution or the War Powers Resolution.[92] That resolution was withdrawn without a vote.[93]

Meanwhile, the president proceeded with the operation, while his lawyers made the case that, as a legal matter, "under these circumstances, the President had constitutional authority, as Commander-in-Chief and Chief Executive and pursuant to his foreign affairs powers, to direct such limited military operations abroad, even without prior specific congressional approval."[94] More specifically, the OLC memo justified the use of force by arguing that the multilateral nature of the oprations and the reduced risk of casualties – made possible by drones and contractors – exempted the operation from any constitutional requirement that Congress declare war or explicitly authorize the use of force. The memo reasoned that a particular use of force is a war "for constitutional purposes" only if there are "prolonged and substantial military engagements, typically involving exposure of U.S. military personnel to significant risk over a substantial period." In contrast, the military action in Libya, the memo argued, was "limited in scope and duration." Again, the so-called limited scope was made possible by the expanded role of drones and contractors substituting for military personnel.

In addition to the constitutional arguments in the OLC memo, other administration lawyers (over OLC's objections) also contended that the Libya campaign did not even fall within the War Powers Resolution's sixty-day limit on engaging in "hostilities" absent congressional authorization. Crucial to their reasoning was the fact that manned planes had moved quickly into a mere supporting role, even though "unmanned forces" continued to carry out strikes up to the very day Qaddafi

---

[92] A joint resolution authorizing the limited use of the United States Armed Forces in support of the NATO mission in Libya, S.J. Res. 20, June 21, 2011, https://perma.cc/5JZ5-TFJH. The resolution also would have prohibited deployment of security contractors. See Kirkpatrick, "A Deadly Mix in Benghazi."
[93] See generally L. Fisher, "Senate Should Protect War Powers on Libya," Roll Call, June 28, 2011, https://perma.cc/6Q72-69QU.
[94] "Authority to Use Military Force in Libya,"

was killed. In other words, so the argument went, because US blood would not be at risk, it didn't count as warfare.[95]

C ISIS

The military engagements in the Balkans and Libya set the stage for the campaign against ISIS in Iraq and Syria. ISIS is a transnational Sunni Islamist insurgent and terrorist group that expanded its control over areas of northwestern Iraq and northeastern Syria beginning in 2013, threatening the security of both countries and drawing increased attention from the international community.[96] The group emerged against the backdrop of the brutal Syrian regime of President Bashir al-Assad, the disintegration of the country into civil war, and the sectarian strife within Iraq, where many segments of the Sunni population felt that they had not received adequate representation within the predominantly Shiite government.[97]

In the face of ISIS military gains in Iraq and Syria, the United States launched an air campaign against the group beginning in August 2014. Drone strikes in this campaign have been significant. During the ten-month period from August 2014 to June 2015, for example, US and coalition forces used combat aircraft, armed unmanned aerial vehicles, and sea-launched cruise missiles to conduct more than 3,700 strikes in Iraq and Syria.[98] Further strikes have continued since that time.

The stated objective of these US strikes has evolved somewhat as circumstances have changed. Initially, the goal was to halt the advance of ISIS forces and reduce threats to American personnel and religious minorities in northern Iraq; subsequently, the objective included support for Iraqi military and Kurdish forces, with the aim of weakening the ability of ISIS to support its Iraq operation from within Syria. Other US strikes have targeted individuals belonging to the "Khorasan Group," an Al Qaeda affiliate whose leadership is based in Pakistan.[99]

Drones have played an even more significant part in the campaign against ISIS than they did in Libya. US Central Command has reported that drone strikes constituted about 15 percent of the overall number of strikes,[100] and drone surveillance capabilities enabled so-called eyes on the ground, particularly in

---

[95] For a discussion of these arguments, see C. Savage, *Power Wars: Inside Obama's Post-9/11 Presidency* (New York: Little, Brown, and Company 2015), pp. 635-654.
[96] C. Blanchard and C. Humud, *The "Islamic State" Crisis and U.S. Policy* (CRS Report No. R43612) (Washington, DC: Congressional Research Service, 2015), https://perma.cc/CV2E-PV9A.
[97] Ibid.
[98] Ibid.
[99] Ibid.
[100] E. Schmitt, "Obstacles Limit Targets and Pace of Strikes against ISIS," *The New York Times*, November 9, 2014, www.nytimes.com/2014/11/10/world/middleeast/trouble-pinning-down-isis-targets-impedes-airstrikes.html?mcubz=2.

denser urban areas. Indeed, demand for drones was so high and drone pilots became so overworked that in 2015 the US Air Force announced a crisis of drone "pilot fatigue."[101]

Contractor involvement has also been significant. While the number of military and security contractors has not swelled to levels witnessed during the previous phase of the conflicts in Iraq and Afghanistan,[102] US Central Command reported in July 2015 that the Department of Defense was employing 41,900 military contractors in the area within Central Command's responsibility.[103] And this figure did not even include the State Department contractors who were providing services such as security to US diplomats, as the State Department does not release data on contractor numbers.[104]

Contractors continue to provide extensive support to the drone operations themselves – performing maintenance on unmanned vehicles, loading bombs into them, analyzing the intelligence feeds they provide, piloting unarmed drones, and in some cases steering the takeoff and landing of armed drones.[105] And although contractors have not officially selected targets, some of their activities, such as intelligence analysis, have, at times, contributed to target selection. Certainly contractors were (and are) performing virtually the entire array of tasks associated with the operation of unmanned vehicles.

As in the Kosovo and Libya conflicts, the combined force of drones and contractors enabled the Obama administration to maintain the appearance of a small military footprint with minimal risk of harm to US troops. Indeed, President Obama repeatedly stressed that he would not deploy ground combat units to fight ISIS, maintaining that US troops could not fix the underlying political problems that ultimately caused the crisis.[106] While the president left the door open for the possibility of small special forces rescue missions and operations to target the ISIS leadership, he ruled out "enduring offensive ground combat operations."[107]

The Obama administration's legal justification under domestic law for the military intervention rested on the preexisting Authorization to Use Force against Al

---

[101] C. Drew and D. Philipps, "As Stress Drives Off Drone Operators, Air Force Must Cut Flights," *The New York Times*, June 16, 2015, www.nytimes.com/2015/06/17/us/as-stress-drives-off-drone-operators-air-force-must-cut-flights.html?_r=1.

[102] "Transforming Wartime Contracting."

[103] *Contractor Support of U.S. Operations in the USCENTCOM Area of Responsibility* (Department of Defense, 2016), https://perma.cc/FFE8-SXFZ.

[104] See J. Risen and M. Rosenberg, "Blackwater's Legacy Goes Beyond Public View," *The New York Times*, April 14, 2015, www.nytimes.com/2015/04/15/world/middleeast/blackwaters-legacy-goes-beyond-public-view.html.

[105] See discussion below at text accompanying nn. 12–23.

[106] Blanchard and Humud, *The "Islamic State" Crisis and U.S. Policy*.

[107] President Barack Obama, Letter to the Congress of the United States, February 11, 2015, https://perma.cc/9SJN-K4X6; Draft Joint Resolution for Congress to Authorize the Limited Use of Force Against the Islamic State of Iraq and the Levant, February 11, 2015, https://perma.cc/FMQ9-BJVR.

Qaeda (and the Taliban) from 2001,[108] as well as the Authorization to Use Force in Iraq.[109] This legal theory, however, required the administration to draw a tenuous link between ISIS and Al Qaeda, two terrorist organizations that were literally fighting each other on the ground in Syria.[110] The administration sought a new authorization to use force, but action on that authorization stalled. Instead, members of Congress seemed content to acquiesce in this legal pretzel logic in order to avoid another vote.

However, even without new authorization from Congress, the precedent now exists for executive branch lawyers to contend that operations against ISIS (or similar operations elsewhere) might not require congressional authorization. To be sure, the Obama administration lawyers who supplied the legal reasoning for the Libya intervention would, in all likelihood, not have carried their arguments that far. Nonetheless, the legal groundwork now exists for the claim that a low risk of troop casualties, made possible by the involvement of contractors and autonomous weapons, could form the basis of military involvement of such "limited nature, scope, and duration" that it would not be big enough to trigger the need for congressional authorization.

The implications of this argument are profoundly unsettling for the historical constitutional tradition of "mutual participation" between the president and Congress in waging war.[111] To be sure, Congress's abdication of its role and the corresponding swelling scope of presidential authority is a continuing storyline from the Korean War to the present.[112] Yet the force of new, increasingly automated military technologies, combined with growing military contracting, has arguably catapulted the phenomenon to another level. Not only have these new technologies and methods of warfare strengthened the executive's *political* hand to use force unilaterally while keeping US military casualties low, but the twin trends of autonomous weapons and privatization have opened the door to *legal* claims that the use of force is so limited that congressional assent is neither constitutionally nor statutorily necessary. In so doing, these twin trends have profoundly upset the very architecture of US governance.

---

[108] Joint Resolution to Authorize the Use of United States Armed Forces Against Those Responsible for the Recent Attacks Against the United States, Pub. L. No. 107-40, § 2(a), 115 Stat. 224 (2001).

[109] Authorization for Use of Military Force Against Iraq Resolution of 2002, Pub. L. No. 107-243, 116 Stat. 1498 (2002).

[110] See, e.g., D. Byman, "Comparing Al Qaeda and ISIS: Different goals, different targets" (prepared testimony before the Subcommittee on Counterterrorism and Intelligence of the House Committee on Homeland Security), April 29, 2015, https://perma.cc/L7UH-KS4Z.

[111] See DaCosta v. Laird, 448 F.2d 1368 (2d Cir. 1971); see also Koh, *The National Security Constitution*.

[112] See J. Ely, *War and Responsibility: Constitutional Lessons of Vietnam and Its Aftermath* (Princeton, NJ: Princeton University Press, 1993). Indeed, during the Korean War, Truman administration officials referred to the conflict as a "police action," despite the deployment of more than five million US troops to the region.

IV THE INTERNATIONAL HUMANITARIAN LAW CRIMINAL ACCOUNTABILITY FRAMEWORK

The combined impact of unmanned weaponry and military privatization not only destabilizes the constitutional and statutory separation of powers arrangements enshrined in US law, but also unsettles the framework of international humanitarian law, also known as the law of armed conflict. In particular, increasingly autonomous weaponry and growing privatization tend to spread responsibility for decision-making across a larger number of actors who do not fit neatly within either a military command structure or an ordered bureaucracy. Thus, as multiple actors work together to gather intelligence, make targeting decisions, and deploy weapons, the authority and responsibility for decisions involving violence are diffused and fragmented. Accordingly, if it turns out that the use of a weapon violates IHL, it is far more difficult even to determine who is responsible. In this new world of frag- ~ented authority, our existing legal and bureaucratic processes are inadequate to provide sufficient mechanisms of accountability. As a result, we face potential difficulties in trying to ensure compliance with IHL norms.

Scholars and policy-makers have debated extensively whether the use of unmanned and autonomous weapons threatens fundamental IHL principles. For example, some have argued that the asymmetric harm such weapons can inflict flouts core principles of humanity that undergird IHL, or that the promise of precision that such weapons offer may be overstated.[113] In contrast, others have contended that such weapons are potentially more precise than human beings who pull triggers and might therefore provide an opportunity for more humane warfare, stripped of the irrationality, hot-blooded decision-making, and emotional toll of the battlefield.[114] Such weapons, according to this argument, might paradoxically lead to better effectuation of the IHL principles that lie at the heart of that body of law, such as distinction and proportionality.[115]

Still others have emphasized that, regardless of whether or not unmanned and autonomous weapons might better implement substantive IHL principles, autonomy itself can pose serious problems for accountability.[116] Perhaps most

---

[113] *See, e.g.,* M. Wagner, "The Dehumanization of International Humanitarian Law: Legal, Political, and Ethical Implications of Autonomous Weapons Systems" (2014) 47 *Vanderbilt Journal of Transnational Law* 1371–424; Docherty, "Losing Humanity."

[114] *See, e.g.,* U.N. Report of the Special Rapporteur, Heyns; M. Waxman and K. Anderson, "Law and Ethics for Autonomous Weapon Systems: Why a Ban Won't Work and How the Laws of War Can," The Hoover Institution, April 13, 2013, https://perma.cc/C4UA-7DJN; R. Arkin, "The Case for Ethical Autonomy in Unmanned Systems" (2010) 9 *Journal of Military Ethics* 332–41, https://perma.cc/6K99-248D; *see also* D. Cohen, "Drones off the Leash," *U.S. News*, July 25, 2013, https://perma.cc/XT35-LB2H.

[115] For an excellent overview of the debate about autonomous weapons and a moderate approach to the issue, *see* J. Vilmer, "Terminator Ethics: Should We Ban 'Killer Robots'?," *Ethics & International Affairs* (March 23, 2015), https://perma.cc/U6QD-TMDE.

[116] *See* Wagner, "The Dehumanization of International Humanitarian Law."

significantly, these systems threaten the framework of individual criminal responsibility that undergirds all of IHL. From the Nuremberg trials of Nazi war criminals to proceedings before the International Criminal Court, a fundamental aspect of international humanitarian law is that perpetrators will be held individually responsible for egregious violations of the law of war that constitute war crimes. While such prosecutions are rare, they serve as core sanctions that ensure compliance with this body of law.

But whether it is individual criminal punishment or some other form of accountability, unmanned autonomous weapons pose problems for this framework. As a number of scholars have pointed out, in the case of truly autonomous systems, who will be responsible for the decision to strike?[117] In 2012, partly due to accountability concerns, the United States issued a temporary ban on fully autonomous systems.[118] Subsequently, policy-makers and scholars gathered at the United Nations in 2015 to discuss these issues.[119]

Meanwhile, a rich debate has also emerged over the impact of military and security privatization on international humanitarian law and policy.[120] For example, some have called for an outright ban on the practice.[121] Others, including myself, have advocated for new means of regulation, such as the use of contracts and public-private accreditation regimes, that attempt to bring the public values embedded in IHL into this newly privatized realm.[122] Without such reforms, existing systems of accountability and oversight will be weak and ineffectual.

Yet few, if any, scholars have addressed the combined force of increasingly autonomous weaponry and privatization. Paired together, these trends significantly enhance the accountability and oversight problems each poses separately. Thus, my

[117] Ibid.; J. Thurnher, "Examining Autonomous Weapon Systems from a Law of Armed Conflict Perspective," in H. Nasu and R. McLaughlin (eds.), *New Technologies and the Law of Armed Conflict* (Den Haag, the Netherlands: T. M. C. Asser Press, 2014), p. 225; M. Sassòli, "Autonomous Weapons and International Humanitarian Law: Advantages, Open Technical Questions and Legal Issues to Be Clarified" (2014) 90 *U.S. Naval War College, International Law Studies* 308; *see also* B. Keller, "Smart Drones," *The New York Times*, March 16, 2013, www.nytimes.com/2013/03/17/opinion/sunday/keller-smart-drones.html.
[118] Department of Defense, *Autonomy in Weapon Systems*, DoD Directive 3000.09, Washington, DC: U.S. Department of Defense, 2012, https://perma.cc/NLG5-ETGS.
[119] United Nations Meeting of Experts on Lethal and Autonomous Weapons Systems (April 13–17, 2015), https://perma.cc/ZC4Q-PMPJ.
[120] *See generally* Avant, *The Market for Force*; L. Cameron and V. Chetail, *Privatizing War: Private Military and Security Companies under International Law* (Cambridge: Cambridge University Press, 2013); Dickinson, *Outsourcing War and Peace*; McFate, *The Modern Mercenary*; Singer, *Corporate Warriors*; International Committee of the Red Cross, *The Montreux Document: On pertinent legal obligations and good practices for States related to operations of military and security companies during armed conflict* (Geneva: International Committee of the Red Cross, 2009), https://perma.cc/6THJ-XV68.
[121] *See, e.g.*, S. Percy, *Mercenaries: A History of a Norm in International Relations* (Oxford: Oxford University Press, 2007).
[122] Dickinson, *Outsourcing War and Peace*; L. Dickinson, "Regulating the Private Security Industry: The Promise of Public/Private Governance" (2013) 63 *Emory Law Journal* 417–54.

goal in this section is to break new ground by scrutinizing how these two phenomena intersect and to assess how that intersection could impact the potential for imposing accountability under IHL when serious violations occur.

Although much could be said about multiple forms of accountability, my focus here will be on individual criminal responsibility. In particular, my contention is that both privatization and the use of increasingly autonomous weapons fragment decision-making and bring it outside the ordinary bureaucratic chain of command. This fragmentation makes accountability harder to impose, because it becomes very difficult to hold individuals responsible for serious harms that occur. Thus, the use of these weapons by a military that has also outsourced multiple functions to private contractors presents distinct problems that add up to more than the combined impact of each of the two trends. In other words, the whole of the problem very likely is worse than the sum of its parts.

A *Autonomous Weaponry and Individual Criminal Accountability*

A growing body of scholarship highlights the difficulties in holding individuals criminally responsible for uses of unmanned and autonomous weapons that lead to significant violations of IHL. For example, Markus Wagner has observed that, if these weapons can effectively make decisions such as engaging in target selection, it will be very difficult to determine which individuals to hold responsible for those decisions, let alone fit the humans' decisions into existing doctrinal frameworks.[123] If a soldier is monitoring the operation of a weapon that goes on a rampage and the soldier fails to stop it, should he or she be held responsible? Or does responsibility lie with the programmer who made it possible for the weapon to act in this way? Or the manufacturer? Or possibly even the military commander who approved the use of the weapon in the first place?

In a seminal article, Rebecca Crootof explained that a significant aspect of the problem flows from the doctrine of international criminal law itself, in particular the intent requirement for most war crimes.[124] War crimes usually must be committed "willfully," which means that the accused must have either acted with the intent to commit a violation or acted recklessly. Thus, if autonomous weapons were used to commit war crimes, a prosecutor might demonstrate the requisite intent by proving that a software engineer had deliberately programmed the weapon to target civilians, or a commander had ordered them to be used in such a manner. But Crootof emphasizes that these are the "easy cases" and instead focuses on the hard case of "whether anyone might be accountable in the more complicated situation where no individual acts intentionally or recklessly, but an autonomous weapon system

---

[123] *See* Wagner, "The Dehumanization of International Humanitarian Law."
[124] *See* R. Crootof, "War Torts: Accountability for Autonomous Weapons" (2016) 164(6) *University of Pennsylvania Law Review* 1347–402.

nonetheless takes action that constitutes a serious violation of international law."[125] Assuming that "no one intended the violation or acted recklessly, no one can be held directly liable."[126] Because of these difficulties, Crootof does not support criminalizing negligence, arguing that civil responsibility in tort is a more appropriate mechanism of accountability in such cases than criminal responsibility.

International criminal law does contain multiple doctrines that provide criminal accountability for participants in war crimes who do not actually pull the trigger. For example, the doctrine of command responsibility permits the imposition of liability on persons with authority over the acts of subordinates in some circumstances. The doctrine of aiding and abetting reaches those who may assist in the commission of a war crime. And the doctrines of complicity and joint criminal enterprise sweep broadly to cover all of those who agree to commit war crimes. These doctrines do not necessarily require all participants to have intended to commit war crimes.

Yet, as Heather Roff points out, these doctrines are premised on the notion that there is at least one individual who *did* possess the requisite intent, an observation that Crootof makes as well.[127] Thus, under the doctrine of command responsibility, a superior can be punished for the war crime of a subordinate if that superior has actual or constructive knowledge of the crime and effectively controls the subordinate. In other words, he or she need not have intended for the crime to be committed, but at least the *subordinate* must have possessed the requisite intent. In the case of a superior in charge of an autonomous weapon, it cannot be fairly said that the autonomous weapon possesses the requisite intent. Thus, merely negligent supervision could not clearly justify criminal punishment under a theory of command responsibility.[128]

---

[125] Ibid., 1377.
[126] Ibid.
[127] H. Roff, "Killing in War: Responsibility, Liability and Lethal Autonomous Robots," in A. Henschke et al. (eds.), *Routledge Handbook of Ethics and War: Just War Theory in the 21st Century* (New York: Routledge Press, 2013); see also Crootof, "War Torts."
[128] Jens David Ohlin has argued that international criminal law can nonetheless provide a framework for accountability with respect to the use of semi-autonomous and autonomous weapons. He observes that the origins of the command responsibility and other doctrines emerging out of the Nuremberg trials are premised on the notion of actions in a bureaucratic context, in which the bureaucracy itself is the instrument of harm. He suggests that the critical element of criminal responsibility of the commander in this context is control, not whether the subordinate possesses the requisite intent. J. Ohlin, "The Combatant's Stance: Autonomous Weapons on the Battlefield," (2016) (92) *Int'l Leg. Stud.* 1–30. The challenge for Ohlin's view is that the concept of effective control in the context of ordinary bureaucracy is not easily applied to those operating autonomous weapons. Even in the case of semi-autonomous weapons, meaningful human control will not always be possible, particularly when systems are complex and involve multiple components. Moreover, Ohlin acknowledges that while the doctrine of command responsibility does not necessarily require commanders to intend harm, at a minimum a showing of recklessness is required. But in many cases involving autonomous or even-semi-autonomous weapons, there may be no human being within the bureaucracy who could be deemed to be reckless.

The problem depends in part on how one defines "autonomous," which is the source of significant debate among scholars and policy-makers. The challenge of criminal accountability is particularly acute for fully autonomous weapons that can engage in independent decision-making. No such weapons are in operation, and the technology to implement this kind of autonomy in weapons systems may still be many years away. Nonetheless, the problem could still arise for a semi-autonomous system, as we will see below.

B *Privatization and Individual Criminal Accountability*

Privatization also undermines criminal accountability for war crimes, but in a slightly different way. When governments turn over military and security functions to armed contractors and authorize them to use force, there is, of course, a risk that those contractors could abuse the responsibility and commit criminal acts. To be sure, under well-established doctrines of international criminal law dating back to the Nuremberg trials of Nazis in the aftermath of World War II, non-state actors can be prosecuted for serious violations of the law of war during an armed conflict. As I have written elsewhere,[129] the mere fact that these contractors are private actors does not insulate them from war crimes prosecution.

Yet the fragmentation of decision-making within bureaucracies has long presented hurdles for such prosecutions. Notions of due process and fundamental justice in criminal proceedings tend to rest on principles of individual criminal responsibility and the opportunity to individually defend oneself in a fair proceeding. Guilt by association or group membership is antithetical to these principles. Yet, while individuals commit war crimes, in modern society they often do so within a much broader organizational context: as members of militaries, civilian governmental departments, private corporate contractors, terrorist organizations, or other groups. Thus, the person who pulls the trigger is often not the decision-maker. Subordinates may sometimes go rogue and act on their own, but at other times, they may be doing the dirty work for an organization's leaders. At the same time, those leaders may not know of the subordinates' specific actions, making it difficult to impose command responsibility. And, in between group leaders and trigger-pullers, other members of an organization may take small steps that facilitate or contribute to the commission of criminal acts; for example, a secretary may sign an order requisitioning the weapons used to target civilians, a budgeting official may allocate funds to purchase such weapons, and so on.

As noted above, international criminal law contains multiple doctrines that allow courts to convict individuals for war crimes and other atrocities even when they contribute to the commission of those crimes due to their role in larger organizations. Nevertheless, applying these doctrines remains fraught with difficulty. Perhaps

---

[129] Dickinson, *Outsourcing War and Peace*.

the least controversial of the doctrines is the theory of command responsibility, whereby a superior with authority can be held responsible for the acts of a subordinate. Because of the intent requirement for both the subordinate and the superior, command responsibility fits most cleanly into traditional liberal conceptions of individual criminal responsibility. International courts and tribunals have deployed the doctrine to hold both military and civilian leaders responsible, and they have determined that formal legal authority is not necessary if the individual superior in question exerts *de facto* control over a subordinate.

Yet, privatization makes it far more difficult to impose even this less controversial version of criminal responsibility on the government actors who supervise contractors. This is because contractors often fall outside the military or bureaucratic chain of command, making it far more difficult to demonstrate that government actors exercised the requisite degree of authority and control to be criminally responsible themselves. For example, the fact that government actors may have generally supervised the large government contracts under which contractors operate would probably not be sufficient to impose command responsibility.

More broadly, the doctrine of command responsibility has never been able to comfortably encompass cases of fragmented responsibility, in which multiple actors together commit a wrong even if no one individual in the group intends the harm. Probably the most controversial and notable effort to address the problem of group-based contributions to war crimes took place during the prosecution of major Nazi war criminals before the International Military Tribunal at Nuremberg (IMT). There, prosecutors tackled the problem head-on by seeking to designate Nazi groups such as the Gestapo and the SS as criminal organizations.[130] Prosecutors had hoped to avoid the difficulty of proving the criminal responsibility of each individual within these organizations by persuading the IMT to deem these organizations criminal. Such a designation would have allowed subsequent proceedings to convict individuals for mere membership in these organizations. Relying on charter provisions that opened the door to organizational accountability,[131] the prosecution essentially argued that the entire Nazi state was a criminal operation: "We shall ... trace for you the intricate web of organizations which these men

---

[130] In addition to charges against twenty-two individual defendants, the prosecution alleged that six organizations constituted criminal organizations due to their role in perpetrated acts of aggression, war crimes, and crimes against humanity, the three crimes within the jurisdiction of the tribunal. The six organizations were the Leadership Corps of the Nazi Party, the Gestapo, the SS, the SA, the Reich Cabinet, and the general staff and high command of the Nazi Party.

[131] The Charter of the Tribunal specifically provided that "[a]t the trial of any individual member of any group or organization the Tribunal may declare (in connection with any act of which the individual may be convicted) that the group or organization of which the individual was a member was a criminal organization." Charter of the International Military Tribunal, August 8, 1945, 59 Stat. 1546, 82 U.N.T.S. 284 § II, art. 9. The charter also specified that, in subsequent proceedings before national courts of states parties to the charter, individuals could be tried for mere membership in organizations deemed criminal by the IMT. Ibid., art. 10.

formed and utilized to accomplish these ends. We will show how the entire structure of the offices and officials was dedicated to the criminal purposes and committed to the use of criminal methods planned by these defendants and their co-conspirators."[132]

The IMT rejected the prosecution's broad theories, however, and took a much more limited approach to the question of the organizations' criminality. Although the court did not throw out the conception of organizational criminality altogether, it defined quite narrowly the groups that could be deemed criminal, and concluded that only three of the six organizations qualified.[133] Moreover, the IMT struggled with the implications of a process that might lead to guilt by association and emphasized that "innocent persons" should not be punished.[134] Thus, the IMT concluded that a group could be deemed criminal only in the limited circumstances when it could be proved that group members were "bound together and organized for a common purpose" and the group was "formed or used in connection with the commission of crimes denounced by the Charter."[135] The IMT further restricted the circumstances in which individuals could be convicted for membership in such associations, essentially crafting a requirement for an individualized finding of *mens rea* for each person accused of membership in a criminal organization.[136] Thus, while the tribunal recognized that group-based liability might be useful because of the challenges of bringing charges against each individual within a large organization,[137] the court was nevertheless reluctant to accept a broad theory of organizational responsibility because of the risks to fundamental due process in criminal prosecutions.

The Nuremberg Tribunal's foray into organizational accountability has largely been viewed as a failure, certainly in comparison to its highly influential development of the substantive principle of individual criminal responsibility for war crimes

---

[132] Trial of the Major War Criminals Before the International Military Tribunal, Nuremberg, November 14, 1945 – October 1, 1946, Second Day, Wednesday, November 21, 1945, Morning Session, p. 104 (1947). For a detailed account of the case against the Nazi organizations at Nuremberg, see D. Fraser, "(De)Constructing the Nazi State: Criminal Organizations and the Constitutional Theory of the International Military Tribunal" (2017) 39(1) *Loyola of Los Angeles International and Comparative Law Review* 117–86.
[133] Judgment of the International Military Tribunal for the Trial of German Major War Criminals, pp. 67–80 (1946).
[134] Ibid., p. 67.
[135] Ibid.
[136] The tribunal noted: "[It] should exclude persons who had no knowledge of the criminal purposes or acts of the organization and those who were drafted by the State for membership, unless they were personally implicated in the commission of acts declared criminal ... in the Charter as members of the organization. Membership alone is not enough to come within the scope of these declarations." Ibid.
[137] For example, the IMT stated that "[w]here an organization with a large membership is used for such [criminal] purposes, a declaration [of criminality] obviates the necessity of inquiring as to its criminal character in the later trial of members who are accused of participating through membership in its criminal purposes and thus saves much time and trouble." Ibid.

and crimes against humanity. Indeed, the tribunal's efforts to wrestle with the problem of organizational accountability highlights what David Luban has described as a core challenge of modern life: "the problem of moral responsibility in bureaucratic settings."[138]

Drawing on the work of Robert Conot and Hannah Arendt, Luban notes that part of the horror of Nazi atrocities stemmed from the organizational setting in which they took place. These were not the acts solely of evil masterminds or bloodthirsty gangsters. Rather, "through the fragmentation of authority and tasks, it was possible to fashion a murder machine."[139] Luban cites Conot as observing that "[t]housands of people were involved, but each considered himself nothing but a cog in the machine and reasoned that it was the machine, not he, that was responsible."[140] In Luban's view, this "bureaucratic irresponsibility" is the "moral plague of modern life," and he refers to Hannah Arendt's insight that bureaucracy is "the latest and perhaps most formidable form of ... dominion" because it is "the rule of an intricate system of bureaus in which no men, neither one nor the best, neither the few nor the many, can be held responsible, and which could be properly called rule by Nobody."[141]

Luban believes that the bureaucratic Nazi machine does not ultimately represent the most difficult version of this problem, however, because many of the participants could be said, at a minimum, to be "willfully" blind to their role. That is, they fit within relatively traditional conceptions of moral culpability and recognized criminal law concepts. To the extent that individual Nazi bureaucrats did not know about the final solution, for example, it was because they deliberately shielded themselves from that knowledge and therefore possessed some degree of intent. As Luban puts it,

> the top Nazis may not have known what the inside of a death camp looked like, but they knew what the code name "Final Solution" meant. Lower level flunkies such as Ivan the Terrible, the notorious executioner at Treblinka, may not have known why they were gassing Jews, but they could not help but know that mass killing was mass murder. Whatever blindness such people allege is willful blindness.[142]

The more difficult situation in which to assign moral culpability and impose criminal legal responsibility, however, is one in which each of the cogs in a bureaucratic machine is truly ignorant of the commission of atrocities. Such a situation may well be presented by the combination of automation and contracting.

---

[138] D. Luban, *Legal Modernism: Law, Meaning, and Violence* (Ann Arbor: University of Michigan Press, 1994), p. 362.

[139] Ibid., p. 363 (quoting R. Conot, *Justice at Nuremberg* [New York: Basic Books, 1983], pp. 210–11).

[140] Ibid. (quoting Conot, *Justice at Nuremberg*).

[141] Ibid. (quoting H. Arendt, *On Violence* [Orlando, FL: Harcourt Publishing, 1970], p. 38).

[142] Ibid., p. 372.

Individual contractors, operating outside of military hierarchies, are each responsible for one aspect of a drone's construction, maintenance, and operation. Unlike the "willful blindness" described by Luban, these individuals may well be entirely ignorant of the larger purpose for which the drones are employed. Thus, automation and contracting together immensely complicate the search for moral and legal accountability within bureaucratic settings.

Subsequent proceedings before international criminal tribunals demonstrate the tensions inherent in the effort to impose criminal responsibility based on group membership, along with the challenges posed by war crimes committed within a bureaucratic setting more broadly. Arguing before these tribunals, prosecutors have used a variety of doctrines, including command responsibility, aiding and abetting, and joint criminal enterprise, to impose criminal responsibilities on individuals implicated in the commission of war crimes and other mass atrocities due to their roles in large, complex organizations. Yet most of these doctrines, as in the proceedings before the Nuremberg Tribunal, have retained a relatively traditional conception of individual criminal responsibility that is based on individual *mens rea* and individual culpability.

Thus, to the extent that such doctrines are predicated on individuals intentionally agreeing to and supporting criminal acts, they reflect a more traditional conception of individual responsibility. And when these doctrines have been applied more broadly and have attempted to sweep in a broader array of actors who may not have intended to commit or may not have truly participated in the underlying crime, critics have charged that these doctrines are antithetical to fundamental principles of individual criminal responsibility. For example, Allison Danner and Jenny Martinez have argued that these doctrines, "if not limited appropriately, have the potential to lapse into forms of guilt by association, thereby undermining the legitimacy and the ultimate effectiveness of international criminal law."[143] With respect to joint criminal enterprise in particular, Danner and Martinez argue that international judges "should require that prosecutors demonstrate that each individual charged under a joint criminal enterprise theory made a substantial contribution to the fulfillment of the common objective of the enterprise" and that any form of the doctrine that tolerates a "reduced *mens rea* should not be used for specific intent crimes."[144] With respect to command responsibility, they argue that "something more than ordinary negligence should remain the touchstone for criminal responsibility."[145]

Although these arguments make perfect sense from the traditional criminal law framework perspective, they may render many acts undertaken within bureaucratic institutions unaccountable, even when those acts represent tiny cogs in a

---

[143] A. Danner and J. Martinez, "Guilty Associations: Joint Criminal Enterprise, Command Responsibility, and the Development of International Criminal Law" (2005) 93(1) *California Law Review* 77–170 at 79.
[144] Ibid., at 79.
[145] Ibid.

bureaucratic machine that is committing grave atrocities. Moreover, in a world of increasingly automated weaponry and increasing military privatization, assigning traditional individual criminal responsibility may only become more difficult over time.

C The Combination of Autonomous Weaponry and Privatization

When an autonomous machine is injected into a bureaucratic system and then private contractors are hired to help operate it, individual criminal responsibility becomes even more challenging and problematic. Lethal force at the robotic hands of the autonomous weapon presents an even grimmer vision of Arendt's "rule by no one," probably far beyond even Arendt's already dystopian imagination. Indeed, when information necessary for targeting decisions is sifted by private contractors, then the fragmentation of decision-making no longer even occurs inside the government, but now falls outside the bureaucracy itself. And when those contractors feed information into an autonomous system that is engaging in the targeting, the problem magnifies exponentially.

The threat is perhaps easiest to perceive when considering fully autonomous weapons systems that move, detect threats, and engage targets based on predetermined criteria. But because no fully autonomous weapons systems are currently operational (and some commentators question whether such technology is imminent) and because many governments have, for now, said they will not use them, I will focus instead on partially autonomous systems. These systems are preprogrammed to detect and strike certain categories of targets, but usually a human being remotely oversees the operation and can override the automated system.

One such example is the Israeli Harpy, which, when launched by a human operator, can detect an enemy radar system and autonomously dive-bomb and strike that target. If the Harpy killed a large number of civilians in a manner that could be said to violate the principle of proportionality, who could or should be held responsible? The human being with the responsibility to override the weapons system? The commander of the territory where the weapons system was deployed? The individuals who set policy for using the technology? The individuals who drafted the targeting criteria? The engineers who designed the weapons system to apply the targeting criteria? Anyone who supplied intelligence that fed into the weapons system and formed the basis for target selection? In the case of even partially autonomous systems, it is difficult to locate a responsible human agent.[146]

Only human beings can be held criminally responsible for actions. But in this case, there is no clear human actor who bears full responsibility. Some of the people in the chain listed above might plausibly be tried, and the governments of the

---

[146] *See* Wagner, "The Dehumanization of International Humanitarian Law"; *see also* Roff, "Killing in War."

United States and the United Kingdom have said that the last human to make a decision regarding the operation of the weapon should be held responsible. Yet arguably this is inadequate, because no one person truly can be said to make that decision. As discussed above, the doctrine of command responsibility allows for superiors in the chain of command to be tried, but that theory is premised on the notion that the human being who actually pulled the trigger could also be tried for the same crime. Yet here there is no human being who can accurately be said to have pulled a trigger.

At its root, the problem is not the mere lack of a human agent. It is that autonomy – even partial autonomy – fragments the decision-making process, spreading the decision across a large number of individuals, thereby making it much harder to hold any one person responsible. And, of course, adding contractors to the mix further diffuses and breaks up the decision-making process. For example, as previously discussed, with many remotely operated weapons systems, such as drones, contractors are involved in supplying intelligence that leads to target selection as well as occasionally calling in targets from the ground. Is the intelligence-gathering that determines the targeting decision itself a targeting decision that implicates criminal responsibility? And does it matter that the contractor is not a state actor? In short, the existing legal doctrine is not equipped to cope with criminal responsibility in the case of automated and outsourced weapons systems.

V CONCLUSION

The purpose of this chapter has been to highlight the profoundly destabilizing impact of unmanned and autonomous weapons systems and privatization on both domestic and international law regulating the use of force abroad. These trends have emerged concurrently, and together they disrupt existing legal frameworks more than each would do separately. On the domestic front, unmanned and autonomous weaponry and privatization potentially alter the balance of power between the president and Congress by buttressing legal arguments that the use of force can be limited in ways that do not require congressional approval. Meanwhile, these trends also work together to put pressure on international humanitarian law, because they fragment and diffuse decision-making, thereby complicating efforts to impose individual accountability. Accordingly, it will be a core challenge for law and governance in the twenty-first century to evolve new legal regimes and new mechanisms of oversight and accountability to respond to the radically changing face of armed conflict.

PART II

Technology and Human Rights Enforcement

Building on recent scholarship and advocacy on the transformation of human rights fact-finding in the digital era,[1] Part II considers the opportunities and challenges presented by the use of new technologies to enforce human rights. In Chapter 6, "The Utility of User-Generated Content in Human Rights Investigations," Jay Aronson addresses the integration of user-generated content in efforts to hold human rights violators accountable. Using a series of case studies, Aronson demonstrates how this evidence can be integrated into human rights investigations and why investigators must be careful when doing so. In Chapter 7, "Big Data Analytics and Human Rights: Privacy Considerations in Context," Mark Latonero analyzes the privacy risks of using large aggregated datasets in human rights monitoring and argues for the development of better normative standards to protect privacy in the process.

While the contributions by Aronson and Latonero are primarily concerned with the collection of information for accountability efforts, the other two chapters in this part address the impact of technological advances on the display and use of information in advocacy contexts. Chapter 8, "The Challenging Power of Data Visualization for Human Rights Advocacy," by John Emerson, Margaret L. Satterthwaite, and Anshul Vikram Pandey, considers the use of new data visualization techniques to communicate and analyze human rights problems. They discuss both the historical evolution of data visuals in advocacy and the risks and benefits of using data visualization in this context. Chapter 9, "Risk and the Pluralism of Digital Human Rights Fact-Finding and Advocacy," by Ella McPherson, draws on field research with human rights organizations to address the way in which these organizations

---

[1] See, e.g., P. Alston and S. Knuckey (eds.), *The Transformation of Human Rights Fact-Finding* (New York: Oxford University Press, 2016), pp. 399–489. For an exploration of related techniques in the humanitarian context, see P. Meier, *Digital Humanitarians: How Big Data Is Changing the Face of Humanitarian Response* (Boca Raton, FL: CRC Press, 2015).

manage risk associated with the introduction of new technologies. She confronts the reality that not all organizations are equally equipped to manage these risks, and she suggests that unless this is addressed, it could have negative impacts on human rights advocacy in the long term.

The chapters in Part II make clear that one of the most significant challenges in regulating the human rights impacts of technology is that the very same characteristics of technology that present the greatest opportunities also create the greatest risks. For example, the increasing availability of low-cost technology to document abuses means more documentation by human rights organizations, bystanders, victims, and even perpetrators. At the same time, more documentation of violations means the generation of greater quantities of data, leading to significant challenges in collecting, sorting, and storing this information. Crucial evidence may also be lost or collected in ways that render it inadmissible in later proceedings to hold perpetrators accountable or unverifiable for use in historical clarification or transitional justice efforts. Video of human rights violations, whether created and shared by bystanders, victims, or perpetrators, can enhance the efficacy of legal tribunals and other accountability mechanisms, but such video also raises a host of legal and ethical challenges regarding ownership of content and concerns associated with making such material public.

Similarly, greater participation in documentation efforts by nonprofessionals could yield democratizing effects for human rights advocacy and bolster the legitimacy of the human rights project, which is often critiqued as elitist.[2] The reduced role for gatekeepers, however, makes verification of information and protection of victims and witnesses much more challenging.[3] As compared to human rights researchers, nonprofessionals engaging in human rights research may have a much higher tolerance for risk, which can have significant implications for victim and witness safety. The persistence of digital information can also frustrate traditional understandings of the right to privacy and undermine efforts to ensure the informed consent of witnesses who share information about human rights abuses.

Even the solutions that technology offers to address some of these challenges can create new ones. As Aronson's chapter makes clear, the ability of human rights practitioners to gather information about victims of human rights violations from user-generated content increases the likelihood that justice and accountability institutions will hear their cases. At the same time, such information gathering can also expose the creator and those portrayed digitally to discovery or harassment by perpetrators and their allies. Similarly, as the chapter by Emerson, Satterthwaite, and Pandey illustrates, data visualization can be a powerful tool for understanding

---

[2] See, e.g., D. Rieff, "The Precarious Triumph of Human Rights," *The New York Times Magazine*, August 8, 1999.

[3] M. Beutz Land, "Peer Producing Human Rights" (2009) 46(4) *Alberta Law Review* 1115–39 at 1126–29; M. K. Land, "Democratizing Human Rights Fact-Finding," in Alston and Knuckey (eds.), *The Transformation of Human Rights Fact-Finding*, pp. 399–424.

and communicating about human rights violations, but it can just as easily obscure or fundamentally misrepresent the details of a complex situation.

Moreover, any democratizing potential that technology might have can be undermined by broad disparities in its distribution, which poses the risk of reinforcing rather than challenging the status quo. Until global power dynamics around technological innovation are changed, these resources will remain unevenly distributed. Some of the most innovative tools, such as satellite imagery, statistical methodology, and sophisticated data analysis techniques, are out of reach for many grassroots organizations, both financially and in terms of expertise. Local groups do not have the resources they need to use technology effectively or safely in their work, and more powerful groups may appropriate the documentation they produce without providing any direct benefit in return. The difficulty of maintaining good digital security is in part a product of poor technological design, but it also reflects preexisting power imbalances and the absence of funders that support the development of technological capacity among small organizations. McPherson argues, in turn, that this unevenly distributed capacity of human rights organizations to manage the risk introduced by new technologies and methodologies is likely to have a negative impact on human rights pluralism – and on human rights.

The chapters in Part II thus question not only the democratizing possibilities of technology, but also its purported objectivity or neutrality. This inquiry can be applied to the design of technology itself. It can also be applied to the activities and processes deployed through or engendered by the physical artifacts of technology, as well as the expertise employed to create it.[4] In the context of big data, for example, this includes the creation of algorithms that make a variety of decisions about information contained in large datasets, including prioritizing or classifying information or creating associations within it.[5] When governments, nongovernmental agencies, or advocacy groups subject data to new forms of analysis, they can introduce algorithmic bias into social and political decision-making. Recent academic and advocacy work has shown the limits of objectivity in data analysis, with respect to both the messiness of real-world data and the fact that an algorithm is just a set of instructions written by humans (who are often prone to bias) to be followed by a computer processor.

Indeed, the destabilizing introduction of new technologies reveals pressures on the idea of "truth" that we often try to ignore. Data visualization techniques, for example, can portray ostensibly "true" material in biased or misleading ways. Much of this indeterminacy is a function of the role of interpretation and perception. Information does not exist in a vacuum, but is constantly interpreted and

---

[4] W. E. Bijker, T. P. Hughes, and T. Pinch (eds.), *The Social Construction of Technological Systems: New Directions in the Sociology and History of Technology.* Cambridge, MA: MIT Press, 2012, p. xiii.

[5] N. Diakopoulos, "Accountability in Algorithmic Decision Making" (2016) 59(2) *Communications of the ACM* 56–62 at 57–58.

reinterpreted by its audience, which "might variously refuse, resist, or recode those materials for their own purposes."[6] In an era of "fake news," with heightened pressure on social media companies to remove "false" material from their sites,[7] understanding the impact of technological developments on concepts of truth is crucial.[8]

All of these potential risks and rewards exist in an environment in which cultural factors, convenience, government aid agencies, technology companies, and human rights funders are encouraging technological solutions to human rights documentation and advocacy problems. Human rights advocates and organizations that might benefit from avoiding new technologies until they develop better mechanisms to cope with risks may feel compelled to adopt new technologies so they can continue to be relevant in the field. In his reflection that concludes this volume (Chapter 13), Enrique Piracés picks up on the theme. He worries that the lure of technological solutions also risks focusing our attention on documentation as an end in and of itself, rather than as part of a larger response. Documentation is important, and in some instances documenting a violation may be the only response possible. There may even be a moral obligation to document, albeit one that must be balanced with security and other concerns. Nonetheless, the seduction of "new technology" should not lead us to overemphasize investigation at the expense of other responses, like transitional justice efforts, legislative reform, or community mobilization.

---

[6] M. McLagan and Y. McKee, "Introduction," in M. McLagan and Y. McKee (eds.), *Sensible Politics: The Visual Culture of Nongovernmental Activism* (Brooklyn, NY: Zone Books, 2012), p. 13.

[7] E. Asgeirsson, "German Social Media Law Threatens Free Speech," Human Rights First, April 10, 2017, www.humanrightsfirst.org/blog/german-social-media-law-threatens-free-speech.

[8] *See, e.g.*, M. P. Lynch, *The Internet of Us: Knowing More and Understanding Less in the Age of Big Data* (New York: Liveright Publishing, 2016).

6

The Utility of User-Generated Content in Human Rights Investigations

Jay D. Aronson

Over the past decade, open source, user-generated content available on social media networks and the Internet has become an increasingly important source of data in human rights investigations. Although use of this data will not always generate significant findings, the sheer volume of user-generated content means that it is likely to contain valuable information. As Craig Silverman and Rina Tsubaki note in the introduction to the *Verification Handbook for Investigative Reporting*, such analysis, at least with respect to journalism, is now "inseparable from the work of cultivating sources, securing confidential information and other investigative tactics that rely on hidden or less-public information."[1]

In this chapter, I will examine how firsthand video recordings of events by citizen witnesses, journalists, activists, victims, and perpetrators are being used to document, prosecute, and find remedies for human rights violations. In this context, I do not focus on why such recordings are made, but rather on the extent to which they provide a particular audiovisual perspective on conflict or human rights violations. I take it as given that most of this material will be biased in what it depicts and what it leaves out, and that all of it requires significant analysis and appraisal before being authenticated and used for investigatory purposes. I focus primarily on video recordings, whether intentionally produced for human rights documentation or not, because of their prevalence in recent human rights investigations, their rich informational content, their verifiability, and their capacity for rapid dissemination via social media. My purpose is to highlight the potential utility, obvious limitations, and significant evidentiary, legal, and ethical dangers of relying on eyewitness videos by examining cases in which they are already being used.

---

[1] C. Silverman and R. Tsubaki, "The Opportunity for Using Open Source Information and User-Generated Content in Investigative Work," in C. Silverman (ed.), *Verification Handbook for Investigative Reporting* (Maastricht, the Netherlands: European Journalism Centre, 2015), http://verificationhandbook.com/book2/chapter1.php.

I THE ROLE OF USER-GENERATED CONTENT IN HUMAN
RIGHTS INVESTIGATIONS

As the nongovernmental organization WITNESS notes in its seminal 2011 report, *Cameras Everywhere*,

> Video has a key role to play, not just in exposing and providing evidence of human rights abuses, but across the spectrum of transparency, accountability and good governance. Video and other communication technologies present new opportunities for freedom of expression and information, but also pose significant new vulnerabilities. As more people understand the power of video, including human rights violators, the more the safety and security of those filming and of those being filmed will need to be considered at each stage of video production and distribution.[2]

Ultimately, WITNESS argues, the ability to access the technology, skills, and expertise needed to analyze these videos will determine "who can participate – and survive – in this emerging ecosystem of free expression."[3]

A wide range of people produce video content and share it through social media, the Internet, semiprivate communication channels like Telegram and Snapchat, or privately via e-mail or physical storage. Conflict events, protests, riots, and other similar events are increasingly being live-streamed as they happen. Some of the creators of this content have been trained in human rights documentation, while others have not. In many cases, damning video will come from the perpetrators themselves, who use the content to boast of their power and accomplishments or seek funding from sympathetic outsiders.

Courts, tribunals, truth commissions, and other fact-finding (or perhaps fact-generating) bodies, as well as journalists and human rights advocates, need to be sensitive to the wide-ranging quality, completeness, and utility of user-generated content. They cannot assume that the content was intentionally created or that the people represented in this material know that their images and activities are being stored, processed, and analyzed for human rights purposes. Extra care must be taken to ensure the privacy, security, and other basic rights of people who produce such content or appear in it. In the case of perpetrator video, they must assume that the content has public relations goals, and they must take care not to spread messages of hate or extremism. Additionally, it is crucial to keep in mind that many war crimes and human rights abuses will continue to leave few electronic traces.[4] Like all other forms of evidence, video is not a magic bullet or panacea that will put an end to atrocities. Nor does it mitigate the need for

---

[2] S. Padania et al., *Cameras Everywhere: Current Challenges and Opportunities at the Intersection of Human Rights, Video, and Technology* (Brooklyn, NY: WITNESS, 2011), p. 16.
[3] Ibid.
[4] A. Azoulay, *The Civil Contract of Photography* (Brooklyn, NY: Zone Books, 2008).

eyewitnesses and victims to provide testimony and for investigators to visit the scenes of crimes and conduct thorough investigations.[5]

Nonetheless, video has potential value at every stage of a human rights investigation, whether that investigation is designed to feed into advocacy or legal proceedings.[6] Most commonly, video generates leads that can be used to start an investigation. It can also provide evidence to establish that a crime or violation occurred, or it can be used to support a particular factual finding, such as whether a particular weapon was used in a conflict or whether pollution from a particular mining site is polluting a water source. Sometimes, it can also link a particular person, group, government, or company to the violation in question.

A great deal of citizen video provides lead evidence or evidence that a crime or violation occurred. For example, it might show the aftermath of an attack or event, such as destroyed infrastructure, suffering victims, dead bodies, inhumane working conditions, or a damaged environment. Evidence linking a perpetrator to a crime or violation is often harder to come by, but the creative investigator can sometimes mine video collections for this information.[7] Videos might show the presence of particular soldiers in a place where a crime has occurred, as demonstrated by uniforms, vehicles with identifiable symbols, or traceable license plates. Further, weapons and munitions (either spent or unexploded) might be traced to a given military unit, government, or manufacturer. Videos posted to the Internet might also include scenes of government officials or other individuals spurring their followers to commit acts of violence against another group. In the nonconflict context, videos might show violations in places of employment, harassment of minority groups, or environmental harm.

As noted above, the information obtained from video can play a central role in criminal proceedings or truth commissions, but it can also contribute to advocacy campaigns aimed at naming and shaming perpetrators or seeking restitution and justice for victims. Video evidence can help combat standard state or corporate responses to allegations of abuse by making it harder to deny the existence of violations. Video can also be used to raise awareness of and elicit sympathy from those who had not previously been aware of the scope or scale of abuses in a particular context. The recent focus on police brutality against black people in the United States is one example of this phenomenon. Although governments and rights violators can deny the content of written reports or question the methodology

---

[5] A. Whiting, "The ICC Prosecutor's New Draft Strategic Plan," *Just Security*, www.justsecurity.org/24808/icc-prosecutors-draft-strategic-plan.

[6] K. Matheson, "Video as Evidence: To be evidence, what does video need?," *New Tactics in Human Rights*, July 20, 2014, www.newtactics.org/comment/7427#comment-7427.

[7] WITNESS offers an excellent guide that explains the elements of linkage evidence and also provides tips to activists and ordinary citizens about filming video in a way that provides information about responsibility for particular actions. WITNESS, "Proving Responsibility: Filming Linkage and Notice Evidence," https://library.witness.org/product/video-as-evidence-proving-responsibility-filming-linkage-and-notice-evidence.

used to generate statistical claims, video and other images require the accused to develop a different response. Visual evidence makes it much harder for violators to engage in the tactic that Stanley Cohen calls "literal denial" and requires them to provide an alternative explanation for their actions or claim that their actions were justified.[8]

In a case of police or security force abuse, for example, it will be more difficult (but not impossible) for the government to claim that the use of deadly weapons was justified when video evidence suggests that a political protester or criminal suspect was not an immediate threat. In the case of a massacre caught on film, the perpetrators will have to convincingly demonstrate that the videos in question were staged or depict another time or place. Video evidence also makes it harder to engage in the standard repertoire of denial techniques used to refute historical claims; i.e., that "it happened too long ago, memory is unreliable, the records have been lost, no one will ever know what [really] took place."[9]

Video evidence also makes it harder (but, again, not impossible) to impugn the credibility of victims or witnesses who testify against perpetrators because the videos that buttress their claims can be authenticated. Using geolocation techniques developed by the video verification community, it is possible to verify the location of a filmed event by matching features using services like Google Earth, Google Street View, and satellite imagery. Additionally, it is sometimes possible to determine the date and approximate time of day using shadows, weather, sun position, and other climatic clues in the footage.[10] Information extracted from video can also be combined with, and corroborated by, forensic evidence (such as autopsy, medical, or pathology reports), other scientific evidence (such as environmental analysis), official records, media reports, intercepted communications, satellite imagery, other open source data, or expert testimony (such as weapons/munitions analysis or political analysis).[11]

II THE CHALLENGE OF VIDEO EVIDENCE IN LAW AND ADVOCACY

Video evidence is, of course, not definitive and presents significant challenges for activists and advocates in legal and other contexts. Audiences might grow tired of viewing traumatic video and become desensitized to its effects. On the other hand, they might become less likely to believe accounts that are not supported by

---

[8] S. Cohen, *States of Denial: Knowing about Atrocities and Suffering* (Cambridge: Polity Press, 2001), p. 7.
[9] S. Cohen, "Government Responses to Human Rights Reports: Claims, Denials, and Counterclaims" (1996) 18 *Human Rights Quarterly* 517–43 at 523.
[10] Amnesty International and Forensic Architecture, "Rafah: Black Friday Report," Methodology section, July 24, 2015, https://blackfriday.amnesty.org/methodology.php.
[11] Ibid.

convincing audiovisual evidence. The persistent creation of falsified media by states, pranksters, and other nefarious actors could also impugn legitimate content.[12] Further, any good defense attorney, government operative, or rights violator can challenge the authenticity of the video or the interpretation of its content (e.g., that victims were indeed dead but were not killed by the claimed perpetrator, or that the victims did not heed warnings to disperse in the moments before the video was shot). They can also offer alternative explanations for what is seen (e.g., that the munitions in question were stolen from the army by a rebel force, or that anti-government forces were dressed up as army personnel). Video evidence cannot always help dispute claims made by perpetrators that their actions were justified on national or internal security grounds, that the victims posed a threat, or that the action captured on video was wrong but the matter has already been dealt with internally.

III THREE CASE STUDIES: SYRIA, UKRAINE, AND NIGERIA

In order to better understand the landscape of video content, this section discusses three recent case studies in which videos were used to support human rights advocacy and accountability. These cases are not fully representative of all of the possible uses of audiovisual evidence in human rights work, but they provide geographically diverse exemplars that, when viewed together, highlight many of the important issues in this domain. They highlight the diverse ways that video content is being used in advocacy and accountability efforts; they also demonstrate the variety of conflict-related information that can be gathered from video, including the distribution of weapons systems, the relative strength of military units or protest movements, and social linkages among various actors in a conflict or political movement. These uses go beyond the kind of secondhand witnessing and visual representation of an event that have traditionally characterized video evidence and raise important questions about the representativeness and reliability of audiovisual accounts. These cases also illustrate the way in which video can become the source of other forms of evidence (e.g., forensic studies of blood stains or wounds that have not been directly examined by an expert, or ballistics reports on weapons that are seen only on video), as well as how it can complement or add credence to expert testimony about physical evidence.

---

[12] C. Silverman, *Lies, Damn Lies and Viral Content: How News Websites Spread (and Debunk) Online Rumors, Unverified Claims and Misinformation* (New York: Tow Center for Digital Journalism and Columbia Journalism School, 2015), http://towcenter.org/research/lies-damn-lies-and-viral-content; *see also* G. King, J. Pan, and M. E. Roberts, "How the Chinese Government Fabricates Social Media Posts for Strategic Distraction, not Engaged Argument," Working Paper, *American Political Science Review*, http://j.mp/1Txxiz1.

A Case Study 1: Syria

The volume of video generated in the ongoing conflict in Syria may very well be the new norm. As of summer 2017, estimates suggest that more than one million conflict-related videos have been filmed there. Put another way, it would take more time to view videos of the conflict than the amount of time the conflict has actually been taking place. Most of these videos can be found on social media sites like YouTube and Live Leak. Many others are circulated through e-mail and nonpublic networks like Telegram. Perpetrators (including armed combatants and Syrian military personnel), journalists, medical professionals, and citizens caught in the middle of fighting regularly post videos, photographs, and text updates of conflict situations on Twitter, Facebook, and other social media sites.

1 Overview of Projects

Several initiatives have turned to this open source material to better understand the situation in Syria. Perhaps the most extensive of these efforts is the Carter Center's Syria Conflict Mapping Project. Initiated in 2013, this project utilizes publicly available information to document the formation of armed groups in the country and the relationships among these groups; conflict events, including ground troop clashes, aerial bombardments, and artillery shelling; and sightings of advanced weaponry. In doing so, the Syria Conflict Mapping Project is able to better understand "the evolution of armed group relations, the geographic areas of control of the various antagonists involved, and the regional and international dimensions of the conflict."[13] The Carter Center is well aware that this information may be false, misleading, or incomplete, so those who run the project hold "regular discussions with conflict stakeholders [including face-to-face meetings in Turkey and within Syria itself, as well as through phone, Internet, or social media conversations] in order to ensure the accuracy of information collected and gain further insights regarding conflict development."[14]

The Carter Center effort is not the only project that that has sought to map the dynamics of the Syrian conflict through user-generated content. Over the last four years, a British citizen and blogger, Eliot Higgins, has been tracking weapons depicted in social media videos in Syria and is now considered to be one of the foremost experts on the conflict. There are also various military, intelligence, and supranational efforts to monitor open source material from the Syria conflict, but their details are less well known.

---

[13] Carter Center, *Syria Countrywide Conflict Report #4* (Atlanta, GA: Carter Center, 2014), p. 3, www.cartercenter.org/resources/pdfs/peace/conflict_resolution/syria-conflict/NationwideUpdate-Sept-18-2014.pdf.

[14] Ibid.

2 Weapons Analysis

Both Higgins and the Carter Center have used social media content to identify weapons used in the Syrian conflict. Beginning in 2012, Higgins and a network of citizen journalists began monitoring hundreds of YouTube channels for content from a diverse array of participants in the conflict in order to document the types of weapons being displayed.[15] Through this tedious and time-consuming work, they were among the first to document the Syrian government's use of Chinese-made cluster munitions and homemade barrel bombs in civilian areas, even though such weapons are considered illegal by most countries and the government denied their existence.[16] Higgins was also among the first to identify the funneling of weapons stockpiles from the former Yugoslavia to certain rebel groups in Syria, which led to the discovery of an arms supply network financed by the Saudi Arabian government and at least tacitly approved by the American government.[17]

The Carter Center's Syria Conflict Mapping Project has also monitored the flow of heavy and sophisticated weapons throughout the country by identifying this weaponry in social media videos. According to the project leader, Chris McNaboe, there are a variety of reasons why armed groups include information about their weapons capability in videos. Many groups boast of their capabilities (including posting videos of their sophisticated weapons) in order to intimidate enemies or convince funders and citizens to join their cause. Others are required by their weapons suppliers (especially governments) to provide proof of their use – e.g., we give you twenty rockets, you post twenty videos of them being used – to ensure that the weapons are not falling into the wrong hands.[18]

In its September 2014 report, the Carter Center analyzed more than 2,500 videos that provided information about weapons, which allowed it to "gain insight into the amounts, networks, timeframes, impacts, and intentions surrounding these efforts."[19]

---

[15] M. Weaver, "How Brown Moses exposed Syrian arms trafficking from his front room," *The Guardian*, March 21, 2013, www.theguardian.com/world/2013/mar/21/frontroom-blogger-analyses-weapons-syria-frontline.

[16] E. Higgins, "Clear Evidence of DIY Barrel Bombs Being Used by the Syrian Air Force," *Brown Moses Blog*, October 27, 2012, http://brown-moses.blogspot.co.uk/2012/10/clear-evidence-of-diy-barrel-bombs.html; E. Higgins, "Cluster Bomb Usage Rose Significantly Across Syria," *Brown Moses Blog*, October 12, 2012, http://brown-moses.blogspot.co.uk/2012/10/cluster-bomb-usage-rises-significantly.html.

[17] E. Higgins, "Evidence of Multiple Foreign Weapons Systems Smuggled to the Syrian Opposition in Daraa," *Brown Moses Blog*, January 25, 2013, http://brown-moses.blogspot.co.uk/2013/01/evidence-of-multiple-foreign-weapon.html; E. Higgins, "Weapons from the Former Yugoslavia Spread through Syria's War," At War: Notes from the Front Line, *The New York Times*, February 25, 2013, http://atwar.blogs.nytimes.com/2013/02/25/weapons-from-the-former-yugoslavia-spread-through-syrias-war/?_r=0; C. J. Chivers and E. Schmitt, "Saudis Step Up Help for Rebels in Syria With Croatian Arms," *The New York Times*, February 26, 2013, p. A1.

[18] C. McNaboe, e-mail to author, June 3, 2015.

[19] Carter Center, *Syria Countrywide Conflict Report #4*, p. 23.

The report argued that that the analysis of three particular weapons – the Croatian RAK-12 rocket launcher, the Chinese HJ-8 antitank guided missile, and the American BGM-71 TOW antitank guided missile – provided valuable clues about the arming of Syrian opposition groups by Saudi Arabia and Qatar with American assistance. The report further notes that all three weapons systems seem to have been distributed on a very limited basis to select groups, but adds that that they did not stay in the hands of the intended recipients.[20]

3 Network Dynamics

One of the most novel applications of user-generated content in the analysis of the Syrian conflict involves documentation of the emergence and shifting allegiances of armed groups. Beginning with the earliest defections from the Syrian Armed Forces and continuing with the formation of new armed groups to fight against Assad's regime, a large majority of anti-Assad fighters and factions announced their intention to defect via highly stylized videos posted online to social media. Indeed, as of June 2015, the Syria Conflict Mapping Project had access to an archive of videos documenting nearly 7,000 armed group formations that had been collected and maintained by the group Syria Conflict Monitor. While there are fighters and armed groups not represented in this collection, either because videos were not located or because the groups formed without such an online announcement, the Carter Center argues that the data gathered from these formation videos present a relatively good picture of the situation on the ground "due to the fact that many of the largest and most capable armed groups operating in Syria have a strong online presence."[21] By counting the number of people in each publicly available formation video, the Carter Center approximated between 68,639 and 85,150 fighters (it is often difficult to establish an exact count because of the low quality of some videos) across the country as of August 2013.[22]

The Carter Center acknowledges that it is very difficult to know what happens with fighters and units after they form, because this information is generally not available on social media. Based on an analysis of connections within the social networks of these groups, though, Carter Center researchers do claim to be able to determine which units are becoming more or less powerful within the opposition. They also make note of which units fade or disband over time based on social media activity. The Carter Center determines relationships among existing armed groups by their connections on social media. The more connections to a particular group,

---

[20] Ibid.
[21] Carter Center, *Regional Conflict Report: Ras al-Ain* (Atlanta: Carter Center, 2013), p. 2, www.cartercenter.org/resources/pdfs/peace/conflict_resolution/syria-conflict/Ras-al-AinReport.pdf.
[22] Carter Center, *Syria Countrywide Conflict Report #1* (Atlanta: Carter Center, 2013), p. 4, www.cartercenter.org/resources/pdfs/peace/conflict_resolution/syria-conflict/nationwidereport-aug-20-2013.pdf.

the more influential it is considered to be. These connections do not necessarily imply cooperation or subordination, though, and the report calls for additional research to understand the exact nature of the linkages.

One finding made by the Carter Center was that as of late 2013, although many groups had claimed allegiance to the Free Syrian Army (FSA, the organization formed by officers who defected from the Syrian Armed Forces at the beginning of the armed conflict) through national-level "military councils," the FSA remained largely a franchise organization, because local-level armed groups had very few direct connections with FSA leadership. "Instead," the report notes, those local groups "have sought to support themselves, and most have established their own patronage networks and leadership structures that have served to increase factionalism of the opposition."[23] This report also noted that social media connections demonstrated the growth of clear divisions among armed groups in particular regions, providing further evidence of factionalism on the ground. At the beginning of 2014, however, the Carter Center reported a coalescing of rebel forces on social media, presumably because of the necessity of banding together to simultaneously fight ISIS and the Syrian government. These new networks seemed to differ from previous military council-type arrangements in that they were much larger and demonstrated "a more credible commitment to integration than previous efforts."[24]

The Carter Center reported that component groups began to dismantle their own organizations' structures in order to better integrate into the larger command structure. "As a sign of this dissolution," the report notes, "component groups of the Islamic Front have been coordinating their imagery and public outreach via their various social media outlets. These groups now coordinate their Twitter hashtags, use a uniform profile picture for all Facebook, Twitter, and YouTube pages, and share each other's posts." Such coordination, however, should not imply unification, which "will prove to be more difficult than coordinating Twitter handles" or creating a few new integrated fighting units.[25] Indeed, many of these armed groups have very different understandings of the conflict, how they should be operating, and what the future of Syria ought to look like. Social media links also seem too tenuous to serve as the basis for legal claims about responsibility for war crimes and human rights abuses, given the multiplicity of reasons why one group might follow or friend another on a social media platform. Such links, however, ought to be considered valuable starting points for investigations into chains of command and strategic alliances.

---

[23] Carter Center, *Syria Countrywide Conflict Report* #2 (Atlanta, GA: Carter Center, 2013), p. 5, www.cartercenter.org/resources/pdfs/peace/conflict_resolution/syria-conflict/nationwideupdate_nov-20-2013.pdf.

[24] Carter Center, *Syria Countrywide Conflict Report* #3, p. 6.

[25] Ibid.

B Case Study 2: Russian Intervention in Ukraine

After gaining public attention for his work on weapons systems in Syria, Eliot Higgins and his colleagues in the citizen journalism community turned their attention to Ukraine in the summer of 2014 after the downing of Malaysia Airlines Flight 17 (MH17) by what was most likely a Russian Buk missile being operated by separatists in eastern Ukraine. Since this initial foray into the Ukraine crisis, Higgins and his collaborators have devoted significant resources to countering Vladimir Putin's claim that the conflict in Ukraine is solely a civil war and that the secessionists are merely disgruntled "people who were yesterday working down in mines or driving tractors."[26] Higgins and his collaborators use social media and satellite imagery to track the flow of weapons and soldiers from Russia over the border into the eastern section of Ukraine, where many ethnic Russians live and where pro-Russian and secessionist feelings are strongest. This content is sourced from media produced and uploaded by Russian soldiers fighting in Ukraine and Ukrainian and Russian civilians on both sides of the war who are "posting photographs and videos of convoys, equipment, and themselves on the Internet" on global sites like Instagram, Facebook, Twitter, and YouTube as well as regional sites like VKontact.[27] They also use commercially available satellite imagery to show the movement of Russian troops and weapons to and over the border with Ukraine. Higgins and his colleagues created the website Bellingcat.com to disseminate their open source findings and share the methods they use. Bellingcat also serves as a sort of virtual community for citizen journalists who analyze open source intelligence.

The Bellingcat team's analyses of publicly available material now holds so much weight in the policy world that the Atlantic Council relied heavily on them in its May 2015 report, entitled *Hiding in Plain Sight: Putin's War in Ukraine*, a direct response to the Russian government's demands for evidence to back up American and European accusations of its involvement in Ukraine. The title refers to the open source information that "provides irrefutable evidence of direct Russian military involvement in eastern Ukraine."[28] This report notes that "the aspect of Russian involvement in Ukraine with the widest breadth of open source information is the movement of heavy military equipment across the border, with hundreds of videos and photographs uploaded by ordinary Russians and Ukrainians who have witnessed direct Russian support of the hostilities in eastern Ukraine."[29] By geolocating reports of individually identifiable weapons (especially tanks, armored trucks, and mobile missile systems not known to be deployed by the Ukrainian Army), the Bellingcat team has been able to trace the flow of military equipment from Russia to separatist

---

[26] Quoted in M. Czuperski et al., *Hiding in Plain Sight: Putin's War in Ukraine* (Washington, DC: Atlantic Council, 2015), p. 7.
[27] Ibid.
[28] Ibid., p. 3.
[29] Ibid., p. 8.

groups in Ukraine. The team does so by analyzing serial numbers; visible markings such as words, phrases, and graphics; paint colors; damage patterns; and other unique identifiers present on the equipment.

The Bellingcat team has also used social media postings by soldiers and citizens to pinpoint the locations of Russian military personnel setting up camps near the border with Ukraine. One such example described in the Atlantic Council report is the Kuziminsky camp, which was "established only forty-six kilometers from the Ukrainian border" in the days after the annexation of the Crimea.[30] This camp did not exist before 2014. Kuziminsky camp, the report notes, "became the site for hundreds of military vehicles, including tanks from the 5th Tank Brigade" of the Russian Army, which is normally stationed far away in Siberia.[31] According to the Atlantic Council report, equipment staged at Kuziminsky camp and elsewhere later turned up in eastern Ukraine, and Bellingcat's analysis of artillery craters created during the conflict (using techniques modified from on-the-ground analysis methods published in US army manuals) suggests that at least some missiles were fired from the Russian side of the border near newly constructed military camps.[32] In a few cases, Russian soldiers actually filmed the launch of these weapons. Damage and craters within Ukraine can be tied directly back to these launch events through analysis of missile trajectories and geolocation of launch sites.[33] The purpose of this shelling, according to the report, was to provide cover for separatists during their offensives.

The report notes that personnel stationed at these camps were decisive in the defeat of the Ukrainian Army at Debaltseve in February 2015. While the Russian government openly acknowledges that hundreds or even thousands of Russian citizens crossed over the border into Ukraine to fight alongside local separatists, it vehemently denied that Russian soldiers had done so. As the Atlantic Council report notes, though, the mounting flow of military casualties back into Russia – revealed through monitoring of the border by the Organization for Security and Co-operation in Europe, reporting by various Western media outlets, and interviews with family members of dead soldiers by the nongovernmental organization Open Russia – contradicts this assertion.[34] Evidence from known Russian soldiers' unauthorized posting of photographs on their social media accounts provides additional evidence of their participation in hostilities within Ukraine.[35]

Perhaps the most damning claim put forth by Bellingcat was that a Russian-supplied Buk missile launcher was used by separatists when they accidentally shot down MH17. Using numerous images and videos from citizens posted to social

---

[30] Ibid., p. 13.
[31] Ibid.
[32] Ibid.
[33] The report provides documentation of Bellingcat's methods on pp. 18–19 and 28–31.
[34] Ibid., p. 17.
[35] Ibid., pp. 25–27.

media sites, Bellingcat investigators were able to trace the movement of a particular Russian Buk launcher (which they call "3x2" because of the identification number on its side) both through Russia to the Ukrainian border in June 2014 and through Ukraine to the likely launch site of the deadly attack in July 2014. The attack itself was established through social media images of a white smoke trail posted soon after the plane was downed; at least one of these images was verified using metadata from the original photograph as well as subsequent review of satellite imagery.[36]

The Russian government went to great lengths to refute claims that it supplied the weapon used to down the plane and instead placed blame on the Ukrainian Army. As early as July 21, 2014, just four days after the MH17 was shot down, the Russian Ministry of Defense published a series of satellite images that purported to show that the actual culprit was a Ukrainian Buk launcher, not a Russian one. These satellite images were used by Russia to claim that a Buk missile launcher and three support vehicles that had previously been seen parked at a Ukrainian military base near Donetsk prior to July 17 were no longer there, and that two Buk missile launchers as well as another Ukrainian military vehicle were stationed near where MH17 was shot down on July 17.

Bellingcat undertook a forensic analysis of the satellite images using a variety of methods, including metadata analysis (which can reveal evidence of manipulation); error-level analysis (which examines compression levels throughout the image – a big difference in compression in a particular area of the image would suggest a modification was made at that location); and reference analysis (which involves comparing an image with other sources of information to determine the extent to which its contents are plausible). In this case, imagery from Google Earth was compared to the images supplied by the Russian Ministry of Defense.[37] Bellingcat investigators determined, based on an examination of vegetation patterns and the growth of an oil leak from one of the trucks located at the site, that the satellite image the Russian Ministry of Defense claimed was taken on July 17, 2014, was actually taken sometime between June 1 and June 18, 2014, and, further, that it was "highly probable" that two large banks of clouds were digitally added to this image using Adobe Photoshop CS5 software to "obscur[e] details that could have been used for additional comparisons with historical imagery."[38]

---

[36] Bellingcat Investigation Team, "Origin of the Separatists' Buk: A Bellingcat Investigation," *Bellingcat*, November 8, 2014, www.bellingcat.com/news/uk-and-europe/2014/11/08/origin-of-the-separatists-buk-a-bellingcat-investigation; D. Romein, "Is This the Launch Site of the Missile That Shot Down Flight Mh17? A Look at the Claims and Evidence," *Bellingcat*, January 27, 2015, www.bellingcat.com/news/uk-and-europe/2015/01/27/is-this-the-launch-site-of-the-missile-that-shot-down-flight-mh17.

[37] Bellingcat Investigation Team, "Forensic Analysis of Satellite Images Released by the Russian Ministry of Defense," *Bellingcat*, May 31, 2015, www.bellingcat.com/wp-content/uploads/2015/05/Forensic_analysis_of_satellite_images_EN.pdf.

[38] Ibid., p. 18.

Bellingcat made similar claims about the digital manipulation of the Russian satellite images purporting to show two Ukrainian Buk missile launchers and another Ukrainian military vehicle in the area where MH17 was shot down on July 17. The report claims that all three locations that show these vehicles on the satellite imagery have a different level of compression (and thus a higher error rate) than other parts of the image. This led them to conclude that the image had been modified in some way in the regions showing the three vehicles. Additional analysis of this image based on a comparison of soil structures (i.e., patterns of vegetation and other markings that result from human activity such as agriculture) with historical Google Earth images suggests that it was taken prior to July 15, 2014. The Bellingcat investigative team ultimately concluded that "the Russian Ministry of Defense presented digitally modified and falsely dated satellite images to the international public in order to implicate the Ukrainian army in the downing of MH17."[39] They could only make this conclusion by analyzing the satellite imagery in tandem with video and photographs found on social media, demonstrating the value that user-generated content can add to investigations of conflict and human rights violations.

C Case Study 3: Nigeria and Boko Haram

Amnesty International recently released two major reports detailing the findings of its investigations into human rights abuses and war crimes committed by Boko Haram and Nigerian security forces in northeastern Nigeria. In both reports, video played a crucial role in demonstrating that the core elements of a crime had occurred. For example, video evidence corroborated Amnesty's claims about violations committed by Boko Haram and Nigerian forces; provided support for its conclusions that those committing the violations were under control of Boko Haram or Nigerian forces; provided evidence that neither force was protecting itself or the public from immediate harm; and provided additional context for Amnesty's conclusions.

The first report, *"Our job is to shoot, slaughter and kill": Boko Haram's Reign of Terror in North East Nigeria*, documents Boko Haram's utter disregard for human rights norms and international laws of war in its quest to impose its own extreme view of Islam on ever larger swaths of the country.[40] Amnesty claims that Boko Haram's actions have led to the deaths of more than 6,800 people, destroyed the homes and livelihoods of hundreds of thousands of people, and forced well over a million people from their homes. The group directly targets all people who do not follow their interpretation of Islam (called "kuffirs," or unbelievers), often killing them dozens at a time simply for not pledging allegiance, or destroying their homes

---

[39] Ibid., p. 42.
[40] Amnesty International, *"Our Job Is to Shoot, Slaughter and Kill": Boko Haram's Reign of Terror in North East Nigeria* (London: Amnesty International, 2015).

and property. Boko Haram also makes life extraordinarily difficult for all who reside in the territories it controls. Christians and members of the Civilian Joint Task Force (JTF) are the most common victims of these attacks, but Muslims who do not subscribe to Boko Haram's rigid interpretation of the Koran and laws of the Caliphate are also targeted. Civilian infrastructure – including schools, religious facilities, transportation systems, and places of business – is often destroyed in its military campaigns. Further, rape and sexual violence, sexual slavery, and forced marriages are all used as tools of war. There are also allegations of Boko Haram forcing children to fight on its behalf.

Many of these atrocities have been captured on video, and property destruction in Gamborou and Bama has been confirmed via satellite imagery. For example, Amnesty reports that after a failed military campaign in September 2014, Boko Haram fighters executed numerous prisoners at one of its facilities in Bama. Boko Haram later released propaganda videos of the killings, which occurred both in the prison and on a bridge in the area. Amnesty describes the first video as follows:

> [A]rmed men are seen offloading approximately thirteen bound men from a truck on a bridge. The detainees are lined up in a row and, one at a time, brought to the railings. The gunmen push each detainee's head between the railings, then they shoot the detainee in the head and tip the body into the river. The video shows eighteen men killed this way and the scene ends with more men being offloaded from a truck. The final scene shows gunmen walking through a small room with bunk beds, checking and then shooting at bodies lying on the floor. It is not possible to tell whether those on the ground were already dead.[41]

In the second video, a continuation of the first, Amnesty reports:

> One of the gunmen turns to the camera and explains that they are executing prisoners: "Our job is to kill, slaughter and shoot, because our promise between god and us is that we will not live with unbelievers. We are either in the grave and with unbelievers on the earth or unbelievers in the grave and us on the earth... There is either Muslim or disbeliever, it is either of the two you will be a Muslim or a non-Muslim. These ones are living under apostate government.[42]

According to Amnesty, local residents, human rights researchers, and satellite imagery confirmed that the locations seen in the videos were in Bama.[43]

In the second report, *Stars on Their Shoulders. Blood on Their Hands*, Amnesty alleges that the Nigerian military committed war crimes (and potentially crimes against humanity) in its noninternational armed conflict against Boko Haram in the northeastern part of the country.[44] The report alleges that Nigerian military

---

[41] Ibid., p. 60.
[42] Ibid.
[43] Ibid., pp. 60–61.
[44] Amnesty International, *Stars on Their Shoulders. Blood on Their Hands: War Crimes Committed by the Nigerian Military* (London: Amnesty International, 2015).

violations included extrajudicial execution of more than 1,200 people, arbitrary arrests of at least 20,000 people, hundreds or more enforced disappearances, the deaths of more than 7,000 people in military custody, and "countless" acts of torture.[45] The Amnesty report relied heavily on classic human rights investigation methods, including more than 400 interviews with "victims, their relatives, eyewitnesses, human rights activists, doctors, journalists, lawyers, and military sources," along with analysis of more than 800 leaked official government documents, including military reports and correspondence between high-level Nigerian military officials and local units stationed in the northeastern part of the country.[46] Amnesty also collected and analyzed "more than ninety videos and numerous photographs showing members of the security forces and their allied militia, the Civilian JTF, committing violations," plus satellite imagery of attack sites and mass graves. Some of these videos were gathered from social media outlets like YouTube and others were acquired by Amnesty International through its network of local researchers and NGO partners.[47]

Using this diverse array of material, Amnesty was able to corroborate victim and eyewitness statements with official government documents and video and photographic evidence, as well as provide context for this visual and textual evidence. Further, Amnesty also "shared its findings with the Nigerian authorities during dozens of meeting[s] as well as 55 written submissions, requesting information and specific action to address the violations."[48] Ultimately, Amnesty also made the report and accompanying evidence available to the Office of the Prosecutor at the International Criminal Court, since Nigeria is a party to the Rome Statute and the ICC has jurisdiction in this case if national courts are unwilling or unable to investigate the allegations and prosecute if appropriate.

As in the MH17 investigation, video and photographic evidence was not used alone in this case. Rather, it was combined with other forms of evidence. Visual evidence provided strong corroboration of evidence drawn from witness interviews that the "young men and boys" subjected to extrajudicial killings in twenty-seven incidents in 2013 and 2014 posed no immediate danger to military personnel or the general public, thus undermining claims that the killings were justified. The victims "were not taking part in hostilities, were not carrying arms, and were not wearing uniforms, insignia or other indications that they were members of Boko Haram."[49] In fourteen of these cases, the report notes, "military forces, sometimes in collaboration with Civilian JTF members, executed a large number of victims, at times dozens or even hundreds in one day."[50]

[45] Ibid., p. 4.
[46] Ibid.
[47] Ibid.
[48] Ibid.
[49] Ibid., p. 37.
[50] Ibid., p. 40.

Video evidence shows that the victims were under the firm control of the military and Civilian JTF members. Victims were shot or had their throats slit in relatively orderly operations without any judicial hearing, filing of formal charges, access to an attorney, or even cursory investigation. Further, evaluation of eyewitness testimony and visual evidence strongly suggests that many additional deaths in detention were caused by "starvation, thirst, severe overcrowding that led to spread of diseases, torture and lack of medical attention, and the use of fumigation chemicals [to kill vermin] in unventilated cells."[51]

Video evidence also played a key role in corroborating allegations of torture and other forms of ill treatment in the context of mass arrests. The report notes that videos show

> soldiers and members of the Civilian JTF beating suspects, making them lie down and walking on their backs, threatening them, humiliating them, tying their arms and making them roll in the mud, and in one case attempting to drown a suspect in a river. Several videos show soldiers loading detainees onto a military truck as if they are sand bags.[52]

A key aspect of Amnesty's argument is that the actions taken by the military and civilian JTFs appear to be similar at multiple locations across the northeastern part of the country, suggesting that there was widespread and systematic effort to target young men and boys regardless of whether they were actually Boko Haram. Most may have just seemed to be sympathetic to the cause, or were simply in the wrong place at the wrong time. The extent to which the policy was formalized cannot, of course, be deduced from video evidence, but the videos do make clear that the acts were not isolated, uncoordinated, or haphazard, but rather followed a similar pattern.[53]

One of the most disturbing instances occurred after Boko Haram fighters let hundreds of detainees out of the prison at the Giwa military barracks on the morning of March 14, 2014. Although the Nigerian military put up little resistance to Boko Haram's initial attack (and, in fact, seemed to have known about the attack in advance and cleared out the night before), they were ruthless in their efforts to seek out and recapture escapees later in the day. The military and civilian JTF conducted house-to-house searches and rounded up more than 600 escapees who had not already fled with Boko Haram fighters. Most of these individuals were extrajudicially executed during the course of the day. These events were recorded in twelve videos, some of which were shot by the perpetrators themselves, obtained by Amnesty International. The videos are described in the report as follows:

> One of the videos, apparently taken by a member of the Civilian JTF with the military commander's camera, shows 16 young men and boys sitting on the ground

---

[51] Ibid., p. 7.
[52] Ibid., p. 36.
[53] Ibid., pp. 37–38.

in a line. One by one, they are called forward and told by the military commander to lie down in front of a pit. Two or three men assist in holding the detainees. Armed soldiers and Civilian JTF members then cut the men's throats with a blade and dump them into an open mass grave. None of the men are tied and they seem to accept their fate without resistance, but the presence of armed soldiers may have prevented them from trying to escape. They may also have been in shock. The video shows five of the men being killed in this way. The fate of the remaining detainees is not shown on the video, but eyewitness accounts confirmed that nine of them had their throats cut while the others were shot dead.

A second video featuring some of the same perpetrators, taken earlier that day at the same location, shows two other detainees digging a grave, guarded by armed men in uniforms. The soldier who, according to witness testimony, is the commander of the group then tells one of the detainees to lie down in front of the pit. Men in civilian clothes who appear to be Civilian JTF members hold his legs and head, while the commander puts his right foot on the man's side, raises his knife, kisses it, shouts "Die hard Commando" and cuts the throat of the restrained young man. All the other soldiers and Civilian JTF members shout "Yes oga [boss], kill him."[54]

These descriptions were corroborated through numerous interviews, by matching uniforms with the battalion said to be involved in the operation (one soldier had the phrase "Borno State Operation Flush" on his uniform), and by the presence of an ID number on a weapon in one of the videos clearly linking it to the battalion in question.[55]

There are, of course, significant limitations to video evidence, like all other forms of evidence in human rights investigations. Lack of records, cover-up efforts by the military, and the challenges of obtaining eyewitness testimony all make it difficult to verify exactly how many extrajudicial executions take place even when events are caught on video. While Amnesty was able to successfully document more than 1,200 killings, the organization notes that it received numerous additional reports that lacked sufficient corroborating detail to be included in its report.[56]

IV DISCUSSION

The three case studies presented above do not represent the entire spectrum of uses of video in human rights advocacy and accountability efforts. They do not illustrate the use of video to protect and promote economic, social, and cultural rights or in

---

[54] Ibid., p. 46.
[55] These videos were first made public by Amnesty International in August 2014. Amnesty International, "Gruesome Footage Implicates Nigerian Military in Atrocities," www.amnestyusa.org/news/press-releases/gruesome-footage-implicates-nigerian-military-in-atrocities.
[56] Amnesty, *Stars on Their Shoulders. Blood on Their Hands*, p. 40.

American or Western European contexts. The three case studies do, however, demonstrate at least some of the ways in which video can be used in human rights and conflict documentation.

Especially in the cases of Ukraine and Nigeria, video is treated as one of many forms of evidence to be corroborated and integrated into a coherent package of evidence. In these two cases, video does not stand on its own in an investigation. The Carter Center takes a somewhat bolder approach, making claims about conflict dynamics – in terms of both the size and influence of particular armed groups and the weapons they possess – based on information extracted from social media networks and social media content.

While it is impossible to definitively test the Carter Center's claim that the online presence of armed groups is roughly similar to their actual presence (with the notable exception of the Islamic State), the Center's analysts have strong personal networks in Syria and in neighboring Turkey and are able to take advantage of human intelligence to point out errors and omissions in their open source intelligence. Even with this confirmation, the Carter Center does not claim that its findings are completely accurate or statistically valid, only that they provide insight that can be used in conflict analysis and in decision-making concerning diplomacy and humanitarian aid. The Carter Center recognizes that definitive quantitative claims cannot be made using social media data without significant additional corroboration and statistical analyses of uncertainty. Like most other forms of evidence, social media is a convenience sample – i.e., data that is relatively easily available to the investigator, as opposed to data collected through some form of systematically randomized or cluster sampling – and they acknowledge the limitations this creates.

The Ukraine case shows that governments are increasingly paying attention to the results of open source investigations, and sometimes go to great lengths (including falsifying data) to call their legitimacy into question. At the same time, it also demonstrates that in an age of open source intelligence, even ordinary people can contribute to the monitoring of conflict and human rights violations. Data about conflict and human rights abuse that was once only available to military or intelligence officials can now be found on the same platforms where cat videos, sports highlights, and other cultural ephemera are routinely shared.

The widespread availability of this content has numerous positive aspects, but it also creates significant challenges. At the most basic level, repeated viewings of audiovisual depictions of death, abuse, and torture can have a variety of negative impacts at the individual and societal levels. Excessive viewing of such content can traumatize the human rights investigator, journalist, or even the casual viewer. I experienced this vicarious or "secondary" trauma firsthand after seeing a particular image of a dismembered child in a video that depicted the aftermath of a shelling in Syria, and I have spoken to numerous colleagues who have experienced it as well. It is increasingly becoming a topic of analysis in both journalism and human rights work.[57] Indeed, much of the impetus for the research my human rights colleagues and I are conducting on semi-automating video analysis using machine learning and computer vision comes from a desire to limit exposure to images of human suffering and brutality.

A corollary to secondary trauma is that the proliferation of horrific content on the Internet may ultimately desensitize the public to what is being portrayed. At a certain point, many people will learn to tune out the reality of what they see because it is too graphic, while a minority of the public will find some perverse pleasure in it in an exploitative and voyeuristic, but not empathetic, way. There are many websites and YouTube channels that present the violence in Syria and other countries accompanied by intense music rather than historical, cultural, or situational context.[58] More work will have to be done to determine the effects of the availability of this content on public opinion, but there is little indication that it led to an upsurge in demand for humanitarian intervention in the conflict.

Even more concerning than the potential negative impacts of video in the human rights domain is the reality that the proliferation of audiovisual evidence of war crimes and human rights abuses has not yet resulted in greater accountability, at either the domestic or international level. Few high-level government or military officials accused of war crimes or crimes against humanity have been forced to admit culpability, provide reparations, or step down from positions of power due to the existence of video evidence. Nor have corporate decision-makers been held to account for the actions of their companies on communities, environments, or workers due to damning video evidence.

But perhaps this is too much to ask of video or any other form of evidence. Achieving justice and accountability requires political will. There is already ample evidence available to convict high-level perpetrators of war crimes and human rights abuses in many violent situations around the world, but no action is taken due to geopolitical or economic realities.

That said, advocacy efforts, including naming-and-shaming campaigns, often lead to important policy change in the short term, and the preservation of relevant content allows for historical clarification and the possibility of justice and accountability in the long term.[59] One need only to look at Guatemala's efforts to bring Efrain Rios Montt to trial for crimes committed during his presidency in 1982–83, Argentina's recent successful conviction of the architects of its Dirty War (including

---

[57] S. Dubberley, E. Griffin, and H. Mert Bal, "Making Secondary Trauma a Primary Issue: A Study of Eyewitness Media and Vicarious Trauma on the Digital Frontline," *Eyewitness Media Hub*, June 15, 2015, http://eyewitnessmediahub.com/research/vicarious-trauma.

[58] For a sampling, search YouTube for "Syria death," among many other combinations of terms. There are many videos with more than one million hits.

[59] J. Aronson, "Preserving Human Rights Media for Justice, Accountability, and Historical Clarification" (2017) **11** *Genocide Studies and Prevention: An International Journal* 82–99.

the dictator at the time and his closest associates), or the Extraordinary African Chamber's recent conviction of Chad's ex-dictator Hissène Habré for crimes against humanity that took place in the 1980s to see that justice can emerge many years, or even decades, after crimes occur.[60] Moving forward, video – properly understood, analyzed, and contextualized – will undoubtedly play an important role in these processes.

[60] "Guatemala Trials before the National Courts of Guatemala," *International Justice Monitor*, www.ijmonitor.org/category/guatemala-trials; "Operation Condor: Former Argentine junta leader jailed," *BBC* News, May 28, 2016, www.bbc.com/news/world-latin-america-36403909; "La Chambre Africaine D'assises Prononce La Perpetuite Contre Hissein Habre," Chambres africaines extraordinaires, May 30, 2016, www.chambresafricaines.org/index.php/le-coin-des-medias/communiqu%C3%A9-de-presse/638-la-chambre-africaine-d%E2%80%99assises-pronounce-la-perpetuite-contre-hissein-habre.html.

7

Big Data Analytics and Human Rights

Privacy Considerations in Context

Mark Latonero[*]

The technology industry has made lucrative use of big data to assess markets, predict consumer behavior, identify trends, and train machine-learning algorithms. This success has led many to ask whether the same techniques should be applied in other social contexts. It is unquestionable that new information and communication technologies bring both benefits and costs to any given domain. And so, when it comes to the human rights context, with higher stakes due to vulnerable populations, the potential risks in applying the technologies associated with big data analytics deserve greater consideration.

This chapter argues that the use of big data analytics in human rights work creates inherent risks and tensions around privacy. The techniques that comprise big data collection and analysis can be applied without the knowledge, consent, or understanding of data subjects. Thus, the use of big data analytics to advance or protect human rights risks violating privacy rights and norms and may lead to individual harms. Indeed, data analytics in the human rights monitoring context has the potential to produce the same ethical dilemmas and anxieties as inappropriate state or corporate surveillance. Therefore, its use may be difficult to justify without sufficient safeguards. The chapter concludes with a call to develop guidelines for the use of big data analytics in human rights that can help preserve the integrity of human rights monitoring and advocacy.

"Big data" and "big data analytics" are catchphrases for a wide range of interrelated sociotechnical techniques, tools, and practices. Big data involves the collection of large amounts of data from an array of digital sources and sensors. Collection often occurs unbeknownst to those who are data subjects. In big data, the subjects are the individuals creating content or emitting data as part of their everyday lives (e.g., posting pictures on social media, navigating websites, or using a smartphone

---

[*] Zachary Gold, JD, a research analyst at the Data & Society Research Institute, contributed research to an early version of this work that was presented as a conference paper.

with GPS tracking operating in the background). This data can be collected, processed, analyzed, and visualized in order to glean social insights and patterns. Behavioral indicators at either the aggregate or individual level can be used for observation, decision-making, and direct action.

Privacy is a fundamental human right.[1] As those in the human rights field increasingly address the potential impact of new information and communication technologies,[2] privacy is of particular concern. Indeed, as G. Alex Sinha states in Chapter 12, since the Edward Snowden revelations in 2013, "perhaps no human rights issue has received as much sustained attention as the right to privacy." Digital technologies have called into question the traditional expectations of privacy, including the right to be free from interference with one's privacy and control over one's personal information, the ability to be left alone, and the right to not be watched without permission. As the technology and ethics scholar Helen Nissenbaum states, "information technology is considered a major threat to privacy because it enables pervasive surveillance, massive databases, and lightning-speed distribution of information across the globe."[3]

The emergence of details about the use of pervasive surveillance technology in the post-Snowden era has only heightened anxieties about the loss of privacy. According to a 2014 report from a UN Human Rights Council (UNHRC) meeting on the right to privacy in the digital age, the deputy high commissioner noted that

> digital platforms were vulnerable to surveillance, interception and data collection... [S]urveillance practices could have a very real impact on peoples' human rights, including their rights to privacy, to freedom of expression and opinion, to freedom of assembly, to family life and to health. In particular, information collected through digital surveillance had been used to target dissidents and there were credible reports suggesting that digital technologies had been used to gather information that led to torture and other forms of ill-treatment.[4]

The UNHRC report gives examples of the risks of surveillance technologies, which expose sensitive information that can produce harms to political freedoms and

---

[1] See Universal Declaration of Human Rights, December 10, 1948, U.N. G.A. Res. 217 A (III), art. 12; International Covenant on Civil and Political Rights, December 16, 1966, S. Treaty Doc. No. 95–20, 6 I.L.M. 368 (1967), 999 U.N.T.S. 171, art. 17.

[2] M. Land et al. demonstrate how the impact of emerging information and communication technologies can be examined from a rights-based framework and international human rights law. M. K. Land et al., "#ICT4HR: Information and Communication Technologies for Human Rights," World Bank Institute, Nordic Trust Fund, Open Development Technology Alliance, and ICT4Gov, November 2012, https://papers.ssrn.com/sol3/papers.cfm?abstract_id=2178484.

[3] H. Nissenbaum, *Privacy in Context: Technology, Policy, and the Integrity of Social Life* (Palo Alto, CA: Stanford University Press, 2010).

[4] United Nations Human Rights Council, "Summary of the Human Rights Council panel discussion on the right to privacy in the digital age," December 19, 2014, www.un.org/en/ga/search/view_doc.asp?symbol=A/HRC/28/39.

physical security. The role of big data analytics in perpetuating anxieties over surveillance will be discussed later in this chapter, after highlighting the importance of understanding the human rights contexts in which big data analytics might transgress privacy norms. The chapter will first take a closer look at what is meant by privacy in relation to the technologies that comprise big data analytics in the human rights context.

I BIG DATA ANALYTICS FOR MONITORING HUMAN RIGHTS: COLLECTION AND USE

Assessing the legitimate application of big data analytics in human rights work depends on understanding what the right to privacy protects. Nissenbaum's framework of "contextual integrity" can provide a way to understand the value of privacy within the human rights context and the situations in which the use of big data might infringe this right. Contextual integrity ties "adequate protection for privacy to norms of specific contexts, demanding that information gathering and dissemination be appropriate to that context and obey the governing norms of distribution within it."[5] Nissenbaum does not go down the perilous path of trying to define privacy in absolute terms or finding a precise legal definition against the thicket of competing legal regimes, and neither will this chapter. Instead, she discusses privacy in terms of an individual's right to determine the flow of information about him- or herself.[6] What individuals care about, Nissenbaum argues, is that their personal information flows appropriately depending on social context. This chapter will employ similar terminology, such that privacy is violated when an individual's information is collected, analyzed, stored, or shared in a way that he or she judges to be inappropriate.

One challenge with examining whether a specific technology may violate privacy is that technology is not a single artifact that exists by itself. Technology is a combination of sociotechnical processes. Thus, it is useful to divide the technologies that comprise big data broadly into two categories: collection and use. As Alvaro Bedoya notes, most questions dealing with data privacy and vulnerable populations focus on how data will be used rather than how it will be collected.[7] Yet collection and use are intrinsically tied together. Disaggregating these categories into their individual processes provides us with a better understanding of potential privacy concerns related to human rights monitoring.

---

[5] H. Nissenbaum, "Privacy as Contextual Integrity" (2004) 79(1) *Washington Law Review* 119–58.
[6] Nissenbaum, *Privacy in Context*.
[7] A. Bedoya, "Big Data and the Underground Railroad," *Slate*, November 7, 2014, www.slate .com/articles/technology/future_tense/2014/11/big_data_underground_railroad_history_says_ unfettered_collection_of_data.html.

A Collection

Discovery, search, and crawling are activities that involve finding data sources that may contain information relevant to purpose, domain, or population. Data sources can be publicly available; for example, Twitter tweets, articles on online news sites, or images shared freely on social media.[8] Other sources, such as Facebook posts, are quasi-public in that they contain data that may be intended to be accessible only to specific members of an online community with appropriate login credentials and permissions. Other data sources, such as the e-mail messages of private accounts, are not publicly searchable. Even the collection of data that is publicly available can violate the privacy expectations of Internet users whose data is being collected. These users have their own expectations of privacy even when posting on sites that are easily accessible to the public. Users may feel that their posts are private, intended only for their friends and other users of an online community.[9]

Scraping involves the actual collection of data from online sources to be copied and stored for future retrieval. The practice of scraping can have an impact on individual Internet users as well. With scraping, users' data may be collected by an entity unbeknownst to them, breaching their privacy expectations or community/social norms.

Classification and indexing involve categorizing the collected data in a structured way so it can be searched and referenced. The data can be classified according to social categories created by the data collector or holder, such as name, gender, religion, or political affiliation. Classification extends the privacy risk to individual Internet users whose data has been collected. The subjects' data, put in a database, is now organized in a way that may not correctly represent those subjects or may expose them if the data were inadvertently released. Placing subjects' personally identifiably data into categories that may be incorrect may cast those in the dataset in a false light.

Storing and retention of large quantities of data is becoming more prevalent as storage become less expensive. This situation means that more data can be kept for longer periods of time by more entities. A post someone may have thought was fleeting or deleted can persist in numerous unseen databases effectively in perpetuity. Storing data for long periods of time exposes users to unforeseen privacy risks. Weak information security can lead to leaks or breaches that reveal personal data to others whom either the collectors or users did not intend to inform. This could expose individuals to embarrassment, extortion, physical violence, or other harms.

---

[8] See Chapter 6. *See also* J. Aronson, "Mobile Phones, Social Media, and Big Data in Human Rights Fact-Finding: Possibilities, Challenges, and Limitations," in P. Alston and S. Knuckey (eds.), *The Transformation of Human Rights Fact-Finding* (Oxford: Oxford University Press, 2015).

[9] In Chapter 12, G. Alex Sinha discusses the need to determine when revealing information to others constitutes a waiver of the human right to privacy.

B Use

Big data analytics involves deploying a number of techniques and tools designed to find patterns, behavioral indicators, or identities of individuals, groups, or populations. Structuring data, performing statistical modeling, and creating visualizations transform otherwise incomprehensible datasets into actionable information.

The threat to privacy from the use of big data analytics is clear. The entity performing the analysis could learn more about a person's life than would be anticipated by a typical citizen, thereby violating the right to determine the flow and use of one's personal information. By combining disparate data sources, these entities may be able to link online identities to real-world identities or find out about a person's habits or personal information.[10] There is also a risk is that the analysis is wrong. Datasets, and the analyses carried out on them, always carry some form of bias, and such analyses can lead to false positives and negatives that decision-makers may later act on. Deploying resources to the wrong place or at the wrong time can cause significant harm to individuals. Even if the analysis is correct in identifying a human rights violation, the victims may be put at greater risk if publicly identified due to community stigma or retaliation by perpetrators.

Access and sharing applies to both data use and collection. The way the data is indexed or classified or the type of data collected can already reveal information considered private. For some human rights issues, personally identifiable information is needed to monitor, assess, and intervene in real time. Unauthorized sharing and access presents a major breach of privacy norms in the human rights context.

II PRIVACY TRADE-OFFS, URGENCY, AND TEMPORALITY

Privacy considerations associated with applying big data analytics in the human rights context has received considerably less attention in the literature than issues like representativeness and validity. A recent volume edited by Philip Alston and Sarah Knuckey discusses new advances in technology and big data that are poised to transform the longstanding activity of human rights fact-finding.[11] In his contribution to that volume, Patrick Ball raises major reservations about using big data for human rights. Ball argues that sampling procedures and the data itself in many big datasets are riddled with biases.[12] Regardless of its "big-ness," data that is not

---

[10] "The 'mosaic theory' describes a basic precept of intelligence gathering: Disparate items of information, though individually of limited or no utility to their possessor, can take on added significance when combined with other items of information." D. E. Pozen, "The Mosaic Theory, National Security, and the Freedom of Information Act" (Note) (2005) 115 *Yale Law Journal* 628–79.
[11] Alston and Knuckey (eds.), *The Transformation of Human Rights Fact-Finding*.
[12] P. Ball, "The Bigness of Big Data," in Alston and Knuckey (eds.), *The Transformation of Human Rights Fact-Finding*, p. 428.

representative leaves out key populations. Incomplete data can hide the very populations most affected by human rights violations. For example, when estimating the number of victims in a conflict based on available data, those who were secretly killed by paramilitary groups and never reported in the media or government records are left out. The models and analyses resulting from such flawed data, Ball argues, can lead to faulty conclusions that could imperil or discredit human rights fact-finding and evidence gathering.

Patrick Meier, on the other hand, suggests an alternative perspective, buoyed by the high-profile applications of big data analytics in humanitarian crises.[13] From crowdsourced maps after the Haiti earthquake in 2009 to millions of tweets related to Hurricane Sandy in New York, such data is important and can help save lives. Meier concedes that big data can be biased and incomplete. Yet data and information on vulnerable populations are almost always lacking in completeness, even more so in the immediate aftermath of a crisis. Thus, big data, for all its flaws, can serve to inform decision-making in real time (i.e., during a crisis event) where comprehensive information does not exist.

One of the core questions that needs to be answered when using big data for human rights purposes is the extent to which the urgency of the need being addressed impacts the decision to use imperfect data and risk privacy violations. Consider, on the one hand, a fact-finding mission to ascertain whether a human rights violation took place in the past. Sometimes the data collection can take months or years in order to produce evidence for use in justice and accountability proceedings or for historical clarification. On the other hand, in a humanitarian response to a disaster or crisis, data collection seeks to intervene in the present, to find those in immediate need. Of course, temporality is not an absolute rule separating human rights and humanitarian domains. Data can be collected both in protracted humanitarian situations and when human rights violations are happening in the present. The issue at hand is whether an urgent situation places unique demands on the flow of personal information, which impacts one's dignity, relationships with others, and right to privacy.

During humanitarian responses to time-bound crises like natural disasters, decisions must often be made between maintaining privacy and responding quickly by using available data that is often deeply personal. For example, in the response to the Ebola crisis, humanitarian organizations deliberated on how big data could be used legitimately.[14] Consider a responder requesting the mobile phone contact list of a person who tested positive for Ebola in an attempt to stop the spread of the disease.

---

[13] P. Meier, *Digital Humanitarians* (Boca Raton, FL: CRC Press, 2015); P. Meier, "Big (Crisis) Data: Humanitarian Fact-Finding with Advanced Computing," in Alston and Knuckey (eds.), *The Transformation of Human Rights Fact-Finding*.

[14] B. Campbell and S. Blair, "How 'big data' could help stop the spread of Ebola," *PRI's The World*, October 24, 2014, www.pri.org/stories/2014-10-24/how-big-data-could-help-stop-spread-ebola.

Further, consider a response organization asking a mobile phone company for the phone numbers and records of all the users in the country in order to trace the network of individuals who may have become infected. That data would need to be analyzed to locate those at risk, and personal information might be shared with other responders without the consent of the data subjects.

The assumption in such an operational decision is that saving the lives of the person's contacts and protecting the health of the broader public outweigh the privacy concerns over that person's personal information. Yet such decisions about trade-offs, particularly in the case of the Ebola response, remain highly controversial due to the potential privacy violations inherent in collecting the mobile phone records of individuals at a national level without the consent of individuals.[15] Even in an urgent humanitarian response context, there is little agreement about when it is appropriate and legitimate to limit privacy rights. Applying real-time data collection and analytics to the investigation of historical human rights violations could raise even more concerns.

When, if ever, is it appropriate to limit privacy rights in order to achieve human rights objectives? Big data can support human rights fact-finding by providing a basis for estimating the probability that an individual falls within a group that has been subjected to past rights violations. Data analytics can also personally identify an individual whose rights are being violated in the present. For example, access to millions of mobile phone records in a large city may reveal patterns of calls by individuals that suggest human trafficking for commercial sexual exploitation. An analysis of these "call detail records" can indicate the movement of people to buildings where exploitation is known to take place at unique times and pinpoint calls with known exploiters. Yet to establish whether a specific individual's rights have been violated, personal information like cell phone numbers, names, and building ownership would need to be identified. Such data can reveal sensitive information (e.g., religion, political affiliation, or sexual orientation) that could lead to physical harm and retribution if shared with the wrong party.

Ultimately, permissible limitations on the right to privacy will vary depending on the urgency and importance of the objective to be achieved. The accompanying graph gives a visual sketch of the potential trade-offs involved in privacy. The right to privacy is not an absolute right, and it may be limited in order to achieve a legitimate objective. The extent to which the right is infringed, however, must be proportional to the objective. Thus, the graph illustrates a scenario involving a decision about the proportional relationship between the urgency of the objective of human security and permissible limits on privacy. In a human rights context, protecting the right to privacy is of high concern since safeguarding an individual's data also protects the right to freedom of expression, association, and related rights. Yet, when there is an

---

[15] S. McDonald, "Ebola: A Big Data Disaster," The Centre for Internet and Society, March 1, 2016, http://cis-india.org/papers/ebola-a-big-data-disaster.

increasing threat to human security and safety, such as the imminent danger of an individual's death due to violent conflict, the concern over privacy may start to decrease. Perhaps a decision must be made about whether to obtain or release an individual's personal mobile phone's GPS coordinates without their consent or permission in order to attempt to locate and rescue that person to save his or her life. There may come an inflection point in the situation where the immediate danger to human security is higher than the protection of an individual's privacy. When there is no longer an immediate threat to human security, there is every reason to uphold the integrity of the individual's privacy right at a high level. Of course, there are a number of additional factors that might influence this analysis and decision. In some cases the release of an individual's data will expose that person to the very forces threatening his or her safety and security or may put another individual directly at risk. Clearly, more foundational work is needed to begin to understand how these trade-offs may be operationalize in practice when making a real-time decision.

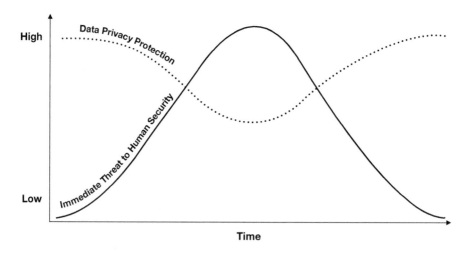

Proportionality of Rights to Data Privacy and Human Security Over Time

Does this mean that we cannot use big data collection for long-term and continuous human rights monitoring? Long-term monitoring of digital data sources for human rights violations holds the promise of providing unprecedented insight into hidden or obscured rights abuses such as labor violations, sexual abuse/exploitation, or human trafficking. The use of big data analytics for monitoring human rights, even by human rights organizations, carries both known and unknown risks for privacy. Indeed, the very nature of data collection and analysis can conflict with normative expectations of privacy in the human right context. It is unclear whether the uncertain benefits of long-term human rights monitoring and advocacy can outweigh the very concrete risks to privacy that accompany the use of big data.

III HUMAN RIGHTS MONITORING AND SURVEILLANT ANXIETIES

The human rights field has a long tradition of employing methods and techniques for protecting privacy while monitoring rights abuses and violations. Determining "who did what to whom" involves identifying victims and alleged violators, doing an accounting of the facts, and collecting relevant information.[16] Ensuring the privacy and confidentiality of sources and victims is a critical concern in human rights fact-finding. A training manual published by the UN Office of the High Commissioner for Human Rights (OHCHR), for example, urges the use of monitoring practices that "keep in mind the safety of the people who provide information," seeking consultation in difficult cases and maintaining confidentiality and security.[17] At the local level, Ontario's Human Rights Commission states that data collection should include informing the public, consulting with the communities that will be affected, using the least intrusive means possible, assuring anonymity where appropriate, and protecting privacy.[18]

In reality, though, ethical data collection protocols that protect privacy in the human rights context, such as obtaining informed consent, are extraordinarily difficult to utilize when deploying big data analytics. Big data analytics often relies on "found" data, which is collected without a user's consent or even knowledge. It also necessarily involves using this information in ways not intended by the individual to whom it relates. Thus, with respect to big data, privacy harms are likely unavoidable.

The use of big data also harms privacy by adding to a growing, although ambiguous, sense of "surveillant anxiety," in which we fear surveillance to the point that it affects our thoughts, behaviors, and sense of self. Kate Crawford describes this anxiety as "the fear that all the data we are shedding every day is too revealing of our intimate selves but may also misrepresent us... [N]o matter how much data they have, it is always incomplete, and the sheer volume can overwhelm the critical signals in a fog of possible correlations."[19] A fact-finding mission that creates privacy

---

[16] UN Human Rights Office of the High Commissioner, *Monitoring Economic, Social and Cultural Rights*, HR/P/PT/7/Rev. 1, 2011, www.ohchr.org/Documents/Publications/Chapter20-48pp.pdf.

[17] UN Human Rights Office of the High Commissioner, *Training Manual on Human Rights Monitoring: The Monitoring Function*, March 21, 1999, www.ohchr.org/Documents/Publications/training7part59en.pdf, § A–C, G, J–K.

[18] Ontario Human Rights Commission, "Count me in! Collecting human rights based data – Summary (fact sheet)," www.ohrc.on.ca/en/count-me-collecting-human-rights-based-data-summary-fact-sheet.

[19] K. Crawford, "The Anxieties of Big Data," *The New Inquiry*, May 30, 2014, https://thenewinquiry.com/the-anxieties-of-big-data/.

risks is not equivalent to surveillance. And certainly not all surveillance technologies violate privacy in a legal sense. Yet any use of big data analytics to monitor human rights creates concerns and anxieties about surveillance, whether or not the surveillance is intended for "good."

Thus, the issue that must be addressed is whether human rights researchers using big data analytics would themselves produce this kind of surveillant anxiety in their data subjects in ways that feel similar to traditional government or corporate surveillance. According to the Special Rapporteur on the right to Privacy, surveillance creates privacy risks and harms such that "increasingly, personal data ends up in the same 'bucket' of data which can be used and re-used for all kinds of known and unknown purposes."[20] Although surveillance for illegitimate purposes necessarily violates privacy, even surveillance for legitimate purposes will do so if the associated privacy harms are not proportional to that purpose. For example, a public health surveillance program that continuously monitors and identifies disease in a population might constitute an appropriate use for a common good shared by many, but could still violate privacy if it produces undue harms and risks. As Jeremy Youde's study of biosurveillance contends:

> [T]he individual human right to privacy had the potential to be eroded thought the increased use of biosurveillance technology by governments and international organizations, such as WHO. This technology requires an almost inevitable intrusion into the behaviours, habits, and interests of individuals – collecting data on individual entries into search engines, Facebook entries in individual travel history and purchases.[21]

In their work on disease surveillance in the United States, Amy L. Fairchild, Ronald Bayer, and James Colgrove document the conflict around public health surveillance during the early days of the AIDS crisis and the struggle to balance the need to collect and share medical records against charges of institutional discrimination against marginalized groups.[22]

Big data collection and analysis by technology companies, sometimes called corporate surveillance, can produce the same kinds of anxieties. And big data collection by either governments or human rights organizations often rely on technologies that serve as intermediaries to the digital life of the public. Both a government agency and a human rights organization may collect data on the lives of millions of individuals from major social media platforms like Facebook. The very same tools, techniques, and processes in the collection and use of big data can be

---

[20] UN Human Rights Council, *Report of the Special Rapporteur on the Right to Privacy, Joseph Cannataci*, U.N. Doc. A/HRC/34/60 (February 24, 2017), p. 9 ("Cannataci Report").
[21] S. Davies and J. Youde (eds.), *The Politics of Surveillance and Response to Disease Outbreaks: The New Frontier for States and Non-State Actors* (London: Routledge, 2016), p. 3.
[22] A. Fairchild, R. Bayer, J. Colgrove, *Searching Eyes: Privacy, the State, and Disease Surveillance in America* (Berkeley: University of California, 2007).

employed to both violate and protect human rights. The collection of one's personal data by governments, corporations, or any number of other organizations using big data analytics may contribute to the constant feeling of being watched and curtail privacy and freedom of expression.

The pressing question is whether the use of big data by human rights organizations is permissible, given the risks to privacy involved. It is fair to say that human rights organizations should be held to the same standards that privacy advocates require of government. Yet the best intentions of human rights organizations using big data are not enough to protect privacy rights or automatically justify privacy violations. Furthermore, any organization collecting, classifying, and storing sensitive human rights data needs to address issues like data protection, secure storage, safe sharing, and access controls. If a human rights organization deploys data collection and analytic tools, how can they incorporate safeguards that responsibly address the inherent risks to privacy and minimize the potential harms?

IV TOWARD GUIDELINES FOR BIG DATA APPLICATIONS IN HUMAN RIGHTS

Because of concerns about privacy, value trade-offs, surveillance, and other potential risks and harms that may befall vulnerable populations, a rigorous assessment of the legitimacy and impact of the use of any big data analytics by organizations in the human rights context is vital. This begs the question of what type of guidelines could help steer such an assessment, particularly given the global proliferation of technologies and the plethora of context-specific harms. As the UN Special Rapporteur on Privacy states, "The nature of trans-border data flows and modern information technology requires a global approach to the protection and promotion of human rights and particularly the right to privacy."[23] Human rights monitoring organizations may need to update their standards for data collection and analysis to take new technologies like big data into account.

New technological applications do not necessarily require entirely new high level principles. For example, the principles of safety, privacy, and confidentiality outlined in the OHCHR Training Manual would still apply to big data, but these principles may need further elaboration and development when they are applied to new information collection regimes. At the same time, new technological challenges need new solutions. For example, confidentiality policies may not be achieved simply through anonymization techniques alone, since the currently accepted fact in computer science is that no dataset can be fully anonymized.[24]

---

[23] Cannataci Report, p. 9.
[24] *See* P. Ohm, "Broken Promises of Privacy: Responding to the Surprising Failure of Anonymization" (2009) 57 *UCLA Law Review* 1701–88; A. Narayanan and E. Felton, "No silver bullet: De-identification still doesn't work," July 29, 2014, http://randomwalker.info/publications/no-silver-bullet-de-identification.pdf.

A major advance in addressing the ethical and responsible use of data is the Harvard Humanitarian Initiative's Signal Code, subtitled "A Human Rights Approach to Information during Crisis," which focuses on the right to information in times of crises. Bridging the humanitarian and human rights contexts, the code states that the right to privacy is fundamental in crisis situations and that "[a]ny exception to data privacy and protection during crises exercised by humanitarian actors must be applied in ways consistent with international human rights and humanitarian law and standards."[25] Other standards in the code include giving individuals the right to be informed about information collection and use as well as the right to have incorrect information about themselves rectified.

Policy work from other fields, such as international development, will also be relevant to creating guidelines for human rights. UN Global Pulse, the data innovation initiative of the UN Secretary General, has published its own "privacy and data protection principles"[26] for using data in development work. These principles recommend that actors "adhere to the basic principles of privacy," "maintain the purpose for data," and ensure "transparency [as] an ongoing commitment."[27] Human rights organizations may also learn lessons from civil society's demands of governments engaging in surveillance. The International Principles on the Application of Human Rights to Communications Surveillance suggests a number of guidelines designed to ensure that government surveillance does not infringe on the right to privacy, including notifying people when they are being watched.[28]

This brings us to the issue of how to ensure the accountability of human rights organizations using big data for interventions or monitoring. Any actor that uses big data to intervene in or monitor human rights, whether a small domestic NGO or a large international organization, should be responsible for the potential risks and harms to the very populations it seeks to help. Yet since human rights organizations often hold governments to account, asking governments to regulate the use of big data by those very organizations will likely provide an avenue for state suppression of human rights monitoring. As such, traditional regulatory mechanisms are unlikely to be effective in this context.

One emerging framework that seeks to regulate any entity engaged in data collection is the EU's General Data Protection Regulation, which comes into

---

[25] F. Greenwood, et al., "The Signal Code: A Human Rights Approach to Information during Crisis," Harvard Humanitarian Initiative, January 2017, http://hhi.harvard.edu/publications/signal-code-human-rights-approach-information-during-crisis.
[26] UN Global Pulse, "Privacy and Data Protection Principles," www.unglobalpulse.org/privacy-and-data-protection; *see also* UN Global Pulse, "Workshop on ICT4D Principle 8: Address Privacy & Security in Development Programs," www.unglobalpulse.org/events/workshop-ict4d-principle-8-address-privacy-security-development-programs.
[27] UN Global Pulse, Unpublished report on data privacy and data security for ICT4D (2015).
[28] Necessary and Proportionate, "International Principles on the Application of Human Rights to Communications Surveillance," May 2014, https://necessaryandproportionate.org/.

force in 2018.[29] This regulation requires all organizations collecting data on EU residents to follow privacy directives about data collection, protection, and consent. For example, individuals in the EU would have the right to withdraw consent given to organizations that collect or process their data. Such organizations would be required to alert the authorities if a data breach of personal information occurred. Any organization, including both companies and nonprofits, could incur heavy fines for noncompliance, such as 4 percent of global revenue.

Another possible avenue may lie in encouraging human rights organizations to engage with technology privacy professionals to assess the possible use of new technologies like big data. Professionals with the appropriate technical, legal, and ethical expertise may be capable of conducting privacy impact assessments of risks that may harm vulnerable populations *before* a new technology is deployed.[30] Or perhaps large donors can require grantees to follow data privacy protections as a condition of ongoing funding.

At the end of the day, though, the lack of certainty about the effectiveness of these approaches speaks to the pressing need for both more foundational research and context-specific assessments in this area. Addressing questions about context would require researchers to include more direct input and participation of the vulnerable communities themselves; for example, through field research. As Nissenbaum suggests, protecting privacy is about upholding the contextual integrity of the underlying norms governing information flow. Understanding norms around data privacy and consent, for example, should necessarily involve the communities where those norms exist.

Applying big data analytics in human rights work reveals tensions around privacy that need to be resolved in order to guide current thinking and future decisions. Unfortunately, a sustained knowledge gap in this area puts vulnerable populations at greater risk. And the anxieties and trade-offs around norms and interventions will be compounded as the human rights field addresses "newer" technologies such as artificial intelligence (AI).[31] Since AI is fueled by big data collection and analytics the same privacy concerns discussed above would apply. Furthermore, AI entails some form of automated decision making, which creates dilemmas over whether only a human rights expert, rather than a computer algorithm, can decide about proportionality and trade-offs between rights in real time. A research agenda should work toward guidelines, principles, and practices that *anticipate* the risks, costs, and benefits inherent in each process involved in emerging technological interventions in the human rights context.

---

[29] General Data Protection Regulation (Regulation [EU] 2016/679), 2–17, http://data.consilium.europa.eu/doc/document/ST-5419-2016-INIT/en/pdf.

[30] See Lea Shaver's chapter (Chapter 2) for an argument that the right to science also requires such an assessment.

[31] See Enrique Piracés's chapter (Chapter 13).

8

The Challenging Power of Data Visualization for Human Rights Advocacy

John Emerson, Margaret L. Satterthwaite, and Anshul Vikram Pandey

I INTRODUCTION

In September 2007, *The New York Times* columnist Nicholas Kristof traveled with Bill Gates to Africa to look at the work the Bill & Melinda Gates Foundation was doing to fight AIDS. In an e-mail to a *Times* graphics editor, Kristof recalls:

> while setting the trip up, it emerged that his initial interest in giving pots of money to fight disease had arisen after he and melinda read a two-part series of articles i did on third world disease in January 1997. until then, their plan had been to give money mainly to get countries wired and full of computers.
>
> bill and melinda recently reread those pieces, and said that it was the second piece in the series, about bad water and diarrhea killing millions of kids a year, that really got them thinking of public health. Great! I was really proud of this impact that my worldwide reporting and 3,500-word article had had. But then bill confessed that actually it wasn't the article itself that had grabbed him so much – it was the graphic. It was just a two column, inside graphic, very simple, listing third world health problems and how many people they kill. but he remembered it after all those years and said that it was the single thing that got him redirected toward public health.
>
> No graphic in human history has saved so many lives in africa and asia.[1]

Kristof's anecdote illustrates the sometimes unexpected power of data visualization: Expressing quantitative information with visuals can lend urgency to messages and make stories more memorable.

Data visualization is the "visual representation of 'data,' defined as information which has been abstracted in some schematic form."[2] The use of data visualization

---

[1] "Talk to the Newsroom: Graphics Director Steve Duenes," *The New York Times*, February 28, 2008, www.nytimes.com/2008/02/25/business/media/25asktheeditors.html.
[2] M. Friendly and D. J. Denis, "Milestones in the history of thematic cartography, statistical graphics, and data visualization," August 24, 2009, www.math.yorku.ca/SCS/Gallery/milestone/milestone.pdf.

can strengthen human rights work when data is involved, and it does something for the promotion of human rights that other methods don't do. Combining data and visuals allows advocates to harness the power of both statistics and narrative. Data visualization can facilitate understanding and ultimately motivate action. And within human rights research, it can help investigators and researchers draw a bigger picture from individual human rights abuses by allowing them to identify patterns that may suggest the existence of abusive policies, unlawful orders, negligence, or other forms of culpable action or inaction by decision-makers. As human rights researchers and advocates look for new ways to understand the dynamics behind human rights violations, get their messages across, and persuade target audiences, they are also expanding the epistemology of advocacy-oriented human rights research. By broadening their evidence base and using new methods, human rights advocates come to know different things – and to know the same things differently.

The use of data visualization and other visual features for human rights communication and advocacy is a growing trend. A study by New York University's Center for Human Rights and Global Justice reviewing all Human Rights Watch (HRW) and Amnesty International reports published in 2006, 2010, and 2014 revealed an increase in the use of photographs, satellite imagery, maps, charts, and graphs.[3] In some cases, data visuals augment existing research and communications methodologies; in other cases, they represent alternative and even novel tools and analytical methods for human rights NGOs.

While data visualization is a powerful tool for communication, the use of data and visualization holds exciting promise as a method of knowledge production. Human rights researchers and advocates are adding new methodologies to their toolbox, drawing on emerging technologies as well as established data analysis techniques to enhance and expand their research, communications, and advocacy. This chapter introduces ways data visualization can be used for human rights analysis, advocacy, and mobilization, and discusses some of the potential benefits and pitfalls of using data visualization in human rights work. After a brief historical review of data visualization for advocacy, we consider recent developments in the "datafication" of human rights, followed by an examination of some assumptions behind, and perils in, visualizing data for human rights advocacy. The goal of this chapter is to provide sufficient grounding for human rights researchers to engage with data visualization in a way that is as powerful, ethical, and rights-enhancing as possible.

II A BRIEF HISTORY OF STATISTICAL GRAPHICS AND ADVOCACY

Visual storytelling has a long and colorful history. Past generations not only created depictions of their reality, but also crafted visual explanations and diagrams to

---

[3] K. Rall, "Data Visualization for Human Rights Advocacy" (2016) 8 *Journal of Human Rights Practice* 171–97.

convey and understand the invisible forces governing the visible world and other realms beyond perception. In *The Book of Trees: Visualizing Branches of Knowledge*, Manuel Lima charts the use of the branching tree as a visual metaphor in charts from Mesopotamia to medieval Europe to the present.[4] In addition to visual storytelling, ancient civilizations developed visual methods to record, understand, and process large numbers. The ancient Babylonians, Egyptians, Greeks, and Chinese used visual systems to record data about vital resources, chart the stars, and map their territories.[5] While visual storytelling was intended for communication, data visualization was used by the ruling powers to interpret and keep tabs on their empires.[6]

Along with the development of modern statistics in the late eighteenth century, there emerged graphical methods of quantitative analysis – new kinds of charts and graphs to visually show patterns in data.[7] The Scottish political economist William Playfair was a mechanical engineer, statistician, and activist who authored essays and pamphlets on the politics of the day and helped storm the Bastille in Paris in 1787.[8] That same year, he published the kind of line, area, and bar charts that we routinely use today for the first time in his *Commercial and Political Atlas*[9] to display imports and exports between Scotland and other countries and territories. He published the first modern pie chart in his 1801 *Statistical Breviary*.[10]

In the first half of the nineteenth century, new technologies helped spread and inspire enthusiasm for both statistics and data visualization.[11] Commercial mechanical devices for counting, sorting, and calculating became popular, and the first successful mass-produced mechanical calculator was launched in 1820.[12] Printing technology, particularly lithography and chromolithography, enabled a more

---

[4] M. Lima, *The Book of Trees: Visualizing Branches of Knowledge* (New York: Princeton Architectural Press, 2014).

[5] F. Cajori, *A History of Mathematical Notations* (Mineola, NY: Dover Publications, 1928), pp. 2–18, 21–29, 43–44; see also *History of Cartography* (Chicago: University of Chicago Press, 1997), vol. 1, pp. 107–147, vol. 2, pp. 96–127.

[6] J. C. Scott, *Seeing Like a State: How Certain Schemes to Improve the Human Condition Have Failed* (New Haven, CT: Yale University Press, 1998), pp. 2–52.

[7] M. Friendly, "The Golden Age of Statistical Graphics" (2008) 23 *Institute of Mathematical Statistics in Statistical Science* 502–35.

[8] I. Spence and H. Wainer, "Who Was Playfair" (1997) 10(1) *Chance* 35–37.

[9] W. Playfair, *The Commercial and Political Atlas: Representing, by Means of Stained Copper-Plate Charts, the Progress of the Commerce, Revenues, Expenditure and Debts of England during the Whole of the Eighteenth Century*, 3rd ed. (New York: Cambridge University Press, 2005).

[10] W. Playfair, *Statistical Breviary; Shewing, on a Principle Entirely New, the Resources of Every State and Kingdom in Europe* (London: Wallis, 1801). For more on Playfair's development of the pie chart, see I. Spence, "No Humble Pie: The Origins and Usage of a Statistical Chart" (2005) 30 *Journal of Educational and Behavioral Statistics* 353–68.

[11] M. Friendly, "The Golden Age of Statistical Graphics" (2008) 23 *Institute of Mathematical Statistics in Statistical Science* 502–35.

[12] P. A. Kidwell, "American Scientists and Calculating Machines – From Novelty to Commonplace" (1990) 12 *IEEE Annals of the History of Computing* 31–40.

expressive range of printing. The Statistical Society of London was founded in 1834 and, by royal charter, became the Royal Statistical Society in 1887.[13] As the psychology scholar Michael Friendly notes, between 1850 and 1900, there was explosive growth in both the use of data visualization and the range of topics to which it was applied.[14] In addition to population and economic statistics, mid-nineteenth-century Paris saw the publication of medical and mortality statistics, demographics, and criminal justice data.[15]

A few examples from this period show how data visualization contributed new findings using spatial and other forms of analysis, and allowed such findings to be made meaningful to a broader public via visual display.

In 1854, Dr. John Snow mapped cholera deaths around the thirteen public wells accessed in the Soho district of London.[16] Using this method, he made a dramatic discovery: there was a particular cluster of deaths around one water pump on Broad Street. His findings ran contrary to the prevailing theories of disease. At the time, the medical establishment believed in the miasma theory, which held that cholera and other diseases, such as chlamydia and the Black Death, were caused by "bad air." Dr. Snow believed that cholera was spread from person to person through polluted food and water – a predecessor to the germ theory of disease. Despite skepticism from the medical establishment, Snow used his map to convince the governing council to remove the handle from the pump, and the outbreak quickly subsided.[17]

Florence Nightingale is known primarily as one of the founders of modern nursing, but she was also a statistician who used data visualization to campaign for improvements in British military medicine.[18] In 1858, she popularized a type of pie chart known as the polar area diagram.[19] The diagram divides a circle into wedges that extend at different lengths from the center to depict magnitude. Nightingale used statistical graphics in her reports to members of Parliament about the condition of medical care in the Crimean War to illustrate how improvements in hygiene could save lives: at a glance, one could see that far more soldiers died of sickness

---

[13] I. D. Hill, "Statistical Society of London – Royal Statistical Society: The First 100 Years: 1834–1934" (1984) 147 *Journal of the Royal Statistical Society. Series A (General)* 130–39 at 131, 137.

[14] See Friendly, "The Golden Age" at 502.

[15] See ibid.

[16] For an extensive discussion of Dr. Snow's advances in graphical reasoning, see E. R. Tufte, *Visual Explanations: Images and Quantities, Evidence and Narrative* (Cheshire, CT: Graphics Press, 1997) pp. 27–37.

[17] The map also shows notable outliers: there were no cholera deaths reported at the neighboring brewery, where, presumably, there were other things to drink. See E. Tufte, *The Visual Display of Quantitative Information*, 2nd ed. (Cheshire, CT: Graphics Press, 2001), p. 30.

[18] I. B. Cohen, "Florence Nightingale" (1984) 250 *Scientific American* 128–37.

[19] S. Rogers, "Florence Nightingale, datajournalist: Information has always been beautiful," *The Guardian*, August 13, 2010, www.theguardian.com/news/datablog/2010/aug/13/florence-nightingale-graphics.

than of wounds sustained in battle.[20] Nightingale persuaded Queen Victoria to appoint a Royal Commission on the Health of the Army, and her advocacy, reports, and the work of the commission eventually led to systemic changes in the design and practices of UK hospitals.[21]

After a long and distinguished career as a civil engineer in France, Charles Minard devoted himself in 1851 to research illustrated with graphic tables and figurative maps.[22] His 1869 visualization of Napoleon's 1812 Russian Campaign shows the march of the French army from the Polish-Russian border toward Moscow and back.[23] The chart was heralded by Minard's contemporaries and is held up by twentieth-century data visualization critics as a marvel of clarity and data density. The visualization displays six different dimensions within the same graphic: The thickness of the main line shows the number of Napoleon's troops; the scale of the line shows the distance traveled; rivers are depicted and cities are labeled; dates indicate the progress of the march relative to specific dates; the orientation of the line shows the direction of travel; and a line chart below the route tracks temperature. Reading the map from left to right and back again, another message beyond the data emerges: As the march proceeds and retreats, the horrific toll of the campaign slowly reveals itself as the troop numbers decline dramatically. The graphic not only details historical fact, but serves as a powerful antiwar statement.

In the United States, data visualization was used to sound the alarm on the frequency of racist violence and its consequences. In 1883, the *Chicago Tribune* began publishing annual data on lynching in the form of a monthly calendar listing victims by date.[24] The journalist and anti-lynching campaigner Ida B. Wells cited the data in her speeches and articles. The annual publication of state and national statistics fed the public's outrage, and on September 1, 1901, the *Sunday Tribune* published a full front-page table of data on 3,000 lynchings committed over 20 years, as well as information about the victims of 101 lynchings perpetrated in 1901 and the allegations against the victims that their killers had cited to rationalize the violence.[25] Rather than focusing on individual cases, the data, table, and narrative exploration presented a powerful picture of the frequency and scale of the crisis: though lynchings occurred in mostly southern states, they were found in nearly every state of the union. The pages appeal to public opinion to support

---

[20] Cohen, "Florence Nightingale," at 132.
[21] Ibid.
[22] V. Chevallier, "Notice nécrologique sur M. Minard, inspecteur général des ponts et chaussées, en retraite" (1871) 2 *Annales des ponts et chaussées* 1–22 (translated by D. Finley at www.edwardtufte.com/tufte/minard-obit).
[23] For an extensive discussion of C. Minard's "space-time-story graphics," see Tufte, *Visual Display*, pp. 40–41.
[24] L. D. Cook, "Converging to a National Lynching Database: Recent Developments and the Way Forward" (2012) 45 *Historical Methods: A Journal of Quantitative and Interdisciplinary History* 55–63 at 56.
[25] Ibid.

change, explicitly calling out the failure of state and local law enforcement and demanding congressional action.

These historic charts and graphs are analytical, but also rhetorical, using visual conventions to identify the dynamics of important phenomena and to communicate findings and make an argument and a persuasive case for policy change. We will investigate some of the promises and pitfalls of coupling visual rhetoric with data below, but first we briefly examine datafication and human rights.

III DATAFICATION AND HUMAN RIGHTS

As a field, advocacy-oriented human rights research traditionally favors qualitative over quantitative research methodologies. Research is typically driven by interviews with victims, witnesses, and alleged perpetrators, usually supplemented by official documents, secondary sources, and media accounts.[26] Additional qualitative methods, including focus groups and participatory observation, are used by some groups, as are forensic methods such as ballistics, crime scene investigations, and exhumations. Quantitative methods such as data analysis and econometrics have been very rare until recently. There are many reasons for the traditional emphasis on qualitative methods. Historically, advocacy-oriented human rights research developed out of legal and journalistic traditions.[27] Ethically, rights advocates are committed to the individual human story. Human rights practice has been defined as "the craft of bringing together legal norms and human stories in the service of justice."[28]

At the same time, researchers in social science, epidemiology, and other fields have long used quantitative methods for research on human rights related issues. Political scientists have developed cross-national time-series datasets to interrogate the relationships between human rights and major social, economic, and political processes.[29] Epidemiologists have studied inequalities in access to health care and disparities in health outcomes between social groups.[30] Research psychologists have

---

[26] M. Langford and S. Fukuda-Parr, "The Turn to Metrics" (2012) 30 *Nordic Journal of Human Rights*, 222–38 at 222.

[27] M. L. Satterthwaite and J. Simeone, "A Conceptual Roadmap for Social Science Methods in Human Rights Fact-Finding," in P. Alston and S. Knuckey (eds.), *The Transformation of Human Rights Fact-Finding* (New York: Oxford University Press, 2016), p. 323.

[28] P. Gready, "Introduction – Responsibility to the Story" (2010) 2 *Journal of Human Rights Practice* 177–90 at 178.

[29] For an examination of several such data sets, *see* M. L. Satterthwaite, "Coding Personal Integrity Rights: Assessing Standards-Based Measures Against Human Rights Law and Practice" (2016) 48 *New York University Journal of International Law and Politics* 513–79.

[30] O. F. Norheim and S. Gloppen, "Litigating for Medicines: How Can We Assess Impact on Health Outcomes?," in A. E. Yamin and S. Gloppen (eds.), *Litigating Health Rights: Can Courts Bring More Justice to Health?* (Cambridge, MA: Harvard University Press, 2011), pp. 306–07.

examined the way human psychology may limit our ability to respond to widespread suffering such as that arising from genocide and mass displacement.[31]

Human rights NGOs are increasingly embracing scientifically based methods of research that involve data and quantification, and they are beginning to use data visualization to reach broader audiences. Using data-driven methods from other fields enables different ways of knowing, of gathering and processing information, and of analyzing findings.

The spread of digital network infrastructure, increased computing speeds, and a decrease in the cost of digital storage have made collecting, sharing, and saving data easy and prevalent. The economic accessibility of mobile technology has made cell phones widely available, even in poor countries.[32] Smartphones have put Internet access and the production of digital content in the hands of the people – generating an enormous swarm of digital exhaust and big data about many populations across the world. This explosion of new data reflects a democratization of sorts, but it also puts a new means of surveillance at the command of state agents,[33] increases the power of private data owners and brokers, and creates pockets of digital exclusion, where communities that do not benefit from the digital revolution are further marginalized.

In this "datified" world, decision-makers seek evidence in the form of data and quantitative analysis. As Sally Merry notes, "quantitative measures promise to provide accurate information that allows policy makers, investors, government officials, and the general public to make informed decisions. The information appears to be objective, scientific, and transparent."[34] While Merry's language suggests that numbers themselves promise to smooth over the messiness of decision-making by appearing scientific, it is, of course, human beings who insist on quantification. Theodore Porter has identified quantification as a "technology of distance" capable of mediating distrust, such as that between governments and citizens.[35] In the human rights context, quantification sometimes functions as a way of disappearing the judgment-laden practices of monitoring and assessment, where governments

---

[31] See, e.g., P. Slovic and D. Zionts, "Can International Law Stop Genocide When Our Moral Intuitions Fail Us?," in R. Goodman, J. Derek, and A. K. Woods (eds.), *Understanding Social Action, Promoting Human Rights* (New York: Oxford University Press, 2012), pp. 100–28.

[32] Globally, the number of mobile cell phone subscriptions reached 97 per 100 people in 2014. Serious inequality remains, with only 56 per 100 in low-income, 96 per 100 in middle-income, and 122 per 100 in high-income countries. See International Telecommunication Union, World Telecommunication/ICT Development Report and database, "Mobile cellular subscriptions (per 100 people)," http://data.worldbank.org/indicator/IT.CEL.SETS.P2.

[33] As Mark Latonero writes, "The very same tools, techniques, and processes in the collection and use of big data can be employed to both violate and protect human rights." M. Latonero, "Big Data Analytics and Human Rights: Privacy Considerations in Context," Chapter 7.

[34] S. E. Merry, *The Seductions of Quantification: Measuring Human Rights, Gender Violence, and Sex Trafficking* (Chicago and London: University of Chicago Press, 2016), p. 3.

[35] M. Power, *The Audit Society: Rituals of Verification* (Oxford: Oxford University Press, 1997), pp. 4–5.

may distrust their monitors as much as monitors distrust officials.[36] In such a context, knowing how many were killed, assaulted, or detained can be seen to satisfy a yearning for objective knowledge in a chaotic and often brutal world.

Metrics are also attractive because they can be weighed against other data and wrapped up into "indicators." Kevin Davis, Benedict Kingsbury, and Sally Merry define indicators as:

> a named collection of rank-ordered data that purports to represent the past or projected performance of different units. The data are generated through a process that simplifies raw data about a complex social phenomenon. The data, in this simplified and processed form, are capable of being used to compare particular units of analysis (such as countries or institutions or corporations), synchronically or over time, and to evaluate their performance by reference to one or more standards.[37]

In the human rights realm, indicators have been developed, *inter alia*, to directly measure human rights violations,[38] assess compliance with treaty norms,[39] measure the impacts of company activities on human rights,[40] and ensure that development processes and humanitarian aid are delivered in a rights-respecting manner.[41] As Merry notes in *The Seductions of Quantification*, indicators are attractive in their simplicity, particularly country rankings that have proven effective in catching the attention of the media and the public.[42] While indicators provide a convenient analysis at a glance, they are complicated and often problematic: Data collected may be incomplete or biased, not comparable between countries, or compromised by encompassing metrics of behavior that may not capture a diversity of values or reasons for the behavior.[43] There may also be a slippage between the norm and

---

[36] For a discussion of this dynamic, see M. L. Satterthwaite and A. Rosga, "The Trust in Indicators: Measuring Human Rights" (2009) 27 *Berkeley Journal of International Law* 253–315 at 253.

[37] K. E. Davis, B. Kingsbury, and S. E. Merry, "Indicators as a Technology of Global Governance" (2012) 46(1) *Law & Society Review* 71–104 at 73–74.

[38] E. Witchel, "Getting Away with Murder: CPJ's 2015 Global Impunity Index spotlights countries where journalists are slain and the killers go free," Committee to Protect Journalists, October 8, 2015.

[39] Office of the United Nations High Commissioner for Human Rights, *Human Rights Indicators, a Guide to Measurement and Implementation* (New York: United Nations, 2012).

[40] D. de Felice, "Business and Human Rights Indicators to Measure the Corporate Responsibility to Respect: Challenges and Opportunities" (2015) 37 *Human Rights Quarterly* 511–55.

[41] For development processes, see T. Landman et al., *Indicators for Human Rights Based Approaches to Development in UNDP Programming: A User's Guide* (New York: United Nations Development Programme, 2006). For a discussion of rights-based indicators for humanitarian aid, see M. L. Satterthwaite, "Indicators in Crisis: Rights-Based Humanitarian Indicators in Post-Earthquake Haiti" (2012) 43(4) *New York University Journal of International Law and Politics* 865–965.

[42] S. E. Merry, *The Seductions of Quantification: Measuring Human Rights, Gender Violence, and Sex Trafficking* (Chicago and London: University of Chicago Press, 2016), p. 16.

[43] Ibid.

the data used to assess the norm, a dynamic in which difficult-to-measure phenomena are assessed using proxy indicators that may become attenuated from the original norm.

In the human rights field, the relationship between the norm and the data can be especially complicated. Human rights data is almost always incomplete and often fraught with bias and assumptions.[44] The imperfect nature of human rights data is a consequence of the challenges facing its collection. There are inherent difficulties in getting a complete or unbiased dataset of anything, but it is particularly challenging when it is in a government's self-interest to hide abuses and obstruct accountability. Marginalized groups may be excluded from the available information as a result of implicit bias, or even by design.[45] For human rights researchers, there may be dangers and difficulties associated with asking certain questions or accessing certain information. Much of the data gathered about civil and political rights violations is collected through case reports by human rights organizations, making it inherently biased by factors such as the organization's familiarity and accessibility, the victims' willingness to report, and the security situation.[46] Data about economic and social rights may seem easier to gather, since there is a plethora of official data in most countries about education, housing, water, and other core rights. This data is not designed to assess rights, however, meaning that it is, at best, proxy data for rights fulfillment.[47]

However, even when there are flaws in the data collection or the data itself, the results can sometimes be useful to researchers and rights advocates. For instance, if the methodology for gathering data is consistent year after year, one may be able to draw certain types of conclusions about trends in respect for rights over time even absent a representative sample. If the data in question was collected by a government agency, it can be strategic for activists to lobby the government using its own data despite the flaws it contains, since such a strategy makes the conclusions that much harder to refute.

Further, a great power of statistics is the ability to work with data that is incomplete, biased, and uncertain – and to quantify bias and uncertainty with some measure of precision. Patrick Ball, Megan Price, and their colleagues at the Human Rights Data Analysis Group have pioneered the application of multiple systems estimation and other statistical methods to work with limited data in post-conflict

---

[44] M. Price and P. Ball, "Big Data, Selection Bias, and the Statistical Patterns of Mortality in Conflict" (2014) 34 (1) *The SAIS Review of International Affairs* 9–20.
[45] For instance, see P. Heijmans, "Myanmar criticised for excluding Rohingyas from Census," *Al Jazeera*, May 29, 2015, www.aljazeera.com/news/2015/05/myanmar-criticised-excluding-rohingyas-census-150529045829329.html.
[46] Brian Root lists other biases affecting human rights data collection in "Numbers Are Only Human," in Alston and Knuckey (eds.), *The Transformation of Human Rights Fact-Finding*, p. 363.
[47] S. McInerney-Lankford and H. Sano, *Human Rights Indicators in Development: An Introduction* (Washington, DC: World Bank Publications, 2010), pp. 16–17.

and ongoing conflict zones.[48] The group is often asked to evaluate or correct traditional casualty counts using their experience with statistical inferences.[49] They have contributed data analysis to both national and international criminal tribunals and truth commissions.

IV CHALLENGES OF DATA

Data is always an abstraction – a representation of an idea or phenomenon. Data is also a product of its collection method, whether it is a recording of a signal, survey, mechanical trace, or digital log. As the scholar Laura Kurgan explains:

> There is no such thing as raw data. Data are always translated such that they might be presented. The images, lists, graphs, and maps that represent those data are all interpretations. And there is no such thing as neutral data. Data are always collected for a specific purpose, by a combination of people, technology, money, commerce, and government. The phrase "data visualization" in that sense, is a bit redundant: data are already a visualization.[50]

Analysts may try to use algorithms and data to limit human bias and preconceptions in decision-making. However, researchers can't help but cast a human shadow on facts and figures; data is affected by people's choices about what to collect, when and how it is collected, even who is doing the collecting. Human rights researchers have begun to call attention to these hidden aspects of data gathering and analysis, examining the rights implications of their elision, and the perils and promise in their use.

For example, data-driven policing based on computerized analyses of arrest and crime data has been advanced as a method for making law enforcement less prone to bias.[51] However, the use of algorithms and visualization in such "predictive policing" often amplifies existing assumptions and historical patterns of prejudice and discrimination, driving police to increase scrutiny of already over-policed neighborhoods.[52] A human rights critique is needed to assess the use of algorithms in predictive policing as well as other practices, like the use of computer scoring to recommend

---

[48] See, for instance, P. Ball et al., "The Bosnian Book of the Dead: Assessment of the Database," Households in Conflict Network Research Design (2007), and M. Price et al., "Full Updated Statistical Analysis of Documentation of Killings in the Syrian Arab Republic," Human Rights Data Analysis Group (2013).
[49] P. Ball, et al. "How Many Peruvians Have Died? An Estimate of the Total Number of Victims Killed or Disappeared in the Armed Internal Conflict between 1980 and 2000," American Association for the Advancement of Science, August 28, 2003.
[50] L. Kurgan, "Representation and the Necessity of Interpretation," *Close Up at a Distance: Mapping, Technology, and Politics* (New York: Zone Books, 2013), p. 35.
[51] A. G. Ferguson, "Policing Predictive Policing" (forthcoming) 94 *Washington University Law Review*.
[52] K. Lum and W. Isaac, "To predict and serve?" (2016) 13 *Significance* 14–19 at 16.

sentencing ranges in overcrowded justice systems.[53] Techniques developed in the algorithmic accountability movement are especially useful here: Audits and reverse engineering can uncover hidden bias and discrimination,[54] which could be assessed against human rights norms.

Metadata describes the origin story of data: the time and place it was created, its sender and receiver, the phone used, network used, IP address, or type of camera. As former US National Security Agency General Counsel Stewart Baker hauntingly put it, "Metadata absolutely tells you everything about somebody's life. If you have enough metadata, you don't really need content."[55] Outside the domain of state security or intelligence, metadata can be useful to human rights researchers and activists as well, for instance, to counter claims that incriminating images or video recordings were falsified or to corroborate that a set of photos were taken in the same place, on the same day, by the same camera. Amnesty International used metadata from photos and videos to corroborate attacks on suspected Boko Haram supporters by Nigerian soldiers, thereby implicating them in war crimes.[56] Building on its experience training grassroots activists to use video for advocacy, the NGO WITNESS worked with the Guardian Project to develop a mobile application called CameraV to help citizen journalists and human rights activists manage the digital media and metadata on their smartphones by automatically encrypting and transmitting media files to a secure server, or, conversely, by deleting and obscuring the metadata when it could put activists at risk.[57]

In addition to concerns about accuracy, rights groups should be cautious in their approach to privacy and ownership of data, and to its analysis and expression through visualization. Over the years, researchers and lawyers have developed a set of best practices to guide the proper collection and use of data, with particular attention to human subjects research.[58] Questions related to the collection of data go to the heart

---

[53] On June 29, 2016, the ACLU filed a lawsuit on behalf of a group of academic researchers, computer scientists, and journalists challenging the US Computer Fraud and Abuse Act. The law creates significant barriers to research and testing necessary to uncover discrimination in computer algorithms. See E. Bhandari and R. Goodman, "ACLU Challenges Computer Crimes Law That Is Thwarting Research on Discrimination Online," American Civil Liberties Union, June 29, 2016, www.aclu.org/blog/free-future/aclu-challenges-computer-crimes-law-thwarting-research-discrimination-online.

[54] See, e.g., C. Sandvig et al., "Auditing Algorithms: Research Methods for Detecting Discrimination on Internet Platforms," presentation at the 64th Annual Meeting of the International Communication Association, Seattle, WA, May 24–26, 2014.

[55] A. Rusbriger, "The Snowden Leaks and the Public," *The New York Review of Books*, November 21, 2013, www.nybooks.com/articles/2013/11/21/snowden-leaks-and-public/.

[56] C. Koettl, "Chapter 7: Using UGC in human rights and war crimes investigations," *Verification Handbook for Investigative Reporting* (Maastricht, the Netherlands: European Journalism Centre, 2015), pp. 46–49.

[57] WITNESS, "Is This for Real? How InformaCam Improves Verification of Mobile Media Files," WITNESS, January 15, 2013, https://blog.witness.org/2013/01/how-informacam-improves-verification-of-mobile-media-files/.

[58] See, for instance, US Department of Health and Human Services Office for Human Research Protections, Informed Consent Checklist (1998), www.hhs.gov/ohrp/regulations-and-policy/

of what constitutes ethical research methods: Did the subjects give informed consent regarding the way their personal data would be used? Does using, collecting, or publishing this data put anyone at risk? Is the data appropriately protected or anonymized? The rules about data continue to evolve and are not without gray areas and open questions. Universities in the United States and many other countries have review processes in place to provide guidance and ensure that critical ethical questions are raised before research is approved. In fact, these ethical questions and review processes are required under US law for research institutions that receive federal funding. However, the "common rule" underlying these processes is widely seen as out of date when it comes to data ethics – especially big data ethics.[59] Ethical discussions and guidelines about data visualization are almost nonexistent, with a 2016 Responsible Data Forum on the topic a very welcome outlier.[60] The forum brought together academics, activists, and visualization practitioners to discuss issues such as the ethical obligation to ensure that data is responsibly collected and stored before being visualized; representing bias, uncertainty, and ambiguity in data visualization; and the role of empathy and data visualization in social change.[61]

V VISUALIZING QUANTITATIVE DATA

With data and data visualizations, physical phenomena like the impact of disease and the movement of troops can become legible – as can systems like economies, relationships, and networks of power. Using data to examine policies, populations, actions, and outcomes over time, individual cases can be seen as instances of widespread and systematic patterns of abuse. However, possession of data does not constitute knowledge. Data requires interpretation, context, and framing. Graphics are a powerful way to help contextualize and frame data, present interpretations, and develop understanding. Through visualization and analysis, correlations and

---

guidance/checklists/index.html, and UK Information Commissioner's Office, Code of Practice on Anonymisation (2012), https://ico.org.uk/for-organisations/guide-to-data-protection/anonymisation/.

[59] A lively debate is currently under way concerning revisions to the Common Rule and federal regulations, with an especially relevant part of that debate centered on ethics in big data research. See J. Metcalf, E. F. Keller, and D. Boyd, "Perspectives on Big Data, Ethics, and Society," The Council for Big Data, Ethics, and Society, May 23, 2016, http://bdes.datasociety.net/wp-content/uploads/2016/05/Perspectives-on-Big-Data.pdf; J. Metcalf and K. Crawford, "Where Are Human Subjects in Big Data Research? The Emerging Ethics Divide" (2016) Big Data & Society 1–14.

[60] See "Responsible Data Forum: Visualization," Responsible Data Forum, January 15, 2016, https://responsibledata.io/forums/data-visualization/; see also F. Neuhaus and T. Webmoor, "Agile Ethics for Massified Research and Visualization. Information" (2013) 15 Communication & Society pp. 43–65.

[61] For more details about the discussion, see M. Stempeck, "DataViz for good: How to ethically communicate data in a visual manner: #RDFviz," Microsoft New York, January 20, 2016, https://blogs.microsoft.com/newyork/2016/01/20/dataviz-for-good-how-to-ethically-communicate-data-in-a-visual-manner-rdfviz/.

patterns of structural violence and discrimination as well as the scope or systemic nature of abuses can become clear.

Data visualization is useful not only for explaining patterns in a dataset, but also for discovering patterns. For its 2011 report *A Costly Move*, HRW used mapping tools to visualize and analyze patterns in a large dataset of more than five million records concerning the transfer of immigrant detainees around the United States.[62] Analyzing twelve years of data, the group found that detainees were transferred repeatedly, often to remote detention centers, a process that impeded their right to fair immigration proceedings. In 2000, HRW and the American Association for the Advancement of Science visualized statistical analyses of extrajudicial executions and refugee flows from Kosovo to Albania in 1999. Instead of random violence, they found distinct surges of activity that suggested purposeful, planned, and coordinated attacks by government forces.[63]

Data is particularly useful to those seeking to understand structural, systemic violations such as abuses of economic, social, and cultural rights. Taking data from development surveys, activists have used data visualization to compare trends in health,[64] education,[65] housing,[66] and other areas against government budgets, tax revenues, and other economic data to paint a picture of progressive realization of rights against "maximum available resources," as outlined in the International Covenant on Economic, Social, and Cultural Rights.[67]

Within the context of communications and advocacy, one powerful characteristic of data visualization is that it is perceived as scientific. In one study, Aner Tal and Brian Wasnick found that including visual elements associated with science, such as graphs, can enhance a message's persuasiveness.[68] Our research group at New York University also found that when viewers did not already hold strong opinions against the subject matter, graphics presented in diagrams or charts were more persuasive than the same information presented in a table.[69] In another study, on people's

[62] B. Root, "Data Analysis for Human Rights Advocacy," School of Data, November 23, 2013, https://schoolofdata.org/author/broot/.
[63] Human Rights Watch, "Chapter 15: Statistical Analysis of Violations," in *Under Orders: War Crimes in Kosovo* (New York: Human Rights Watch, 2001), pp. 345–68.
[64] Center for Economic and Social Rights (CESR), "Visualizing Rights: Guatemala Fact Sheet" (2008), www.cesr.org/sites/default/files/Guatemala_Fact_Sheet.pdf.
[65] Ibid.
[66] CESR, "Visualizing Rights: Cambodia Fact Sheet" (2009), www.cesr.org/sites/default/files/cambodia_WEB_CESR_FINAL.pdf.
[67] A. Corkey, S. Way, and V. Wisniewiki Otero, *The OPERA Framework: Assessing Compliance with the Obligation to Fulfill Economic, Social and Cultural Rights* (Brooklyn, NY: Center for Economic and Social Rights, 2012), p. 30.
[68] A. Tal and B. Wansink, "Blinded with Science: Trivial Graphs and Formulas Increase Ad Persuasiveness and Belief in Product Efficacy" (2014) 25 *Public Understanding of Science* 117–25.
[69] A. V. Pandey et al., "How Deceptive Are Deceptive Visualizations?: An Empirical Analysis of Common Distortion Techniques," presentation at the 33rd Annual ACM Conference on Human Factors in Computing Systems, Seoul, Republic of Korea, April 18–23, 2015).

ability to remember visualizations, Michelle Borkin and colleagues found that specific visual elements of a given presentation affected its memorability. Memorable graphics had a main visual focus, recognizable objects, and clear titles and annotations.[70]

The persuasive power of charts and graphs may come at a cost: In a report full of text, numbers crucial to making a rights case are a prime target for attack and dispute.[71] The currency of human rights work is credibility, and researchers and program staff at human rights organizations carry the additional burden of having to take special care to protect their credibility and the incontrovertibility of the evidence they present. If a number in a report is convincingly challenged, the rest of the report may be called into question. Charts and numbers are also easily taken out of context, with readers understanding representations of data as statements of fact. This is all the more reason to interrogate the methodology and unpack conditions of production of specific datasets before they are used in visualizations. The powerful impact and memorability of data visualization come with a responsibility to put this knowledge to use with care and attention to the potential ethical pitfalls.

Effective data visualization can make findings clear and compelling at a glance. It provides readers with an interface to navigate great quantities of data without having to drill down into the various data points. This can obscure the fact that visualization is only a part of working with data – and often only a small part. The lead-up to the creation of a data visualization can be the key to its usefulness. Acquiring, cleaning, preparing, and analyzing data very often make up the bulk of the work. When exploring a visualization, the sometimes tedious and decidedly unsexy data work that has been done behind the scenes is not always immediately visible. And given its persuasive power, data visualization in polished and final form may gloss over issues with data collection or analysis. This may be especially true with human rights visualization, where analysis includes normative judgments about the fit between a given dataset and the legal standards at issue.

As noted above, the data used in visualization is subject to bias and the underlying assumptions around data collection and processing. In addition to this, the presentation and design of visualization is also susceptible to distortion and misinterpretation. In a series of experiments performed by our research group at NYU, empirical analysis of common distortion techniques found that these techniques did indeed mislead viewers. Distortion techniques include using a truncated y-axis (starting at a number greater than zero when illustrating percentages) or using area to represent

---

[70] M. A. Borkin et al., "Beyond Memorability: Visualization Recognition and Recall" (2016) 22 *IEEE Transactions on Visualization and Computer Graphics* 519–28.

[71] B. Root, "Numbers Are Only Human: Lessons for Human Rights Practitioners from the Quantitative Literacy Movement," in Alston and Knuckey (eds.), *The Transformation of Human Rights Fact-Finding* (referring to a 196-page report on sexual assault in Washington, DC; roughly five pages of statistical analysis bore the brunt of the negative criticism).

quantity (such as comparing areas of circles.)[72] While manipulation of the facts or deception of the reader is usually unintentional in the human rights realm, accidentally misleading visualizations can affect the clarity of the message and could damage advocacy efforts.[73] As data and visualization command attention, they can also become the focus of criticism. If a misleading visualization is called to account, it could distract from the credibility of the rest of a given project's research and advocacy, and perhaps even damage the reputation of the organization.

These risks must be borne in mind as advocates have the opportunity to analyze and visualize the increasing quantities of data made available online as a result of open government efforts. While the call for open sharing of scientific data long predates the Internet, connectivity has spurred an explosion in the use, production, and demand for high-quality data, particularly data collected by government agencies. Governments are sharing great quantities of data online, and making them accessible via Freedom of Information or other "sunshine" requests. Open government data has been used to uncover and analyze patterns of human rights abuse in criminal justice data,[74] inequalities in wage data,[75] unequal burdens of pollution,[76] and the impacts of climate change in environmental data.[77] Data created to track human development, like that collected by international demographic and health surveys, has proven to be fruitful for human rights analysis as well.[78] Under the title "Visualizing Rights," the Center for Economic and Social Rights (CESR) has published a series of country fact sheets that use publicly available data for analysis and visualization to convey patterns of discrimination and the failure to fulfill rights obligations, e.g., the rights to health, food, and education in Guatemala,[79] or with regard to poverty, hunger, and housing in Egypt.[80] The CESR briefs are designed to

---

[72] Experiments have shown that people perceive position more effectively than area. A line next to another line half its length is more easily understood as *double* than are circles or squares compared with area doubled. See W. Cleveland and R. McGill, "Graphical Perception: Theory, Experimentation, and Application to the Development of Graphical Methods" (1984) 79 *Journal of the American Statistical Association* 531–54.

[73] Pandey et al., "How Deceptive are Deceptive Visualizations?"

[74] J. Fellner et al., *Nation Behind Bars: A Human Rights Solution* (New York: Human Rights Watch, 2014).

[75] P. Overberg and J. Adamy (reporting), L. T. Vo and J. Ma (interactive), A. V. Dam and S. A. Thompson (additional development), "What's Your Pay Gap?," *The Wall Street Journal*, May 17, 2016, http://graphics.wsj.com/gender-pay-gap/.

[76] S. A. Perlin, D. Wong, and K. Sexton, "Residential Proximity to Industrial Sources of Air Pollution: Interrelationships among Race, Poverty, and Age" (2001) 51 *Journal of the Air & Waste Management Association* 406–21.

[77] E. Rosten and B. Migliozzi, "What's Really Warming the World?" *Bloomberg*, June 24, 2015.

[78] E. Felner, "Closing the 'Escape Hatch': A Toolkit to Monitor the Progressive Realization of Economic, Social, and Cultural Rights" (2009) 1 *Journal of Human Rights Practice* 402–35.

[79] CESR, "Visualizing Rights: Guatemala Fact Sheet" (2008), www.cesr.org/sites/default/files/Guatemala_Fact_Sheet.pdf.

[80] CESR, "Visualizing Rights: Egypt Fact Sheet" (2013), www.cesr.org/sites/default/files/Egypt.Factsheet.web_.pdf.

be read at a glance by a busy audience of policy officials and individuals who staff intergovernmental human rights mechanisms.

VI VISUALIZING QUALITATIVE DATA

A core output of traditional human rights research is the fact-finding report, which tends to rely on qualitative data such as the testimony of witnesses and survivors of human rights violations.[81] Visualization can provide useful context for the broader rights messages in such reports. For instance, to provide visual and spatial context for the findings presented in human rights reporting, it is not uncommon to include a timeline of events,[82] a map of the areas affected,[83] or a map of towns the researcher visited.[84]

Qualitative visualization for human rights generally falls into the category of visual storytelling. Techniques like breaking down an explanation into stages and walking the audience through these stages can elucidate the narrative, building an understanding of the sequence of events or the layers of information. Examining Amnesty International and HRW reports, the 2016 NYU study found that the use of visual features nearly tripled between 2006 and 2014, and that the majority of visual features used were qualitative.[85] For example, the study found that the number of reports using satellite images increased from one in 2006 to four in 2010 to seventeen in 2014. Most maps included in reports during this period displayed geographic information (such as places visited by researchers) and were only rarely used for quantitative display or analysis (such as displaying numbers of refugees).

Some of the changes in human rights reporting were made possible by advances in technology and newly available data. In the 1990s, Global Positioning System (GPS) and high-resolution satellite imagery became available for civilian use.[86] Since then, high-quality satellite imagery has become increasingly accessible from vendors and through free applications like Google Earth and other web-based mapping tools. Human rights groups have used GPS and

---

[81] M. Langford and S. Fukuda-Parr, "The Turn to Metrics" (2012) 30 *Nordic Journal of Human Rights* 222–38 at 222.

[82] A particularly elaborate interactive timeline is Human Rights Watch, "Failing Darfur, Five Years On," www.hrw.org/sites/default/files/features/darfur/fiveyearson/timeline.html.

[83] Human Rights Watch, "DR Congo: M23 Rebels Committing War Crimes," September 11, 2012, www.hrw.org/news/2012/09/11/dr-congo-m23-rebels-committing-war-crimes.

[84] See map of prisons visited in E. Ashamu, *"Prison Is Not for Me": Arbitrary Detention in South Sudan* (New York: Human Rights Watch, 2002).

[85] K. Rall et al., "Data Visualization for Human Rights Advocacy" (2016) 8 *Journal of Human Rights Practice* 171–97 at 179, 183.

[86] On GPS, see L. Kurgan, "From Military Surveillance to the Public Sphere," in *Close Up at a Distance: Mapping, Technology, and Politics* (New York: Zone Books, 2013), pp. 39–40. On satellite imagery, see C. Lavers, "The Origins of High Resolution Civilian Satellite Imaging – Part 1: An Overview," *Directions Magazine*, January 13, 2013, www.directionsmag.com/entry/the-origins-of-high-resolution-civilian-satellite-imaging-part-1-an-ov/303374.

satellite imagery to present vivid pictures of changes brought about by events such as mass violence, secret detention, extrajudicial executions, internal displacement, forced evictions, and displacement caused by development projects.[87] Satellite imagery has proven especially powerful in showing the visual differences before and after an event.[88] It can show the creation or destruction of infrastructure by marking changes in the landscape designed to hide underground weapons development, or migrations of people by tracking changes in the contours of refugee camps. Satellite images provide local activists with a way to contextualize and document a bigger picture than can be seen from the ground, and enable human rights researchers outside of the country to survey places that are difficult or dangerous to access.[89] These techniques are especially crucial for closed states and in emergency contexts, though researchers based outside of the countries of interest should avoid relying solely on geospatial analysis, since it may not include local voices or context. Integrating local voices with satellite imagery provides both the "near" and the "far" and paints a more complete picture of the situation on the ground.[90]

Network graphs are another visual tool that can help illuminate human rights reporting and narrative. Network graphs are a special kind of visualization showing relationships between and among entities. A family tree is one simple example of a network graph. Networks relevant to human rights investigations include networks of corruption, formal and informal chains of command, the flow of resources among industries and the government agencies charged with regulating them, and relationships between military and paramilitary groups. Visualization serves as a useful shorthand, a way to illustrate complex networks that would be cumbersome to describe in text. For example, for its 2003 report on violence in the Ituri region of the Democratic Republic of Congo, HRW used a network graph to illustrate the web of training, funding, and alliances among national governments, national militaries,

---

[87] American Association for the Advancement of Science, Geospatial Technologies Project, www.aaas.org/page/geospatial-technology-projects. The AAAS has played a leading role in developing geospatial analysis as a human rights documentation tool.

[88] For example, see Amnesty International and Zimbabwe Lawyers for Human Rights, *Zimbabwe – Shattered Lives: The Case of Porta Farm* (London: Amnesty International, International Secretariat, 2006).

[89] The Committee for Human Rights in North Korea has published a series of reports that rely heavily on analysis of satellite images of prison camps and other sites of abuse in the Democratic People's Republic of Korea. See HRNK publications at www.hrnk.org/publications/hrnk-publications.php.

[90] Combining "near" and "far" is a powerful storytelling technique, mixing testimonies, individual stories, or data points of personal interest to the reader (the near view) with the overview of large-scale trends and data abstraction. This creates an empathetic entry point into the larger story, and locates and contextualizes it. Scott Klein touches on this a bit more in "The Design and Structure of a News Application," *ProPublica*, https://github.com/propublica/guides/blob/master/design-structure.md, as does Dominikus Baur in "The superpower of interactive datavis? A micro-macro view!," *Medium*, April 13, 2017, https://medium.com/@dominikus/the-superpower-of-interactive-datavis-a-micro-macro-view-4d027e3bdc71.

and local paramilitary groups.[91] The graph clarifies the complicity of the national governments in local atrocities. The 2007 Global Witness report *Cambodia's Family Trees* uses a network graph to illustrate the connections and relationships among more than sixty individuals and family members, companies, the military, and government agencies in a deeply entrenched web of corruption around illegal logging.[92]

Graph theory, the study of networks and their properties, can be used to model the spread of information or influence along social networks. One of Google's early innovations was analyzing the network structure of the Internet – i.e., determining which pages are linked to from other pages – in order to rank web pages by relevance. Graph theory algorithms that weigh connections among entities to gauge their importance have proven useful to help navigate millions of pages in document dumps such as WikiLeaks and the Panama Papers. Network analysis and visualization can help make these large sets of data navigable and give researchers and the public a starting point toward understanding connections between parties. Like statistics, network analysis is a tool of social scientists that is increasingly being used by human rights researchers. As noted in Jay Aronson's chapter (Chapter 6), the Carter Center has used network analysis of social media postings by armed groups in Syria to estimate chains of command and track emerging and shifting alliances among groups.[93]

At the nexus of qualitative and quantitative analysis, Forensic Architecture is an international collaboration that is researching incidents around the world through crime scene reconstructions of human rights violations. Founded by the architect Eyal Weizman, Forensic Architecture uses diverse sources, including photos, cell phone audio and video, satellite imagery, digital mapping, and security camera and broadcast television footage, to painstakingly reconstruct the scene of a violation as a virtual three-dimensional architectural model. The team looks at traces and clues in the data sources. For instance, ascertaining the time of an incident from time stamps on digital metadata and even the fall of shadows in imagery and footage allows the team to establish a sequence of events and uncover falsifications or omissions in recordings. The reconstructions go so far as to adjust the virtual camera lens to match the parallax distortion of video, allowing for analysis of things like line of vision or the position of a munitions impact. The spatial data and architectural model become the nexus that stitch together the reconstruction to determine just what happened at a given point in time and space, how it happened, and who was responsible.[94]

---

[91] A. Van Woudenberg, *Covered in Blood: Ethnically Targeted Violence in Northern DRC* (New York: Human Rights Watch, 2003), p. 24.
[92] Global Witness, *Cambodia's Family Trees: Illegal Logging and the Stripping of Public Assets by Cambodia's Elite* (London: Global Witness, 2007), pp. 48–49.
[93] Carter Center, "Carter Center Makes Dynamic Syria Conflict Map Available to Public," Press Release, March 8, 2016, www.cartercenter.org/news/pr/syria-030916.html.
[94] See, for instance, Forensic Architecture's interactive report "The Killing of Nadeem Nawara and Mohammad Mahmoud Odeh Abu Daher in a Nakba Day Protest Outside of Beitunia on May 1, 2014," http://beitunia.forensic-architecture.org.

Recent developments in machine learning have also made possible a kind of qualitative data analysis of imagery by computers: the use of computer vision algorithms to detect patterns and recognize objects depicted in digital image data. This can include faces or pictures of weapons in massive bodies of social media images, or feature detection in footage from camera-enabled drones, closed-circuit video surveillance, or satellite imagery. Applying these techniques to human rights research, the Event Labeling through Analytic Media Processing (E-LAMP) project at Carnegie Mellon University combines computer vision and machine learning for conflict monitoring by searching through large volumes of video for objects (weapons, military vehicles, buildings, etc.), actions (explosions, tank movement, gunfire, structures collapsing, etc.), written text, speech acts, human behaviors (running, crowd formation, crying, screaming, etc.), and classes of people such as soldiers, children, or corpses.[95] Project ARCADE is a prototype application that uses computer vision to analyze satellite imagery in order to automate the detection of bomb crater strikes and determine their origin.[96] In these instances, after its algorithmic processing, the source imagery is often annotated with visual indicators that are more readily interpreted by humans and the image data is made understandable through a layer of visualization. Such applications are likely to become more common in the human rights field, though, as noted above, machine learning is only as good as its input and the assumptions embedded in it.

VII TECHNICAL DECISIONS CONVEY MEANING

Given the vital role that data visualization can play in analyzing data and delivering a persuasive human rights message, there is great temptation and good reason to incorporate it into the process of human rights research and its outputs, such as reports and other advocacy products. However, the power of data visualization must be harnessed with care. The techniques of data visualization are a form of knowledge production, with constraints and connotations associated with forms and their interpretation. The meanings conveyed by seemingly technical decisions must be unpacked and considered when designing human rights visualizations.

The keystone technique in the field of data visualization is "encoding" to visually identify a specific aspect of a dataset.[97] Visual encoding associates visual properties like location, color, shape, texture, and symbol with data properties like time,

---

[95] J. D. Aronson et al., *Video Analytics for Conflict Monitoring and Human Rights Documentation: Technical Report* (Pittsburgh: Carnegie Mellon University Center for Human Rights Science, 2015), www.cmu.edu/chrs/documents/ELAMP-Technical-Report.pdf.

[96] Rudiment and the Centre for Visual Computing, ARCADE: ARtillery Crater Analysis and Detection Engine, https://rudiment.info/project/arcade/.

[97] W. Cleveland and R. McGill, "Graphical Perception: Theory, Experimentation, and Application to the Development of Graphical Methods" (1984) 79 *Journal of the American Statistical Association* 531–54.

category, and amount. More than one variable and visual encoding can be combined within the same representation, such as x-axis for time, y-axis for magnitude, and color for category.[98] Variables need not be strictly visual, either. Data can be encoded using different aspects of sound (tone, volume, pitch) or touch (height, location, texture).[99] A great power of visualization is the ability to combine multiple encodings within the same visual space, enabling rich exploration. Some kinds of data, like geographic data, can also act as a bridge between other kinds of data. For instance, poverty rates in a particular geographic area can be compared against resources and services available in the same geographic area.[100] A standard collection of chart styles has emerged as conventions, and these chart styles come preloaded in popular data tools like Microsoft Excel. As one's audience becomes familiar with certain chart forms, their legibility is reinforced. As the typographer Zuzanna Licko once noted, "You read best what you read most."[101] The pie chart, bar chart, and line chart have become visual conventions. However, the form of a visualization and its interface are not neutral; it constitutes a choice, a way of structuring experience and knowledge. Form imposes its own assumptions and arguments, and the most familiar and common forms of data visualization may not always be suitable for data presentation and can, in some cases, obscure or even distort findings. In the human rights context, it is important to reflect on design decisions and visual conventions – especially when designing research-based products for diverse and often international audiences.

Visual tone can also carry connotations. For instance, does a visualization make beautiful something monstrous and tragic? This is an especially important query in a field marked by its "affect-laden conception of humanity."[102] Presenting visualizations in a "neutral" style could downplay the assumptions behind the data and its collection and analysis. Seemingly less weighty design decisions can also affect the way a map or other graphic is perceived; for example, design decisions that affect color contrast and legibility can obscure a graphic's meaning or, at their worst, create a misleading visualization.

Color can also carry cultural weight. Consider the color orange, which is associated with Hindu nationalists in India and with Unionism and the Orange Order in Northern Ireland, and was used by groups that participated in the 2004–05 "Orange Revolution" in Ukraine. Depending on how color is used in a given visualization,

---

[98] Ibid.
[99] D. Carroll, S. Chakraborty, and J. Lazar, "Designing Accessible Visualizations: The Case of Designing a Weather Map for Blind Users," in C. Stephanidis and M. Antona (eds.), *Universal Access in Human-Computer Interaction, Design Methods, Tools, and Interaction Techniques for eInclusion*, vol. 8009 of *Lecture Notes in Computer Science* (Berlin: Springer, 2013), pp. 436–45.
[100] T. Bedi, A. Coudouel, and K. Simler (eds.), *More than a Pretty Picture: Using Poverty Maps to Design Better Policies and Interventions* (Washington, DC: World Bank, 2007).
[101] Zuzanna Licko, interview by R. VanderLans, in *Emigré*, 1990, 15, p. 43.
[102] L. A. Allen, "Martyr Bodies in the Media: Human Rights, Aesthetics, and the Politics of Immediation in the Palestinian Intifada" (2009) 36 *American Ethnologist* 161–80.

such associations can invoke secondary cultural meanings, particularly when representing geopolitical data across borders.

VIII ACCESS AND INCLUSION

Particularly for the sake of advocacy, outreach, and transparency, it is important for human rights information to be accessible and inclusive. Though a strength of data visualization is the ability to invite readers to engage with analysis and communications, an ongoing challenge is access and inclusion. How can human rights researchers and NGOs make their visualizations accessible to the populations to whom their data is relevant? Or work with communities to help them access the tools and expertise necessary to generate their own data visualizations?

One challenge for interactive digital visualization is the physical constraints of the screen, particularly of small mobile screens. The popularity of smartphones and widespread mobile Internet access have overtaken desktop and broadband access in much of the world.[103] While the increasing ubiquity of access is promising, the smaller screen size poses a challenge to making complex data visualizations legible and interactive.

The physical attributes of a visualization and its interaction can profoundly affect how it is accessed and understood by users of different abilities. Limited motor control, color blindness, color vision deficiency, restricted vision, or blindness can affect how a visualization is read. The visualization community has made great strides toward awareness of color blindness and color vision deficiency. But while tools are available to check the accessibility of color use, there is still far more work to be done to make visualizations accessible to blind users. Can visual information be accessed through other means as well, such as accessible HTML, that can be processed by automated screen-reader software? Can the visualization be navigated with a keyboard instead of only by a mouse?

In addition to visual encodings, data visualization relies on culturally coded visual metaphors: an "up" symbol or bigger size means "more," time moves from left to right, clusters indicate similarity, big is important, etc. As a result, reading, interpreting, and understanding data visualizations require a certain degree of cultural and visual literacy. In the case of quantitative data visualizations, numeracy is key. The success of a data visualization for both analysis and advocacy relies not only on the visualization itself, but on its accessibility to the reader. For its advocacy work to promote better health, the design team behind the now ubiquitous Nutrition Facts label tested more than thirty variations of charts and other formats before settling on the organized table we know so well. They found that graphs, icons, and pie charts

---

[103] M. Kende, *Global Internet Report 2015: Mobile Evolution and Development of the Internet* (Reston, VA: Internet Society, 2015).

were more complicated for consumers than they'd originally thought, requiring a relatively high degree of visual literacy to understand.[104]

IX CRITICAL MAPPING

Mapping is a particularly popular form of data visualization being used in human rights research and advocacy today. Critical cartography is a set of practices and critiques based on the premise that maps are not neutral; it holds that visual design decisions about what to include or exclude, what boundaries to show, etc., are political expressions about space and power. Authors choose the data to include or exclude and decide how to highlight it. Marginalized populations may be excluded from maps for a variety of reasons, or maps may privilege spaces of commerce over spaces of community. Commercial vendors may not consider it profitable to digitize the streets and addresses of villages in developing countries. State statistical agencies with limited resources must inevitably prioritize their activities and focus.

Even when focusing on specific evidence, decisions about what to include or exclude and how to represent visual elements can carry political implications; histories can be contentious, particularly where nationalism and national identities are woven into narratives of conflict. The drawing of maps can raise human rights issues in the act of visualization itself. For example, borders and place names can be particularly contentious. The government of China, for instance, takes border demarcations very seriously – confiscating maps that, through their visuals, "violate the country's positions on issues such as Taiwan, islands in the South China Sea or territory in dispute with India" or that reveal "sensitive information."[105] Lawmakers in India also considered a draft bill threatening fines and jail time for depicting or distributing a "wrong or false" map of its borders.[106] Human rights researchers need to take political sensitivities and significance into account when they engage in visualization involving maps – both the visual expression and where sensitive interactive media is hosted.

Mapping for social justice purposes has a long history, from the historical examples above to the current use of digital mapping. A powerful example of "counter-mapping," the Detroit Geographic Expedition was formed after the 1967 Detroit riot to conduct and publish research on racial injustice in Detroit and offered free college courses on geography and urban planning for inner-city African American students. The group critically challenged plans put forward by the

---

[104] J. Emerson, "Guns, butter and ballots: Citizens take charge by designing for better government" (2005) January–February *Communication Arts* 14–23.
[105] N. Thomas and M. Martina, "China tightens rules on maps amid territorial disputes," Reuters, December 16, 2015, www.reuters.com/article/us-china-maps-idUSKBN0TZ1AR20151216.
[106] Agencies, New Delhi, "7-year jail, Rs 100 crore fine for wrong depiction of India map," Times of India, May 5, 2016, timesofindia.indiatimes.com/india/7-year-jail-Rs-100-crore-fine-for-wrong-depiction-of-India-map/articleshow/52133221.cms.

Board of Education, visualized inequities of Detroit's public spaces for children's play, and mapped traffic fatalities of children along commuter routes, all of which pointed to patterns of spatial and racial injustice in the built environment.[107]

To claim the power traditionally held by governments and companies, grassroots organizations and individuals are also using tools for digital mapmaking. For example, in 1993, Daniel Weiner and Trevor Harris worked with communities in the central lowlands of South Africa to develop participatory applications of GIS in support of the redistribution of natural resources in the post-apartheid transition.[108] By far the largest current open mapping collaboration is OpenStreetMap, a free editable map of the world. An ecosystem of tools has been developed around OpenStreetMap data, including some specifically designed for supporting humanitarian responses to crises.[109] Other projects are using mapping to capture local knowledge in order to assert claims of land ownership. Inspired by their work with indigenous communities in the Amazon, the organization Digital Democracy developed a method for contributing data to OpenStreetMap without having continuous access to the Internet.[110] The organization Hidden Pockets found that sexual and reproductive health services in Delhi were absent from Google's map, so set out to track this data and create its own publicly available map.[111]

X MOBILIZATION AND OUTREACH

Data visualization can be a powerful vehicle for collaboration and mobilization in human rights outreach and advocacy. It can be used to interface with activist members, donors, and allies. For instance, using visualization, one can illustrate the impact of one's findings or recommendations to present a compelling vision of what is possible. Using data visualization not only to describe systemic abuses, but also to render a concrete vision of the future and project an alternative vision and message of hope, can be a powerful way to mobilize supporters. Visually mapping the activities of supporters also provides participants with a visual overview of activities and creates a virtuous cycle of feedback as well as a sense of both transparency and solidarity. Making feedback visible can be an effective way of engaging participants to build solidarity and momentum.

Data visualization for advocacy can also be participatory in public space, inserted into the world. For example, in February 2009, a series of large-scale projections

---

[107] C. D'Ignazio, *The Detroit Geographic Expedition and Institute: A Case Study in Civic Mapping* (Cambridge, MA: MIT Center for Civic Media, 2013).
[108] D. Weiner and T. M. Harris, "Community-Integrated GIS for Land Reform in South Africa" (2003) 15 URISA Journal 61–73.
[109] See Humanitarian OpenStreetMap Team, https://hotosm.org.
[110] J. Halliday, "OpenStreetMap without Servers [Part 2]: A peer-to-peer OSM database," Digital Democracy, June 9, 2016, www.digital-democracy.org/blog/osm-p2p/.
[111] S. Bagchi, "Feminist mapping initiative tries to reclaim Delhi, one dot at a time," *FactorDaily*, June 21, 2016, https://factordaily.com/mapping-delhi-hidden-pockets/.

were displayed at sites across the center of Bristol, England. Using a powerful video projector, organizers displayed on building facades the line of the future water level anticipated due to climate change.[112] On a smaller scale, in 1991, the artist Félix González-Torres created an emotionally powerful series of portraits of friends with HIV/AIDS using piles of candy to match their body weight. Viewers were encouraged to take a piece of candy as they passed each portrait, thereby reducing the weight of the pile, performing, in essence, the wasting and loss of weight caused by the illness before each person's death, and quietly implicating themselves.[113]

XI TECHNICAL SUSTAINABILITY

Data visualizations are often presented in paper copies of reports and briefs, but they also figure prominently in web-based communications by human rights organizations, thus making an understanding of that technology essential to maintaining best practices with respect to data and data visualizations. A growing number of human rights visualizations have also moved beyond the presentation of a single view of a given dataset, and instead allow users to explore data in online applications. However, online databases sometimes require ongoing maintenance, particularly when they rely on external services such as maps or timelines hosted by third parties. Amnesty International launched a series of interactive data sites in 2007 with "Eyes on Darfur," followed by "Eyes on Pakistan" in 2010, "Eyes on Nigeria" in 2011, and "Eyes on Syria" in 2012, to map human rights abuses in those regions. Whereas the 2007 "Eyes on Darfur" project hosted satellite imagery on Amnesty's own web server, the Syria and Nigeria sites used Google Maps to display points of interest. Google Maps provides a low-cost, easy-to-use, web-based interface to map information, which is attractive for human rights organizations with budgetary constraints. However, as Amnesty International's experience illustrates, reliance on a third-party service provider comes with a long-term cost: By 2016, Google had updated its interface and both "Eyes on Syria" and "Eyes on Nigeria" no longer functioned. Amnesty would be required to update the back-end code for these sites in order to continue to plot information on Google Maps. Though human rights and humanitarian crises continue in these countries, the "interactive evidence" in Amnesty International's visualization has become inaccessible.

Periodic updating of interactive sites is, in fact, essential to maintaining their accessibility and relevance. While "Eyes on Darfur" continues to function, its use of Flash[114] to display information makes it inaccessible on tablet and mobile devices,

---

[112] Watermarks project, "Visualizing Sea Level Rise," February 2009, www.watermarksproject.org.

[113] F. Gonzalez-Torres, *Untitled (Portrait of Ross in L.A.)*, 1991, candies individually wrapped in multicolor cellophane, endless supply, Art Institute of Chicago.

[114] Flash is a multimedia software package the runs primarily in web browsers. Flash enables web browsers to stream audio or video, play animation and interactive games, and interact with rich applications. It has been criticized for its poor accessibility and a series of high-profile security

which have become popular since the site's development in 2007 and do not support Flash. The Pakistan site featured "a geocoded database of more than 2,300 publicly reported incidents occurring between 2005 and 2009."[115] A collection of rights-related incidents of that scale represents a treasure trove of possible human rights cases and a powerful baseline by which to judge reports of ongoing abuse. As of 2016, however, eyesonpakistan.org is no longer online and its database is no longer accessible. The apparent failure to prioritize or plan for the demands of changing technology and the lack of ongoing technical support means it is no longer possible for human rights researchers and advocates to use the Syria, Nigeria, or Pakistan data.

One way to mitigate the obsolescence that plagued the Amnesty sites is to make both the data and the source code of a visualization readily available for download by users. This would allow visitors to access the raw data regardless of the technical implementation of the interactive interface. Though still rare among human rights NGOs, this is a growing practice among news organizations. The evolving field of data journalism offers one view of things to come. "Computer-assisted reporting" and "data-driven journalism" use spreadsheets and databases to help find patterns in data by using statistical methods and other techniques from the social sciences. The findings of the investigations are often presented through interactive news applications and data visualizations that can engage readers with data-driven reporting in a richer way, beyond the constraints of print. News organizations are increasingly posting these projects, their data, and tools to code-sharing websites like GitHub for others to download, use, and modify.[116] Like human rights organizations, journalists face limited resources and technical overhead, but news media are, thus far, more readily embracing the use of data analysis to drive investigations and visualization for effective storytelling. Any effort to share the data behind human rights reporting or visualization will need to carefully grapple with crucial ethical and security challenges, including confidentiality and anonymization, consent, and potential misuse of the data by abusive governments or other opponents.

---

vulnerabilities. Released in 2007, the Apple iPhone does not have the ability to run Flash, nor does the Apple iPad, released in 2010. In November 2011, Adobe announced that it would no longer support Flash for mobile browsers. The popularity of Flash has continued to decline and in July 2017, Adobe announced it will end development and distribution of Flash in 2020. See C. Warren, "The Life, Death and Rebirth of Adobe Flash," *Mashable*, November 19, 2012, http://mashable.com/2012/11/19/history-of-flash/ and Adobe Corporate Communications, "Flash & the future of interactive content," July 25, 2017, https://theblog.adobe.com/adobe-flash-update/

[115] Amnesty International, "Pakistan: Millions Suffer in Suffer in [sic] Human Rights Free Zone in Northwest Pakistan," June 10, 2010, www.amnestyusa.org/news/press-releases/pakistan-millions-suffer-in-suffer-in-human-rights-free-zone-in-northwest-pakistan.

[116] See this list of GitHub accounts of various news organizations: https://github.com/silva-shih/open-journalism.

XII CONCLUSION

Human rights researchers and advocates are adding new methodologies to their toolbox, drawing on emerging technologies as well as established data analysis techniques to enhance and expand their work. Data visualization holds exciting potential, bringing new techniques of knowledge production, analysis, and communication to human rights research and advocacy. Organizations are increasingly recognizing the power of data visualization to support human rights analysis and arguments to help make a memorable and persuasive case for change.

Enabled by digital technology and engagement with data, effective visualization is a powerful tool for understanding social problems and their potential solutions. While journalism and academic disciplines, including the social sciences, are using data visualization for both analysis and communication, the human rights field is just beginning to tap its potential. As interest grows, human rights organizations will need to struggle with the ethical and practical considerations of producing data visualizations.[117] More research is still needed on the effective use of data visualization and human rights. Used in a principled way, however, data visualization can benefit human rights researchers and advocates, and those whose rights are in danger. It can help researchers identify patterns and trends; clarify a call to action; make data analysis compelling, understandable, and interactive; rally supporters; and perhaps even visualize the effects of activism itself.

---

[117] J. Emerson, "Ten Challenges to the Use of Data Visualization in Human Rights," *Social Design Notes*, February 9, 2016, http://backspace.com/notes/2016/02/ten-challenges.php.

9

Risk and the Pluralism of Digital Human Rights Fact-Finding and Advocacy

Ella McPherson

I INTRODUCTION[1]

The rise of information and communication technologies (ICTs) has captivated many human rights practitioners and scholars. Particular interest, mine included, is focused on the potential of using ICTs to support the pluralism of human rights fact-finding and advocacy.[2] In theory, now anyone with a cell phone and Internet access can document and disseminate evidence of human rights abuses. But what happens when this theory is put into practice?[3] What happens when ICTs are adopted in empirical realities shaped by unique contexts, distributions of resources, and power relations?[4] I will argue that, while the rise of ICTs has certainly created new opportunities, it has also created new risk – or negative outcomes – for human rights practitioners. This risk is silencing, and unequally so.

In this chapter, I focus on human rights fact-finding and advocacy from the perspective of practitioners at human rights NGOs, while acknowledging that the range of practices and actors involved in human rights work is much broader.[5] These practices form a communication chain: information moves from witnesses on the ground to human rights practitioners during fact-finding, who gather and evaluate this information for evidence of violations. This evidence is then packaged and communicated to audiences such as journalists, policy-makers, and publics as

---

[1] This work was supported by the Economic and Social Research Council (grant no. ES/K009850/1) and the Isaac Newton Trust.
[2] I lead a project at the University of Cambridge called "The Whistle," which is a digital app we are developing to facilitate human rights reporting and verification. See www.thewhistle.org.
[3] I draw on my ongoing digital ethnography of human rights practices in the digital age for examples of empirical realities.
[4] R. Mansell, "The Life and Times of the Information Society" (2010) 28(2) *Prometheus: Critical Studies in Innovation* 165–86 at 173.
[5] K. Nash, *The Political Sociology of Human Rights* (Cambridge: Cambridge University Press, 2015).

advocacy work designed to impel change through persuasion.[6] At each stage, we can think of successful communication as speaking to and being heard and understood by intended audiences.[7] In other words, it is about audibility (or its equivalent, visibility). Unsuccessful communication, in contrast, involves either audibility to *un*intended audiences or *in*audibility to intended audiences. Successful communication can also have unsuccessful outcomes for the actors involved, as when a message is audible to intended audiences but is misunderstood or turns out to be erroneous, or when a message is received and interpreted as the communicator intended but turns out to be deceptive.

The success of communication matters, of course, for human rights practitioners' ability to generate accountability for individual cases of human rights violations. It also matters for a value at the core of human rights: pluralism, or the successful communication of a variety of voices. Three types of pluralism are of concern in this chapter. The first is the pluralism of human rights actors vis-à-vis the state or non-state actors they wish to hold to account. The second is the pluralism of individual human rights actors within the human rights world, which, as with all worlds, has hierarchies corresponding to the distribution of power.[8] The third is the pluralism of access by the subjects and witnesses of violations to the mechanisms of human rights accountability, which, of course, cannot act on a violation without first hearing about it.[9]

The chapter begins by outlining how risk is entwined with communication in the digital age. Rather than considering risk in isolation, we can think of it as manifesting via "risk assemblages," or dynamic combinations of actors, technologies, contexts, resources, and risk perceptions.[10] In the subsequent two sections, I detail selected types of risk for human rights communication resulting from new combinations of actors and technologies involved in digital fact-finding and advocacy. For fact-finding, these include the risk of surveillance, which has consequences for participants' physical security, and the risk of deception, which has consequences for their reputational integrity. For advocacy, these include the risk of mistakes,

---

[6] Human rights fact-finding is also used to produce evidence for courts, where the uptake of ICTs is also an important area for inquiry, but beyond the scope of this chapter.
[7] M. Madianou, L. Longboan, and J. C. Ong, "Finding a Voice through Humanitarian Technologies? Communication Technologies and Participation in Disaster Recovery" (2015) 9 *International Journal of Communication* 3020–38 at 3022; A. T. Thrall, D. Stecula, and D. Sweet, "May We Have Your Attention Please? Human-Rights NGOs and the Problem of Global Communication" (2014) 19(2) *The International Journal of Press/Politics* 135–59 at 137–38.
[8] W. Bottero and N. Crossley, "Worlds, Fields and Networks: Becker, Bourdieu and the Structures of Social Relations" (2011) 5(1) *Cultural Sociology* 99–119 at 105.
[9] E. McPherson, "Source Credibility as 'Information Subsidy': Strategies for Successful NGO Journalism at Mexican Human Rights NGOs" (2016) 15(3) *Journal of Human Rights* 330–46 at 331–32.
[10] D. Lupton, "Digital Risk Society," in A. Burgess, A. Alemanno, and J. O. Zinn (eds.), *The Routledge Handbook of Risk Studies* (Abingdon, UK: Routledge, 2016), p. 302.

which can in turn risk reputational integrity, and the risk of miscalculations, which can jeopardize precious resources. In the following section, I explain how this materialized risk combines with risk perceptions to create a silencing double bind. Human rights practitioners may be silenced if they don't know about risk – and they may silence themselves if they do. This silencing effect is not universal, however, but disproportionately affects human rights practitioners situated in more precarious contexts and with less access to resources.[11] This has consequences for the three types of pluralism outlined above. The chapter finishes by outlining four ways of loosening the risk double bind: educational, technological, reflexive, and discursive approaches to working with risk.

II COMMUNICATION, MEDIATION, AND RISK IN THE DIGITAL AGE

As communicators, we all do a number of things to increase the odds that our communications are successful. We establish the identities of our communication partners through clues we gather from their appearance and bearing. We supplement our messages with cues such as facial expressions or emoticons to guide interpretation, and we look for cues from our audiences that they have heard and understood us.[12] We gather information about our interlocutors' context – the time and place in which they are communicating – and supplement our messages with information about our own contexts (often referred to as metadata). We adjust our production and reception of content to these clues, cues, and contextual information. Still, even with all of these aids, communication can be unsuccessful, and this risk is exacerbated by the mediation of communication over ICTs.

Mediation is the extension of communication across time and/or space using technology. The closer we are in time and space to our communication partners, the easier it tends to be for us to establish their identities, observe and provide cues, and understand context. Easiest of all is face-to-face communication. By introducing "temporal and spatial distances," mediation makes all of this more difficult, as we are no longer in the same environment.[13] It is not, however, just this distance that increases the risk of unsuccessful communication, but also the introduction of intermediaries. These intermediaries are not neutral, but rather introduce new technical features and new actors with new motives, as well as new ways for existing actors to intervene with communications.

The technical features of ICTs can diminish, augment, or alter the clues, cues, and contextual metadata associated with a communication. Furthermore, the complexity of these technical features may make it difficult to understand just what has

---

[11] U. Beck, *Risk Society: Towards a New Modernity* (London: SAGE Publications, 1992), p. 23.
[12] J. B. Thompson, *The Media and Modernity: A Social Theory of the Media* (Stanford, CA: Stanford University Press, 1995), pp. 83–85.
[13] Ibid., p. 22.

happened. For example, many social media user profiles allow people to communicate with pseudonyms or assumed identities. Twitter's character limit squeezes nuance out of tweets, though users have introduced the use of hashtags as an abbreviated interpretation cue. In another example, YouTube automatically dates videos according to the day it is in California at the time of upload, no matter where the upload took place. This metadata is widely misunderstood and has contributed to disputes about the veracity of videos.[14]

A significant proportion of new actors behind ICTs are commercial, governed by profit motives. These motives can shape technical features, like Facebook's "like" and "share" buttons, which are designed to keep eyeballs on timelines peppered with advertisements. The motives of commercial communication platforms may not necessarily align with the motives of communicators. As discussed further below, this is particularly evident in the algorithms that determine visibility on social media and thus who is seen by whom. These algorithms may make certain communications either more or less visible than their producers intended. The phenomenon of commercial intermediaries controlling public visibility is nothing new – think of the gatekeeping role of mainstream news organizations. What is new is the lack of transparency and accountability when visibility decisions are made by a black-box algorithm instead of a human journalist.[15] Just as the technical complexity of ICTs obscures these algorithms and the commercial motives underpinning them, it also hides third-party actors. These include political actors who have a vested interest in human rights communication. The market for digital surveillance is thriving, and hardware and software that allow us to communicate over time and space also create opportunities for eavesdropping. In sum, at the same time as communicators using ICTs usually can glean less about their interlocutors and eavesdroppers than in a face-to-face situation, they must also know more about intermediary technologies that are both complex and opaque.[16] Mediation thereby increases the risk of unsuccessful communication and its attendant consequences.

Alongside many other professional worlds of communication, human rights practitioners are considering and adopting new ICTs. This use of ICTs and the mediation they engender supplements other forms of communication, creating a new "interaction mix" characterized by renewal, as fresh technologies proliferate and slightly stale ones become obsolete.[17] In terms of human rights fact-finding, this

---

[14] R. Mackey, "Confused by How YouTube Assigns Dates, Russians Cite False Claim on Syria Videos," *The New York Times*, August 23, 2013, http://thelede.blogs.nytimes.com/2013/08/23/confused-by-how-youtube-assigns-dates-russians-cite-false-claim-on-syria-videos/.
[15] Z. Tufekci, "Algorithmic Harms Beyond Facebook and Google: Emergent Challenges of Computational Agency" (2015) 13 *Journal on Telecommunications and High Technology Law*: 203–18 at 208–09.
[16] E. McPherson, "Social Media and Human Rights Advocacy," in H. Tumber and S. Waisbord (eds.), *The Routledge Companion to Media and Human Rights* (London: Routledge, 2017), pp. 281–83.
[17] Thompson, *The Media and Modernity*, p. 87.

new mix has been described as enabling a new generation of methodologies.[18] Traditionally, the gold standard of fact-finding has been the face-to-face interview between civilian witnesses and human rights practitioners. Often facilitated by trusted networks cultivated over time, the interview allows for the co-production of information between the witness and the practitioner. This witness testimony and the accompanying analysis done by human rights practitioners are the cornerstones of the weighty, precisely worded, and highly documented orthodox human rights report.[19] These reports, in turn, underpin human rights advocacy, which practitioners traditionally – though not exclusively – communicated to targets via the mainstream media.[20]

Human rights communication has therefore always been mediated, whether information is passed through a trusted network of witnesses or shaped to attract the attention of journalists covering human rights violations.[21] It has also always entailed risk. One only has to dip into the multitude of reports on the conditions of human rights practice to see this – or to consider practitioners' risk-mitigation tactics, ranging from security training to robust and transparent methodologies to publicity strategies.[22] But the new mix of human rights fact-finding and advocacy in the digital age has brought about new risk assemblages shaped by technologies, actors, contexts, resources, and risk perceptions.[23] Over the next three sections of this chapter, I outline elements of these new risk assemblages and explain how they can hinder successful communication, with implications for the pluralism of human rights communication.

III DIGITAL FACT-FINDING AND COMMUNICATION RISK

Human rights practitioners have adopted ICTs for fact-finding in a variety of ways, including using high-technology information sources like satellite images, drone

---

[18] P. Alston, "Introduction: Third Generation Human Rights Fact-Finding," in *Proceedings of the Annual Meeting* (Washington, DC: American Society of International Law, 2013), pp. 61–62. For a recent overview of ways that ICTs are being adopted in human rights practice, see E. McPherson, *ICTs and Human Rights Practice* (Cambridge: University of Cambridge Centre of Governance and Human Rights, 2015).

[19] P. Alston and C. Gillespie, "Global Human Rights Monitoring, New Technologies, and the Politics of Information," *European Journal of International Law* (2012) 23(4) 1089–123 at 1108–09.

[20] M. Powers, "NGO Publicity and Reinforcing Path Dependencies: Explaining the Persistence of Media-Centered Publicity Strategies" (2016) 21(4) *The International Journal of Press/Politics* 492–94.

[21] McPherson, "Source Credibility as 'Information Subsidy'," at 333–35.

[22] S. Hopgood, *Keepers of the Flame: Understanding Amnesty International* (Ithaca, NY: Cornell University Press, 2006), pp. 90–92; A. M. Nah et al., "A Research Agenda for the Protection of Human Rights Defenders" (2013) 5(3) *Journal of Human Rights Practice* 401–20 at 413.

[23] S. Hankey and D. Ó Clunaigh, "Rethinking Risk and Security of Human Rights Defenders in the Digital Age" (2013) 5(3) *Journal of Human Rights Practice* 535–47 at 539.

videos, big data, and statistics as well as open source social media content.[24] Given our concern with communication, here I focus on practitioners' use of digital information that documents human rights violations and has been produced and transmitted by civilian witnesses – "civilian" in contrast with professional to highlight their inexpert status, and "witness" as someone who is purposively communicating experienced or observed suffering.[25] Civilian witnesses can be spontaneous or solicited.[26] In the digital age, spontaneous witnesses might use their smartphones to document violations that they then share with broader audiences via social media or messaging apps; sometimes this information is gathered, curated, and connected to human rights NGOs by networks of activists. Solicited witnesses may be answering a human rights NGO's open call for information made via a digital crowdsourcing project or a digital reporting application.

Digital information from civilian witnesses affords human rights practitioners a number of fact-finding advantages. First, the images and video civilian witnesses produce can provide much more detailed evidence than witness interviews that rely on memory.[27] Second, consulting civilian witnesses can tap wells of knowledge, particularly expertise relating to local contexts unfamiliar to foreign practitioners. Third, a wider incorporation of civilians via ICTs can fire up public enthusiasm about human rights and thus receptivity to advocacy.[28] Fourth, and most important for our concern with pluralism, these new sources can support the variety and volume of voices speaking and being heard on human rights. They supplement interviewing's traditional co-production of information between witnesses and practitioners with both the more autonomous production of spontaneous digital witnesses and new forms of co-production via solicited digital witnesses.[29] If these

---

[24] Amnesty International, Benetech, and The Engine Room, *DatNav: New Guide to Navigate and Integrate Digital Data in Human Rights Research* (London: The Engine Room, 2016). See also M. Latonero, "Big Data Analytics and Human Rights: Privacy Considerations in Context," Chapter 7 in this volume. Open source social media content includes perpetrator propaganda videos. It also includes content originally posted without a witnessing purpose but later repurposed by others, such as the use of a geolocated selfie for corroboration of a military vehicle's movements because it happens to capture that vehicle driving past in the background.

[25] E. McPherson, "Digital Human Rights Reporting by Civilian Witnesses: Surmounting the Verification Barrier," in R. A. Lind (ed.), *Produsing Theory in a Digital World 2.0: The Intersection of Audiences and Production in Contemporary Theory* (New York: Peter Lang Publishing, 2015), vol. 2, p. 206; S. Tait, "Bearing Witness, Journalism and Moral Responsibility" (2011) 33(8) *Media, Culture & Society* 1220–35 at 1221–22.

[26] McPherson, *ICTs and Human Rights Practice*, pp. 14–17.

[27] C. Koettl, *Citizen Media Research and Verification: An Analytical Framework for Human Rights Practitioners* (Cambridge: University of Cambridge Centre of Governance and Human Rights, 2016), p. 7.

[28] M. Land et al., *#ICT4HR: Information and Communication Technologies for Human Rights* (Washington, DC: The World Bank Group, 2012), p. 17; M. Land, "Peer Producing Human Rights" (2009) 46(4) *Alberta Law Review* 1115–39 at 1120–22.

[29] J. Aronson, "The Utility of User-Generated Content in Human Rights Investigations," Chapter 6 in this volume.

witnesses are situated in closed-country contexts or rapidly unfolding events, they might otherwise be inaccessible to human rights practitioners.[30] Indeed, fact-finding in a number of recent cases has hinged on evidence documented digitally by civilian witnesses. For example, Amnesty International's research into a 2017 shooting at an Australian refugee detention center in Papua New Guinea used refugees' photos and videos to challenge both governments' official version of events, which was that Papua New Guinea Defence Force soldiers fired bullets into the air rather than into the center.[31]

These opportunities are all made possible by ICTs' mediation of communication over time and place. Of course, this mediation, and the intermediaries it requires, also introduces risk. Below, I outline two possible manifestations of communication risk and their consequences arising from the introduction of new technologies into fact-finding, associated new commercial actors, and new opportunities for existing actors to interfere with communications. The first is the risk of surveillance, in which the communication is audible to unintended recipients and generates concomitant risk for the physical security of civilian witnesses and human rights practitioners. The second is the risk of deception, in which the producer of a digital communication engineers the recipient's misinterpretation of that communication. Misinterpretation creates follow-on risk to the reputational integrity of human rights practitioners and their NGOs. These are familiar categories of risk in the human rights domain but manifested, as explained below, in new ways. Both are made possible by the technical complexity of mediating ICTs, which allows eavesdroppers to hide and deceivers to manipulate metadata.

A *Surveillance and Physical Security*

Surveillance, understood broadly as monitoring information about others for purposes including management and control, is a risk that civilian witnesses and human rights practitioners have always faced.[32] Surveillance of their identities, networks, and activities is a key tactic deployed by state adversaries in a "cat-and-mouse" game over truth-claims.[33] Human rights practitioners who pioneered the use of ICTs may have had a momentary advantage in this battle by using these technologies to transmit information quickly and widely. Many state actors, however, have caught up quickly and even surpassed human rights actors in their strategic use of ICTs.

---

[30] Alston and Gillespie, "Global Human Rights Monitoring, New Technologies, and the Politics of Information," at 1112–13.

[31] "In the Firing Line: Shooting at Australia's Refugee Centre on Manus Island in Papua New Guinea," Amnesty International, May 14, 2017 www.amnesty.org/en/documents/document/?indexNumber=asa34%2f6171%2f2017&language=en.

[32] D. Lyon, *Surveillance after Snowden* (Cambridge: Polity Press, 2015), p. 3.

[33] Hankey and Ó Clunaigh, "Rethinking Risk and Security of Human Rights Defenders in the Digital Age," at 538.

The surveillance opportunities ICTs afford center on a metadata paradox. ICTs can both reveal and conceal communication metadata; the first facilitates mass surveillance, while the second facilitates spyware.

ICTs are built to collect metadata on their users, often without users understanding just how significant their data trails are. Many ICT companies routinely collect users' metadata for reasons ranging from marketing to legal compliance.[34] This profit-driven surveillance produces information about communications that also meets the surveillance imperatives of states. The US National Security Agency, for example, infamously has a bulk surveillance program that collects telecommunications metadata. Activists worry that this program has set a standard for other governments in terms of the permissible level of spying on their citizenries – as exemplified by the Egyptian government's 2014 request for tenders for a mass social media surveillance system.[35] In addition to its implications for the rights to privacy and freedom of opinion and expression, this form of surveillance is a particular concern for individuals communicating information critical of retaliatory states.[36] Even if the content of these communications remains private, metadata can reveal connections between civilian witnesses and human rights practitioners and, through social network analysis, identify individuals as human rights practitioners.[37]

While mass surveillance depends on ICTs' revelation of communication metadata, spyware depends on its obfuscation, afforded by ICTs' complexity. Spyware hides in victims' communications equipment to track and share information about their activities.[38] In order to get spyware into target devices in the first place, victims must be deceived into installing it. This often happens through a wolf-in-sheep's-clothing tactic called social engineering, where messages containing spyware are disguised through the manipulation of metadata and content. For example, a human rights practitioner in the United Arab Emirates received unsolicited text messages containing a link that appeared to document evidence of prison torture. Had he clicked on the link, this practitioner's iPhone would have been infected with commercial spyware priced at around $1 million – an indication that a powerful

---

[34] "Metadata," Privacy International, www.privacyinternational.org/node/53.
[35] "Egypt's plan for mass surveillance of social media an attack on internet privacy and freedom of expression," Amnesty International, June 4, 2014, www.amnesty.org/en/latest/news/2014/06/egypt-s-attack-internet-privacy-tightens-noose-freedom-expression/; S. Kelly et al., "Tightening the Net: Governments Expand Online Controls," Freedom House, 2014, https://freedomhouse.org/report/freedom-net/2014/tightening-net-governments.
[36] *Report of the Special Rapporteur on the Promotion and Protection of the Right to Freedom of Opinion and Expression, David Kaye*, U.N. Doc. A/HRC/29/32 (May 22, 2015).
[37] Amnesty International, Benetech, and The Engine Room, *DatNav*, p. 23; S. Bender-de Moll, *Potential Human Rights Uses of Network Analysis and Mapping* (Washington, DC: AAAS Science and Human Rights Program, 2008), p. 4.
[38] M. Schwartz, "Cyberwar for Sale," *The New York Times*, January 4, 2017, www.nytimes.com/2017/01/04/magazine/cyberwar-for-sale.html.

actor was behind the attack.[39] In another case, a Mexican human rights practitioner received a text message purporting to share news about the investigation into the 2014 disappearance of forty-three students. He fell for it, clicking on the link and infecting his phone with malware believed to have been sold to the Mexican government by an Israeli cyberwarfare company.[40]

Digital security risk turning into physical security risk is unfortunately becoming more and more common for human rights practitioners and civilian witnesses.[41] If surveillance makes fact-finding communication audible to an unintended audience, its participants may not be aware this has happened until they experience related harassment and attacks. Security risk may spread through practitioners' and witnesses' networks, which are rendered visible by smartphone contacts and social media friends and followers lists. Furthermore, the mediation of digital fact-finding over time and space can make it difficult for practitioners who have learned of a threat to locate and warn civilian witnesses.[42] Human rights practitioners can and do use security tools – such as technologies supporting encryption, anonymity, and the detection of spyware – to counteract the corporate/state surveillance nexus. These technologies are threatened, however, by laws curtailing their use.[43] Furthermore, powerful discourses, such as "nothing to hide, nothing to fear," which have been propagated by state actors and picked up by the media, align the use of these technologies with criminality and threats to national security.[44]

B Deception and Reputational Integrity

Human rights practitioners' use of digital information from civilian witnesses generates another category of risk: susceptibility to misinterpretation through deception. By dint of their accusations of violations, human rights practitioners often engage in battles over truth-claims with their adversaries. Though the manipulation of truth-claims with an intent to deceive has always been a feature of these battles, human rights practitioners may be more exposed to them in the digital age for several reasons. First, ICTs afford a greater number and variety of sources of information,

---

[39] B. Marczak and J. Scott-Railton, "The Million Dollar Dissident: NSO Group's iPhone Zero-Days Used against a UAE Human Rights Defender," The Citizen Lab, August 24, 2016, https://citizenlab.org/2016/08/million-dollar-dissident-iphone-zero-day-nso-group-uae/.

[40] A. Ahmed and N. Perlroth, "Using Texts as Lures, Government Spyware Targets Mexican Journalists and Their Families," *The New York Times*, June 19, 2017, www.nytimes.com/2017/06/19/world/americas/mexico-spyware-anticrime.html?_r=0.

[41] S. Kelly et al., "Silencing the Messenger: Communication Apps Under Pressure," Freedom House, 2016, https://freedomhouse.org/report/freedom-net/freedom-net-2016.

[42] Amnesty International, Benetech, and The Engine Room, *DatNav*, p. 61.

[43] A. Crowe, S. Lee, and M. Verstraete, "Securing Safe Spaces Online: Encryption, Anonymity, and Human Rights," Privacy International, 2015, www.privacyinternational.org/sites/default/files/Securing%20Safe%20Spaces%20Online_0.pdf.

[44] H. Abelson et al., "Keys under Doormats: Mandating Insecurity by Requiring Government Access to All Data and Communications," (2015) 1(1) *Journal of Cybersecurity* 69–79.

many of whom are outside of the trusted networks that human rights organizations traditionally consult. Deceptive actors can camouflage themselves among this broader pool of sources. Second, unlike in a traditional face-to-face interview, human rights practitioners using spontaneous or solicited digital information from civilian witnesses are not present at the moment of production. As such, they cannot rely on their direct perceptions of identity clues, communication cues, and contexts to verify civilian witnesses' accounts.[45] Instead, they must use digitally mediated content and metadata as a starting point, which can be distorted and manipulated. Third, this information is often in image or video format that appears to be amateur. This lends it an aura of authenticity – rooted, perhaps, in a "seeing is believing" epistemology – that may belie manipulation.[46]

Deception through truth-claims manipulation can be divided into at least three categories: outright staging of content, doctoring of content, and doctoring of metadata.[47] Staging of content involves packaging fakery as fact, as with the viral YouTube video "SYRIA! SYRIAN HERO BOY rescue girl in shootout." This video, which claimed to document children dodging bullets while running through a dusty Syrian street, was actually a cinematographic project by a Norwegian director that was filmed in Malta.[48] Doctored content, in turn, uses real rather than staged content but relies on digital editing tools such as Photoshop to alter the images. For example, one human rights practitioner received images via WhatsApp from a source who claimed that they were evidence of torture during detention. These included a picture of a person who seemed, at first glance, to have a bruised face. Additional investigation, however, revealed that this was a highly edited version of an older picture involving changes to its color balance to create the illusion of bruises.[49]

Human rights practitioners report that it is the last of these three forms of deception – the doctoring of metadata – that is by far the most prevalent.[50] This involves scraping videos or images from one context and repackaging them as evidence of violations in another context. Examples include reposting YouTube videos with new descriptions, as in the case of one video depicting the water cannoning of a man shackled to a tree while other men watch and laugh. This video appeared on YouTube multiple times with at least three different sets of metadata entered in the video description. One version claimed to depict Venezuelan armed forces assailing a student, another stated that it was Colombian

---

[45] Diane F. Orentlicher, "Bearing Witness: The Art and Science of Human Rights Fact-Finding" (1990) 3 *Harvard Human Rights Journal* 83–136 at 114.
[46] P. Brown, *"It's Genuine, as Opposed to Manufactured": A Study of UK News Audiences' Attitudes towards Eyewitness Media* (Oxford: Reuters Institute for the Study of Journalism, 2015), http://reutersinstitute.politics.ox.ac.uk/publication/its-genuine-opposed-manufactured.
[47] Amnesty International, Benetech, and The Engine Room, *DatNav*, p. 35.
[48] McPherson, "Digital Human Rights Reporting by Civilian Witnesses," pp. 193–94.
[49] Koettl, *Citizen Media Research and Verification*, pp. 27–28.
[50] Ibid., p. 16.

special forces and a farmer, and a third portrayed the scene as Mexican police and a member of a civil defense group.[51]

Though some instances of deception may be malevolent, other instances may be backed by the best of intentions. For example, civilian witnesses may use images from one event to illustrate another, similar event that was not recorded. Nevertheless, using any kind of manipulated information as evidence creates a follow-on risk to the reputations of human rights practitioners and their organizations. For these, credibility is a fundamental asset, not only for the persuasiveness of their advocacy, but also for garnering donations and volunteers, influencing policy-making, and motivating mobilization.[52] Credibility is also a human rights organization's Achilles' heel, as it can be damaged in an instant with the publication of truth-claims that others convincingly expose as false.[53] Though the verification of information has always been a cornerstone of human rights work as a truth-claim profession, information mediated by ICTs is challenging established verification practices. This is not only because of the new sources and formats of information ICTs enable, but also because verifying digital information requires expertise that, though increasingly standardized, is still emergent.[54]

IV DIGITAL ADVOCACY AND COMMUNICATION RISK

As with fact-finding, human rights practitioners are incorporating ICTs into their advocacy strategies, venturing far beyond websites into formats including apps, livestreaming, and virtual reality. Because human rights practitioners are paying particular attention to mainstream social media platforms to supplement their traditional advocacy practices, I focus on that medium here.[55] Practitioners are communicating advocacy messages via social media to directly target policy-makers, either publicly or via private messages, and to attract the attention of the mainstream media.[56] They are also using social media to mobilize publics for a variety of reasons, including fundraising, creating visibility for an issue, and building networks

---

[51] M. Bair and V. Maglio, "Video Exposes Police Abuse in Venezuela (Or Is It Mexico? Or Colombia?)," *WITNESS Blog*, February 25, 2014, http://blog.witness.org/2014/02/video-exposes-police-abuse-venezuela-mexico-colombia/.

[52] L. D. Brown, *Creating Credibility* (Sterling, VA: Kumarian Press, 2008), pp. 3–8; S. Cottle and D. Nolan, "Global Humanitarianism and the Changing Aid-Media Field: Everyone Was Dying for Footage" (2007) 8(6) *Journalism Studies* 862–88 at 872; M. Gibelman and S. R. Gelman, "A Loss of Credibility: Patterns of Wrongdoing Among Nongovernmental Organizations" (2004) 15(4) *Voluntas: International Journal of Voluntary and Nonprofit Organizations* 35–81 at 372.

[53] Koettl, *Citizen Media Research and Verification*, p. 6.

[54] McPherson, "Digital Human Rights Reporting by Civilian Witnesses," pp. 199–200.

[55] "Incorporating Social Media into Your Human Rights Campaigning," New Tactics in Human Rights, 2013, www.newtactics.org/conversation/incorporating-social-media-your-human-rights-campaigning.

[56] Powers, "NGO Publicity and Reinforcing Path Dependencies," p. 500.

between publics and subjects of violations in a show of global solidarity.[57] Many NGOs undertake advocacy over social media through institutional accounts operated by individuals. Though dedicated communications professionals operate these accounts at some organizations, at others the arrangement is more ad hoc, undertaken by existing staff according to interest or availability.

The use of social media affords human rights practitioners a number of advocacy advantages. It can allow them to amplify messages and reach advocacy targets without depending on the mainstream media, whose human rights coverage may be circumscribed by commercial imperatives, censorship, and norms of newsworthiness.[58] The range of communication formats supported by social media enables development of new and captivating ways to represent human rights information, such as data visualization.[59] Additionally, the quantification metrics built into social media platforms, such as numbers of likes and shares, allow human rights practitioners to track engagement with their messages.[60] They can then incorporate these numbers into their campaigns targeted at policy-makers to quantify public support for their advocacy aims.[61] A wide variety of human rights advocacy communications over social media exists, such as the 2013 Thunderclap campaign created by EDUCA, an NGO based in Oaxaca, Mexico, to raise awareness about ongoing human rights violations there. Thunderclap is a digital platform that allows users to coordinate their supporters' automatic participation in a onetime, synchronized mass social media posting of a particular message.[62] EDUCA surpassed its goal of 100 supporters, and its Thunderclap – timed to coincide with the October 23 UN Universal Periodic Review of Mexico's human rights record – reached more than 58,000 people via social media.[63]

---

[57] McPherson, *ICTs and Human Rights Practice*, pp. 28–32; R. Stewart, "Amnesty International's head of comms on why interactive social campaigns could help find a solution to the refugee crisis," *The Drum*, February 7, 2017, www.thedrum.com/news/2017/02/07/amnesty-international-s-head-comms-why-interactive-social-campaigns-could-help-find.

[58] Alston and Gillespie, "Global Human Rights Monitoring, New Technologies, and the Politics of Information," pp. 1112–13; E. McPherson, "How Editors Choose Which Human Rights News to Cover: A Case Study of Mexican Newspapers," in T. A. Borer (ed.), *Media, Mobilization, and Human Rights: Mediating Suffering* (London: Zed Books, 2012), pp. 96–121.

[59] J. Emerson et al., "The Challenging Power of Data Visualization for Human Rights Advocacy," Chapter 8 in this volume.

[60] D. Karpf, *The MoveOn Effect: The Unexpected Transformation of American Political Advocacy* (New York: Oxford University Press, 2012), pp. 36–37.

[61] E. McPherson, "Advocacy Organizations' Evaluation of Social Media Information for NGO Journalism: The Evidence and Engagement Models" (2015) 59(1) *American Behavioral Scientist* 124–48 at 134–39.

[62] Thunderclap is free, but the platform does decide whether or not to approve campaigns, and the extent of campaign visibility can depend on users' purchase of premium plans. "Take your message even further," Thunderclap, 2017, www.thunderclap.it/pricing.

[63] EDUCA, "Thunderclap: TÚ PUEDES EVALUAR A EPN EN DH," October 23, 2013, www.thunderclap.it/projects/5687-t-puedes-evaluar-a-epn-en-dh.

Again, however, the advantages of social media for advocacy are accompanied by risk, and here I detail two types of communication risk and their consequences. Both stem from the introduction of new technologies into advocacy, which in turn introduces new actors with new motives. Human rights practitioners are accustomed to considering the motives of intermediaries and their intended audiences in shaping their advocacy strategies. For example, they cater to the "media logic" of mainstream media outlets, tailoring the tone and theme of their content as well as building their identities as credible sources to meet journalists' exigencies.[64] The use of social media intermediaries, however, requires them to shape advocacy messages in light of new "social media logics."[65] These are also commercially driven motives, manifested in new technical features. Like journalists and journalism, these technical features can be inscrutable to human rights practitioners and incompatible with human rights advocacy – but in different ways.[66] Conducting advocacy via these intermediaries thus introduces new facets to existing risk. This risk includes audibility to *un*intended audiences, which I refer to as mistakes that can have reputational consequences. The second variety of risk addressed below is *in*audibility to intended audiences, or advocacy miscalculations that waste resources.

A *Mistakes and Reputational Integrity*

An advocacy-related mistake involves something happening that the communication's producer does not wish to happen. Social media's facilitation of mediation to publics, in combination with technical features that both speed up and obscure the dynamics of this mediation, introduce new ways of making mistakes. Analog means of communicating with publics had areas of friction, such as the effort required to set up a press conference.[67] This friction, no doubt, was frustrating during crises in need of immediate response, but it also allowed room for reflexivity and proofing. Digital communication to publics, in contrast, requires only the click of a button. As such, the pace of communication is much faster on social media, as is the pressure to produce at speed. Proofing becomes the friction, and there may not always be time for this to be done as thoroughly as one would like. Furthermore, the technical complexity of social media can make proofing difficult to do. This is particularly the case with respect to ensuring that the right communication is audible to the right audience, as audiences are both blurred and obscured by social media. Mistakes can

---

[64] S. Waisbord, "Can NGOs Change the News?" (2011) 5 *International Journal of Communication* 142–65 at 149–51.
[65] J. van Dijck and T. Poell, "Understanding Social Media Logic" (2013) 1(1) *Media and Communication* 2–14.
[66] McPherson, "Social Media and Human Rights Advocacy."
[67] S. Gregory, "Human Rights Made Visible: New Dimensions to Anonymity, Consent, and Intentionality," in M. McLagan and Y. McKee (eds.) *Sensible Politics: The Visual Culture of Nongovernmental Activism* (New York, Cambridge, MA: Zone Books, 2012), p. 552.

thus be about erroneous content, but they can also involve the transmission of private information to publics or of information intended for one "imagined audience" or communication context to another.[68] The consequences of these mistakes are also caught up with mediation, as ICTs allow endless possibilities of repetition and amplification over time and space.[69]

Here, I develop the example of individual practitioners managing multiple Twitter accounts, each with its own profile and audience. Rather than involving an error in the advocacy itself, the associated mistake results from having social media open as an advocacy channel. Twitter's phone apps easily allow users to switch between accounts, requiring nothing more than holding down the profile icon in the iPhone version. Of course, this also means it is easy to slip between accounts erroneously or forgetfully. When one account is personal and one is institutional, this can create some sticky situations.

One such situation arose in response to a 2014 tweet by Amnesty International about the police shooting in Ferguson, Missouri: "US can't tell other countries to improve their records on policing and peaceful assembly if it won't clean up its own human rights record." Six minutes later, the Center for Strategic and International Studies (CSIS), a major public policy think tank, replied, "Your work has saved far fewer lives than American interventions. So, suck it." CSIS scrambled to quickly explain the tweet as the work of an intern who had access to the CSIS Twitter account but thought he was logged into his personal account instead when he wrote the message. In the context of a flurry of media stories, CSIS's senior vice president of external relations described himself and his colleagues as "distressed," and CSIS quickly sent out an apology tweet to Amnesty. Amnesty followed this by tweeting: ".@CSIS and @amnesty have kissed and made up. Now back to defending human rights!"[70]

Though this example is relatively lighthearted, more serious mistakes can have more serious consequences. One human rights practitioner told me about a mistake made on his organization's Facebook feed when an image of a private meeting was erroneously published. A furious phone call from an important participating organization ensued, creating what the practitioner described as "a terror effect within the organization" about using social media. At the time of our interview, the resulting policy at this NGO was that every social media post made on the institutional account must first be approved by the executive director.

---

[68] A. E. Marwick and D. Boyd, "I Tweet Honestly, I Tweet Passionately: Twitter Users, Context Collapse, and the Imagined Audience" (2011) 13(1) *New Media & Society* 114–33; Thompson, *The Media and Modernity*, pp. 143–44.

[69] Thompson, *The Media and Modernity*, p. 141.

[70] B. James, "Think Tank Apologizes for Intern's 'Suck It' Tweet to Amnesty International," *Talking Points Memo*, August 19, 2014, http://talkingpointsmemo.com/livewire/csis-amnesty-international-suck-it-tweet; M. Roth, "Think Tank Blames Intern for Tweet Telling Amnesty International to 'Suck It'," *MTV News*, August 20, 2014, www.mtv.com/news/1904747/csis-intern-amnesty-international/.

Serious mistakes can jeopardize an organization's reputational integrity, particularly with respect to credibility and professionalism. The relative permanence of information published on social media, as well as the unpredictability of its circulation, means a mistake cannot be undone but must instead be overcome. Repairing a damaged reputation, which may involve performing credibility over time and rebuilding social capital, can divert precious resources from human rights NGOs' core aims.[71] Even if the mistake is quickly forgiven, it can – as Amnesty's last tweet above highlights – detract from the message and work of the organization. Because of the risk of mistakes that accompanies the use of social media, adopting this technology can result in slower and more resource-intensive practices than expectations might suggest.

B Miscalculation and Resources

A communication miscalculation means that one's message is inaudible to one's intended audience. Of course, the risk always exists that one's audience either does not hear or does not listen to the message. This is exacerbated by mediation, not only because distance makes it more difficult to perceive audience cues about attention, but also because the intermediary may do things to the message to make it less audible. In the case of social media, this includes evaluating messages automatically with timeline algorithms to determine how visible they should be, and to whom.

Human rights practitioners are in good company with respect to not knowing exactly how these algorithms make decisions about message visibility. The algorithms that govern social media timeline visibility are considered proprietary trade secrets, and these algorithms in turn may be governed by deep learning, in which the algorithm adapts autonomously based on the information to which it is applied.[72] Furthermore, these algorithms may have thousands of moving parts that are updated weekly or even daily.[73] Deciphering these algorithms – which are black boxes to just about everybody, even possibly to those who design them – is a far cry from building a trusting relationship with a journalist.[74]

Practitioners do know that these algorithms prevent organizations from reaching all of their fans or followers with their posts. "Organic," or unpaid, reach may be only 10 percent of a potential audience, and only a small proportion of those reached

---

[71] Cottle and Nolan, "Global Humanitarianism and the Changing Aid-Media Field" at 871–74.
[72] N. Koumchatzky and A. Andryeyev, "Using Deep Learning at Scale in Twitter's Timelines," Twitter, May 9, 2017, https://blog.twitter.com/engineering/en_us/topics/insights/2017/using-deep-learning-at-scale-in-twitters-timelines.html; Tufekci, "Algorithmic Harms Beyond Facebook and Google."
[73] W. Oremus, "Twitter's New Order," Slate, March 5, 2017, www.slate.com/articles/technology/cover_story/2017/03/twitter_s_timeline_algorithm_and_its_effect_on_us_explained.html.
[74] W. Knight, "The Dark Secret at the Heart of AI," MIT Technology Review, April 11, 2017, www.technologyreview.com/s/604087/the-dark-secret-at-the-heart-of-ai/; McPherson, "Source Credibility as 'Information Subsidy'."

will engage with the post by liking, sharing, or clicking on a link.[75] Facebook does shed some light on how this organic reach is determined, stating in its support materials for nonprofits that the post's timing and its relevance to particular audience members matter.[76] Twitter reveals that it ranks a tweet for relevance on a number of criteria, including how much user interaction it has already generated and how much past interaction exists between the producer and the potential recipient; in other words, visibility returns to the already visible and to the already networked.[77] Still, ambiguity remains for the organic visibility of individual posts. Greater certainty is available, however – at a price: mainstream social media platforms allow users to buy access to larger and targeted audiences. Social media advocacy is therefore a "free-to-play, pay-to-win game."[78]

Human rights practitioners encounter further elements of social media logic that generate communication risk. One is social media platforms' community standards, which outline the grounds for removal of content that might alienate users. Graphic images and videos fall into this category. The problem for human rights advocacy as well as fact-finding is that the documentation of certain categories of violations necessarily involves depictions of violence – though practitioners think through the ethics of such representations very carefully.[79] Like the determination of timeline visibility, content moderation is an opaque decision-making process.[80] Practitioners know that whether or not a graphic video or image stays on social media can depend on a number of factors, including how it is explained by whoever posts it (Facebook allows graphic images and videos to stay up if they are "in the public interest," but not if they are "for sadistic pleasure"), if it is reported by another user, what the content moderator employed by the platform to review content decides, and – as recently happened with the video livestreamed by Diamond Reynolds immediately after police shot her boyfriend – even "technical glitches."[81]

---

[75] M. Collins, "It's time for charities to stop wasting money on social media," *The Guardian*, March 11, 2016, www.theguardian.com/voluntary-sector-network/2016/mar/11/charities-wasting-money-social-media.

[76] Facebook, "Measurement & Tracking," Nonprofits on Facebook, 2017, https://nonprofits.fb.com/topic/measurement-tracking/.

[77] Koumchatzky and Andryeyev, "Using Deep Learning at Scale in Twitter's Timelines."

[78] L. Karch, "Is Social Media a Time-Waster for Nonprofits?" *Nonprofit Quarterly*, March 17, 2016, https://nonprofitquarterly.org/2016/03/17/is-social-media-a-time-waster-for-nonprofits/.

[79] M. Bair, "Navigating the Ethics of Citizen Video: The Case of a Sexual Assault in Egypt" (2014) 19 *Arab Media & Society* 1–7; Gregory, "Human Rights Made Visible," p. 555.

[80] S. T. Roberts, "Commercial Content Moderation: Digital Laborers' Dirty Work," in S. Umoja Noble and B. M. Tynes (eds.), *The Intersectional Internet: Race, Sex, Class, and Culture Online* (New York: Peter Lang Publishing, 2016), pp. 148–49.

[81] "Community Standards," Facebook, 2017, www.facebook.com/communitystandards#violence-and-graphic-content; A. Peterson, "Why the Philando Castile police-shooting video disappeared from Facebook – then came back," *The Washington Post*, July 7, 2016, www.washingtonpost.com/news/the-switch/wp/2016/07/07/why-facebook-took-down-the-philando-castile-shooting-video-then-put-it-back-up/.

A third way in which social media logics can introduce advocacy miscalculations is the content culture they cultivate by rewarding certain types of content with visibility – a culture that contrasts sharply with the traditional registers of human rights advocacy.[82] Facebook, for example, counsels nonprofits that "formal language can feel out of place" and that "placing blame ... typically doesn't lead to high engagement."[83] It may also be that certain types of human rights and certain types of victims are more aligned than others with the logics of social media virality, which is co-constructed by the predilections of algorithms and networked humans.[84] This was the topic of much public contemplation following the 2015 circulation of an image of three-year-old Syrian refugee Alan Kurdi's body washed up on a Turkish beach. Many critically attributed this image's viral spread to Alan's resemblance to a Western child, and thus his relatability to Western social media users.[85] Furthermore, the competition for audience attention on social media has fueled the rise of "clickbait" headlines, which feature a "curiosity gap." These headlines give away just enough to pique someone's attention, but require that person to click on a link to get the full story.[86] An interviewee from a human rights NGO that works with migrants and refugees joked about why this popular format is not an option for her organization's advocacy practices: "We are not going to be like, you know, 'This man got to the border, and you would never believe what happened next!' You can't do that, because it makes you sound ... your credibility is gone. So we don't do that." The content culture that is rewarded on social media, then, may also be at odds with what the target audiences of human rights advocacy want to hear from practitioners – if the audience even pays attention to social media advocacy in the first place.[87]

Using social media allows human rights practitioners to directly address advocacy targets, but whether those targets hear or listen to those advocacy messages is often an open question. The risk of such advocacy miscalculation generates follow-on risks to an NGO's resources. These include wasted time, since maintaining a social media presence – including designing content, building and interacting with networks, and developing advertising strategies – demands significant person-hours. This is also a waste of money, as is targeted advertising that falls on deaf ears. Social media's relative novelty has meant a steep learning curve for human rights

---

[82] McPherson, "Social Media and Human Rights Advocacy," pp. 281–83.
[83] "Grab People's Attention," Nonprofits on Facebook, 2016, https://nonprofits.fb.com/topic/grab-peoples-attention.
[84] van Dijck and Poell, "Understanding Social Media Logic," p. 7.
[85] See, e.g., C. Homans, "The Boy on the Beach," *The New York Times*, September 3, 2015, www.nytimes.com/2015/09/03/magazine/the-boy-on-the-beach.html.
[86] D. Thompson, "Upworthy: I Thought This Website Was Crazy, but What Happened Next Changed Everything," *The Atlantic*, November 14, 2013, www.theatlantic.com/business/archive/2013/11/upworthy-i-thought-this-website-was-crazy-but-what-happened-next-changed-everything/281472/.
[87] Powers, "NGO Publicity and Reinforcing Path Dependencies" at 498.

practitioners, and risk to advocacy communications can be diminished with expertise. At the same time, however, mastery remains somewhat of a mirage, due not only to the inaccessible element of social media logics, but also to the ICT sector's state of permanent renewal. Users regularly encounter new platforms as well as new features within the platforms they use, which appear seemingly overnight as tweaks to commercially driven systems designed to hold our attention.

So far, I have outlined the communication risk posed by digital fact-finding and advocacy related to new technologies and new actors; in the next section, I put these findings into conversation with contexts, resources, and risk discourses to show how risk's silencing effect is not universal, but rather can map onto existing inequalities.

V RISK ASSEMBLAGES, PLURALISM, AND INEQUALITY

Returning to the three types of pluralism introduced earlier in the chapter, it is clear that the manifested forms of risk outlined above have silencing effects on the first category – the pluralism of the human rights world vis-à-vis the world of power it aims to hold to account. New mediating technologies, with commercially driven technical features that complicate communication, fuel new communication cultures and allow new spaces for adversaries to intervene. Surveillance, through its consequences for physical security, can stop human rights practitioners from speaking. The susceptibility of practitioners to deception and mistakes, with the repercussions for reputations, may deafen advocacy targets to their communications. Advocacy miscalculations may prevent advocacy targets from hearing those communications at all. In order to understand the effects of communication risk on the other types of pluralism, however, we must further develop our understanding of these new risk assemblages. We must also think about context, resources, and risk discourses.

As materialized risks are always embodied, an individual practitioner's context and resources matter in understanding how risk impacts the second type of pluralism, namely pluralism within the human rights world. Context – or individuals' "social risk positions" ranging from their political environments to their positions within relevant social hierarchies – influences exposure to risk.[88] In turn, the resources individuals have at hand influence their ability to mitigate risk. Key here is the resource of expertise, such as digital literacy about computer security, knowledge of digital verification practices, and facility with social media. Also relevant are the resources that can be used to secure expertise, including money, of course, but also the social capital and reputations that can connect practitioners to expertise and convince experts to share it. The same resources can be used to secure physical and digital safeguards. The upshot is that risk curtails pluralism within the human rights world by silencing practitioners unequally.

[88] Beck, *Risk Society*, p. 23.

Inequalities in contexts and resources intersect with the types of risk enumerated above in a variety of ways. The risk of surveillance depends greatly on the proclivities of a practitioner's political opponents for purchasing surveillance technologies and enacting pro-surveillance legislation. It also depends on the practitioner's networks, whose resistance to surveillance is only as strong as their weakest links; one member falling prey to malware can unwittingly expose all her communication partners.[89] Security literacy is crucial. As a practitioner at an organization that trains human rights reporters on digital security once told me, "A lot of them don't know that Facebook is where a lot of people who would target human rights defenders go shopping." Security literacy is expensive, in terms of money and time, and it is daunting; therefore, it is more accessible to some than to others.[90]

Deception via the manipulation of truth-claims is also a risk that human rights practitioners experience differently. Like surveillance, this risk is conditional on the political context, since some governments are particularly inclined to engage in information wars. Associated reputational risks are not isolated, but rather may have repercussions for a practitioner's networks. This is because human rights organizations build their credibility in part through networks of association with credible peers; one organization's loss of credibility allows opponents to tar its network with the same brush.[91] Some organizations can weather a hit on their credibility better than others. As human rights organizations also build credibility through performance over time, a more well-established NGO would have more reputational capital to counterbalance an instance of susceptibility to deception or a mediation-related mistake.[92]

The risk of advocacy mistakes and miscalculations can be mitigated by human rights organizations' in-house social media expertise and consequently the money required to acquire this expertise. Funds also allow human rights organizations to buy visibility for their social media communications through targeted advertisements. Those with fewer resources to dedicate to social media advocacy are, unfortunately, more likely to waste resources by engaging in this practice. This is evident in the results of a recent study, which found that, of 257 sampled human rights NGOs, the richest 10 percent had 92 percent of the group's total Twitter followers, 90 percent of their views on YouTube, and 81 percent of their likes on Facebook. The study also found that social media advocacy does not seem to help NGOs set the agenda in the mainstream media – further evidence that unsuccessful digital communication can curtail the greater pluralism that using ICTs could bring, both within the human rights world and vis-à-vis the world of power.[93]

---

[89] Kelly et al., "Tightening the Net."
[90] Hankey and Ó Clunaigh, "Rethinking Risk and Security of Human Rights Defenders in the Digital Age" at 542.
[91] M. Land, "Peer Producing Human Rights," at 1136; Gibelman and Gelman, "A Loss of Credibility" at 376.
[92] McPherson, "Source Credibility as 'Information Subsidy'" at 337.
[93] Thrall, Stecula, and Sweet, "May We Have Your Attention Please?" at 143.

A major purpose of the first and second forms of pluralism is to support the third form: the pluralism of civilian access to human rights mechanisms. Civilians cannot access accountability without their voices – their accounts – being heard. If the NGOs representing them are silenced, they too may be silenced. So, communication risk restricts civilian access to the mechanism of human rights unequally as well. As this effect maps onto context and resource distributions, this means that civilians in more precarious contexts with relatively few resources – in other words, those who might most need human rights mechanisms – are more likely to be silenced. The networked nature of human rights NGOs, which are characterized by solidarity, information exchange, and international communication, goes some way to counteract this effect, as another organization may be able to pick up the communication chain.[94] Still, while ICTs do create human rights communication channels where none existed before, we must be alert to the possibility that they do not level inequalities of audibility, but rather extend them.

So far, this chapter has looked at materialized risk, but risk perception is just as important for understanding the human rights practitioner's lived experience of risk.[95] It is also just as important for understanding how the risk accompanying use can impact the pluralizing potential of ICTs. As evident from interviews with some human rights practitioners, in which they qualified their view of ICTs with words such as "terrified" and "scary," knowing about risk can be distracting and even debilitating. The more complex the risk assemblage, the stronger this effect, as it is more difficult to understand and predict the risk. This knowing but not knowing *exactly* brings its own anxieties.[96]

Risk perception is not necessarily accurate, in part because risks are hard to estimate and because the idea of them can be overwhelming. Furthermore, as explained below, the specter of risk associated with a practice may have been conjured on purpose to prevent people from undertaking that practice; it may be a discourse deployed in the pursuit of power.[97] A full exploration of risk perception, which is outside the confines of this chapter, would consider the practices individuals adopt in anticipation of these risks and would investigate how these practices affect pluralism. For example, some human rights practitioners are renouncing digital communication methods for a return to analog.[98] Some are slow to adopt digital information from civilian witnesses for fact-finding.[99] As mentioned above,

---

[94] M. E. Keck and K. Sikkink, *Activists beyond Borders: Advocacy Networks in International Politics* (Ithaca, NY: Cornell University Press, 1998).
[95] Nah et al., "A Research Agenda for the Protection of Human Rights Defenders" at 405–06.
[96] Beck, *Risk Society*, pp. 22, 54.
[97] D. Lupton, "Introduction: Risk and Sociocultural Theory," in D. Lupton (ed.), *Risk and Sociocultural Theory: New Directions and Perspectives* (Cambridge: Cambridge University Press, 1999), pp. 4–5.
[98] Hankey and Ó Clunaigh, "Rethinking Risk and Security of Human Rights Defenders in the Digital Age" at 542.
[99] Amnesty International, Benetech, and The Engine Room, *DatNav*, p. 8.

practitioners introduce protracted review systems for social media communications and pay for the visibility of their social media messages, and thus the success of their digital communications depends on the resources of time and money. Risk perception can also silence, and unevenly so. Furthermore, as practitioners weigh up pluralism versus security in deciding whether or not to communicate digitally, erroneous risk perception can swing the balance too far to security.

What we have here, then, is a risk double bind – risk is bad for pluralism if you know about it, and it is bad for pluralism if you don't. If the latter, human rights practitioners are more likely to fall prey to communication risk. If the former, risk perception can prevent them from communicating digitally in the first place. This creates its own follow-on risk, like missing vital pieces of evidence or being dismissed as Luddites in the context of a broader pro-technology zeitgeist that has enthused donors. Though this double bind can make practitioners feel caught between paralysis and propulsion, it is not impervious to resistance. Next, I offer four approaches to loosening the silencing risk double bind.

VI LOOSENING THE SILENCING RISK DOUBLE BIND

The silencing risk double bind, constructed in part by commercial and political actors, threatens to squeeze the pluralism potential from human rights practitioners' adoption of ICTs. Political adversaries of the human rights world benefit directly from this silencing effect. The commercial actors of social media companies profit from human rights practitioners shaping their communications to social media logics, which can have silencing consequences. Human rights practitioners – as well as all those involved in the human rights and technology space, such as scholars, technologists, and donors – can, however, counteract these forces. In outlining four approaches to loosening the risk double bind, this chapter moves beyond the techno-pessimistic enumeration of materialized risk, and its potential contribution to silencing risk perception, toward a techno-pragmatic position. The four approaches, which work best in tandem, support the development and adoption of ICTs for human rights pluralism. The first pair of approaches, involving education and technology, are about mitigating materialized risks, while the second pair, involving reflexivity and discourse, relate to the construction and perception of risk. As human rights practitioners know very well, risk is an unavoidable element of their work; the aim is not to eliminate it, but to work alongside risk without it getting in the way.

A *The Educational Approach*

Knowing about risk without overblowing it involves understanding the origins of risk as well as mitigation strategies. Education projects for digital literacy – particularly around data literacy, security training, and social media advocacy – are proliferating apace with interest in digital human rights practices. For example, The Engine Room, Benetech, and Amnesty International recently published *DatNav: How to Navigate Digital Data for Human Rights Research*, which has since been translated into Spanish and Arabic.[100] Amnesty International's Citizen Evidence Lab walks practitioners through techniques for verifying digital truth-claims.[101] New Tactics in Human Rights hosts online conversations about using social media for advocacy, among other topics.[102] These educational resources are targeted at human rights practitioners, who share them through their networks of knowledge exchange.

That said, education has its limits. Expertise in digital fact-finding and advocacy can mitigate the materialization of some risk, but to the extent that the use of ICTs remains inscrutable – due, for example, to black-box algorithms or to an ever-shifting security terrain – some risk always remains. Furthermore, it is difficult to inform diffuse arrays of civilian witnesses about risk, which puts the burden of responsibility for digital security more squarely on the shoulders of human rights practitioners.[103]

B *The Technological Approach*

The technological pathway out of the risk double bind involves using ICTs built to address the risks engendered by digital communications. If human rights practitioners adopt these technologies to communicate with civilian witnesses, they go some way toward protecting those witnesses as well. For example, human rights practitioners are increasingly communicating via messaging applications, like WhatsApp, that are relatively impervious to surveillance. Many are consulting Security in-a-Box, developed by Front Line Defenders and the Tactical Technology Collective to introduce communities of users to digital security tools in seventeen languages.[104]

Of course, introducing technical fixes to digital communication risk may instead compound this risk, even if the technical fixes are done with the best of intentions. This is because the adoption of new technologies escalates the technological "arms race" between human rights practitioners and adversary state actors.[105] A case in point was the 2014 arrest for treason of human rights bloggers in Ethiopia, in which their use of Security in-a-Box was presented as evidence against them.[106] This potential for the inadvertent escalation of risks is one reason why the latter two

---

[100] Ibid.
[101] C. Koettl, "About & FAQ," Citizen Evidence Lab, 2014, www.citizenevidence.org/about/.
[102] "Using Social Networking for Innovative Advocacy," New Tactics in Human Rights, 2016, www.newtactics.org/conversation/using-social-networking-innovative-advocacy.
[103] McPherson, "Digital Human Rights Reporting by Civilian Witnesses," pp. 197–98.
[104] "Security-in-a-Box: Digital Security Tools and Tactics," https://securityinabox.org/en.
[105] Hankey and Ó Clunaigh, "Rethinking Risk and Security of Human Rights Defenders in the Digital Age" at 540.
[106] "Tactical Tech's and Front Line Defenders' statement on Zone 9 Bloggers," Tactical Tech, August 15, 2014, https://tacticaltech.org/news/tactical-techs-and-front-line-defenders-statement-zone-9-bloggers.

approaches, the reflexive and the discursive, are vital complements to the educational and technological approaches.

C *The Reflexive Approach*

Reflexive and discursive approaches call for critical perspectives on risk that unsettle taken-for-granted interpretations and practices. Reflexivity requires considering one's own role in making and perceiving risk, as well as the ways in which broader power relations shape risk assemblages.[107] It is all too easy to think about risk being an individual problem, when actually it is a socially constructed phenomenon.[108] For example, human rights practitioners are told to strengthen passwords, adopt encryption, and be vigilant about social engineering – or risk being hacked or surveilled. This is despite the fact that these risks emerge from the confluence of a multitude of commercial, criminal, and political actors.[109] Our tendency to individualize risk is to the benefit of these powerful actors. A broader view of risk that sheds light on these actors' roles redresses deniability and supports accountability in the determination of risk responsibility.[110] Furthermore, this view helps to safeguard individuals by painting a more comprehensive picture of risk and how and why it occurs.

Reflexivity about one's own roles and responsibilities in constructing risk is also important. Asking individuals to participate in a human rights technology project is also asking them to take on risk. This risk may be difficult to anticipate, in part because the context and resources where technologies are developed – usually the Global North – do not match the context in which the technology is being deployed. For example, digital security experts convinced one NGO to change its operating system, but the new operating system was not compatible with the NGO's printer. The NGO's employees had to bring files on memory sticks to printers at local Internet cafés. The memory sticks got lost in the process, which created a greater security risk than the original risk the operating system change was implemented to address.[111]

Practitioners in this sector must also be reflexive concerning their assumptions about civilian witnesses' participation in digital human rights fact-finding and these witnesses' knowledge of associated risk. Some civilian witnesses are driven to

---

[107] J. Kenway and J. McLeod, "Bourdieu's Reflexive Sociology and 'Spaces of Points of View': Whose Reflexivity, Which Perspective?" (2004) 25(4) *British Journal of Sociology of Education* 525–44 at 527.

[108] Deborah Lupton, *Risk*, 2nd ed. (London: Routledge, 2013) p. 21.

[109] Ulrich Beck, "The digital freedom risk: Too fragile an acknowledgment," openDemocracy, January 5, 2015, www.opendemocracy.net/can-europe-make-it/ulrich-beck/digital-freedom-risk-too-fragile-acknowledgment.

[110] Beck, *Risk Society*, p. 33.

[111] Z. Rahman, "Technology tools in human rights," The Engine Room, 2016, www.theengineroom.org/wp-content/uploads/2017/01/technology-tools-in-human-rights_high-quality.pdf.

document human rights violations by the somewhat idealized goal of speaking truth to power. For others, however, bearing witness may instead be a life-or-death matter, a matter of local or global politics, an exercise of identity, a function of resources – or simply a response to digital solicitations by human rights practitioners.[112] Some are accidental witnesses, while others are activists. Tailoring risk assessment to individual risk profiles and providing support for risk-bearing may require difficult, on-the-ground work that outweighs the mediation benefits of ICTs. Furthermore, practitioners may consider that soliciting digital information from civilian witnesses is too risky for certain contexts. Again, reflexivity is important, as practitioners need to consider whether they are or should be making silencing decisions on behalf of civilian witnesses. While an accidental witness may not have had an opportunity to think through risk, an activist witness's drive to digitally communicate documentation of violations may be underpinned by extremely sophisticated risk calculations.

D *The Discursive Approach*

The discursive pathway out of the risk double bind also involves focusing on the social construction of risk – this time by being aware of the possibility that actors communicate risks in order to control the behavior of others. In other words, risk perception can be a discourse used to protect or pursue power. The discursive approach to loosening the risk double bind involves identifying those who might benefit from risk discourses in order to assess how well perception corresponds to materialized risk.[113] For example, state actors may visibly surveil or punish digital activists not only to quell those individuals, but also to create a broader chilling effect on online human rights reporting.[114] As Amnesty International's secretary general stated following the UK government's 2015 admission that its agencies had been intercepting Amnesty's communications, "How can we be expected to carry out our crucial work around the world if human rights defenders and victims of abuses can now credibly believe their confidential correspondence with us is likely to end up in the hands of governments?"[115]

These discourses don't just serve political purposes; they can have commercial benefits, too. For example, tales of criminal and terrorist use of the "dark web" may

---

[112] M. Loveman, "High-Risk Collective Action: Defending Human Rights in Chile, Uruguay, and Argentina" (1998) 104(2) *American Journal of Sociology* 477–525; S. Madhok and S. M. Rai, "Agency, Injury, and Transgressive Politics in Neoliberal Times" (2012) 37(3) *Signs: Journal of Women in Culture and Society* 645–69 at 661.
[113] Lupton, "Introduction: Risk and Sociocultural Theory," pp. 4–5.
[114] K. E. Pearce and S. Kendzior, "Networked Authoritarianism and Social Media in Azerbaijan" (2012) 62(2) *Journal of Communication* 283–98.
[115] "UK surveillance Tribunal reveals the government spied on Amnesty International," Amnesty International, July 1, 2015, www.amnesty.org/en/latest/news/2015/07/uk-surveillance-tribunal-reveals-the-government-spied-on-amnesty-international/.

arouse public suspicion about human rights practitioners' use of it in fact-finding, but they also sell newspapers.[116] Risk perceptions also create profit for the security sector, a major industry in which digital security is a growth niche.[117] The discursive approach to risk perceptions is particularly important, since, given the technical complexity of ICTs, most human rights practitioners must rely on external expertise to assess actual risk and appropriate responses.[118] Circling back to the education approach, incorporating this external knowledge must involve interrogating its motives.

VII CONCLUSION

Techno-optimism has surfaced in the human rights world, as in many others, based in part on the perceived benefits of ICTs for the pluralism of human rights communication. These benefits have been realized in a number of cases, but the application of ICTs has also materialized risk. As human rights practitioners consider whether and how to incorporate ICTs into their practices, this chapter has sought to outline some types of risk they may face and associated consequences for human rights pluralism. This risk, I argue, is a product of ICTs' affordance for mediation, or communication across time and place. This mediation, and the technical features it requires, alters the identity clues, interpretation cues, and contextual information communicators draw upon in order to increase the likelihood that their communication is successful.[119]

Furthermore, the use of ICTs introduces new intermediary actors to the human rights communication chain, and the technical complexity of ICTs makes these actors and their impact on communication more difficult to identify and assess.[120] Of particular note here are new commercial actors with profit motives. To be sure, human rights reporters have interacted with commercial motives before in their communication practices, such as in considering the marketability of newsworthiness decisions.[121] Never before, however, have commercial actors been so influential over and yet so hidden in mediation.[122] Cases in point are the commercial-political surveillance nexus, the lucrative gray market for spyware, and the proprietary, revenue-maximizing algorithms of social media platforms. Incorporating ICTs into human rights fact-finding and advocacy contributes to new risk assemblages for human rights practitioners.

---

[116] Beck, *Risk Society*, p. 46.
[117] Ibid., pp. 23, 46.
[118] Ibid., pp. 53–55.
[119] Thompson, *The Media and Modernity*, pp. 83–5.
[120] Beck, *Risk Society*, p. 22.
[121] McPherson, "Source Credibility as 'Information Subsidy'" at 333–35.
[122] Tufekci, "Algorithmic Harms Beyond Facebook and Google" at 208–09.

The types of risk outlined here are by no means the only ones that ICTs introduce or exacerbate for the human rights world. Others include the risk to human rights practitioners of secondary trauma brought on by exposure to images and videos of violations, or the retraumatization of individuals featured in advocacy material, particularly if the material is re-mediated and re-mixed.[123] The types of risk detailed here, however, have particular consequences for human rights pluralism. In digital fact-finding, human rights practitioners face surveillance risk that can imperil their physical security and deception risk that can jeopardize their reputational integrity. In digital advocacy, they encounter the risk of mistakes that have negative repercussions for reputations, as well as the risk that miscalculation poses for their resources. Some of these materialized risks and their repercussions silence human rights practitioners and civilian witnesses, while others deafen intended audiences to human rights communication. The perception of these risks can also be silencing, leading to a risk double bind in which both knowing and not knowing about risk can curtail human rights communication.

Acknowledging the silencing risk double bind throws into relief the importance of thinking about risk not in isolation, but rather as socially constructed. These social contexts produce values and connect individuals that could end up on opposite sides of a risk trade-off. In deciding whether or not to speak in the face of risk, human rights practitioners are choosing between the value of pluralism and the value of security. In so doing, they are also choosing between types of follow-on risk: the risk of physical harm and harm to reputations and resources if they choose pluralism, and the risk of ongoing human rights violations if they choose security. This means they are also making choices between risk populations.

The silencing risk double bind can feel unstoppable, part of the "juggernaut" of rapidly advancing technological change – with its associated complexities, inscrutable interconnections, and risk – that characterizes contemporary societies.[124] Yet, silencing is not inevitable. This chapter proposes four approaches to loosening the risk double bind: the educational and technological, which can limit materialized risk, and the reflexive and discursive, which can stay the construction of risk and erroneous risk perceptions. For practitioners, technologists, donors, and scholars, these approaches are useful heuristics for assessing risk. These heuristics also support human rights practices that allow successful digital communication to coexist with risk rather than be dictated by it.

The net impact of ICTs on the pluralism of the human rights world vis-à-vis the world of power it aims to hold to account is difficult to determine. What we can

---

[123] Bair, "Navigating the Ethics of Citizen Video" at 3; S. Dubberley, E. Griffin, and H. M. Bal, "Making Secondary Trauma a Primary Issue: A Study of Eyewitness Media and Vicarious Trauma on the Digital Frontline" Eyewitness Media Hub, 2015, http://eyewitnessmediahub.com/research/vicarious-trauma; Gregory, "Human Rights Made Visible."

[124] Beck, "The Digital Freedom Risk"; A. Giddens, *The Consequences of Modernity* (Cambridge: Polity Press, 1990), p. 139.

establish, however, is that materialized and perceived risk curtail pluralism unevenly within the human rights world. This dampening effect is stronger for human rights practitioners in more perilous political and social contexts and with less expertise and associated resources. It is not only particular organizations that are more affected by the materialized and perceived risk of digital human rights fact-finding and advocacy, but also the particular populations and particular human rights that they represent. For a world fundamentally concerned with pluralism, this momentum toward the use of technology creates risk for human rights enforcement in general, as it may be reinforcing inequalities around who speaks and gets heard on which human rights.

PART III

Beyond Public/Private

States, Companies, and Citizens

This final part considers the role of the actors – states, companies, and citizens – whose conduct has a bearing on the promotion and protection of rights. It breaks with the traditional binary frame of "public/private" to consider the impact that individual conduct has on the enjoyment of rights, as well as the obligations of states to support the capacity of individuals to protect their own rights. By considering not only states and non-state actors but also individuals, this part creates a more complete foundation for developing effective responses to violations of rights such as privacy and freedom of expression.

In Chapter 10, "Digital Communications and the Evolving Right to Privacy," Lisl Brunner focuses first on the obligations of states. She considers the evolution of the right to privacy under international law. Brunner notes several areas where this law needs further development, including with respect to the flow of data across borders, as well as the ways in which current efforts to protect personal data are – and are not – responding to those gaps. In Chapter 11, Rikke Jørgensen then examines the responsibilities of non-state actors in respecting and protecting rights and the role of states in regulating them. She notes that core civil and political rights are being exercised in a commercial domain that is owned and operated by private actors, and that user expressions and personal information are the raw material that drives the business models of Internet and social media companies. Jørgensen notes the existence of a governance gap as applied to these companies, which are subject primarily to moral, not legal, obligations to respect rights. This gap is particularly troubling because current approaches to corporate social responsibility focus on how these companies respond to pressure from the state to violate rights but neglect the extent to which the enforcement of their own terms of service negatively affects privacy and freedom of expression.

Finally, in Chapter 12, G. Alex Sinha addresses the role of individuals in ensuring their own safety and security online. Sinha makes the case for greater attention to what human rights law has to say about waiver of privacy rights, since many of the actions we take in our everyday online communications can undermine our privacy. Drawing a bright line between public and private based on whether information has been shared with, or is visible to, another is increasingly out of sync with modern patterns of communication. The ease with which digitized information can be obtained, shared, and collated today exponentially increases the privacy impacts of even ostensibly "public" information.[1] Further, given the role of private individual conduct in the protection of rights such as privacy, states may have greater obligations than ever before to equip individuals to protect their own rights. Sinha's research persuasively illustrates the challenges of protecting privacy online and the way in which these challenges may force us to choose between protecting our privacy and participating in democratic culture.

The contributions in this part provide two essential insights into the impact of states, corporations, and individuals in regulating the effects of technology on human rights. First, they illustrate many of the ways in which human rights law, particularly with its binary emphasis on states and non-state actors, needs to adapt to the technological changes that have taken place over the past few years. The chapters by Brunner (Chapter 10) and Sinha (Chapter 12), in particular, identify ways in which current approaches to privacy need further development in order to respond to issues created by new technologies. Data protection law, while important, may only protect limited aspects of what a right to privacy entails, and any human right to privacy must also answer the question of when these rights are waived.

Second, viewing the relationships among states, companies, and citizens reveals significant governance gaps in responding to the impacts of new technologies. An essential element of state duties is the duty to protect individuals from interference in enjoying their fundamental human rights by actors such as corporations. In the context of the Internet, states have not paid sufficient attention to this obligation. As Brunner (Chapter 10) and Jørgensen (Chapter 11) both discuss, states, particularly in Europe, have taken action to protect the right to privacy from infringement by companies. But, as Jørgensen observes, they have not acted as effectively to protect freedom of expression, particularly when private companies enforce their terms of service in ways that are detrimental to human rights.

There are, of course, many obstacles to effective state enforcement of the rights to privacy and to freedom of expression, including the state's own desire to limit criticism levied against it or its core ideology. In such circumstances, one of the most effective responses we can advocate for is to require all states to ensure that their citizens can protect themselves; for example, by allowing access to the encryption and anonymity tools Sinha describes in Chapter 12. Such tools, along with the remedies that Jørgensen notes, are required by international law and may be a way to increase respect for freedom of expression and privacy even in the face of intransigent states and companies.

---

[1] D. G. Johnson, P. M. Regan, and K. Wayland, "Campaign Disclosure, Privacy and Transparency" (2011) 19 *William & Mary Bill of Rights Journal* 959–82 at 960.

10

Digital Communications and the Evolving Right to Privacy

Lisl Brunner

I INTRODUCTION[1]

The meaning of the human right to privacy is evolving in response to developments in communications technology and an increasingly connected world in which data transits national boundaries imperceptibly. Although governments have had the capacity to access and store unprecedented quantities of digital communications data for some time, high-profile terrorist attacks and expanding transnational criminal activity have provided a strong motive to continue and expand these activities. When Edward Snowden revealed the global scope of existing communications surveillance capacity, states and civil society organizations turned to international law to seek clarity on how the right to privacy protects individuals, preserves legitimate state interests, and addresses the realities of the large-scale collection of data across traditional borders.

The tribunals and experts who interpret international human rights law have developed a rich body of standards on the right to privacy in communications, with European institutions leading the way. These standards address much of the present-day collection and use of digital communications, but significant gaps still exist. Until recently, there were few clear norms regarding the bulk collection of communications data, the responsibility of private companies to respect privacy rights, and the rules and protections that apply when communications data crosses borders.

This chapter explores the evolution of the right to privacy as it is established in international human rights law, and the ways in which human rights law is beginning to bridge these gaps. The first part provides an overview of the right to privacy and highlights developments in the digital age that international human rights law

---

[1] All opinions expressed in this chapter are those of the author alone and should not be attributed to any organization. Lisl would like to thank Sarah St. Vincent for her thoughtful comments on prior versions of this chapter.

must urgently address. The second part outlines the scope and meaning of the right to privacy in communications as it appears in international human rights treaties and in interpretations of these treaties by international tribunals and experts. The chapter then examines how European institutions are interpreting data protection law in a way that seeks to bridge some of the gaps in privacy protection that have formed in international human rights law. The chapter concludes by describing the incipient steps that UN and European institutions are taking to address the privacy challenges presented by the seamless flow of data across borders.

II THE EVOLUTION OF THE RIGHT TO PRIVACY AND ITS PRESENT CHALLENGES

A *The Protection of Privacy in Human Rights Law*

The right to privacy has a broad scope. Scholars note that there is no universal conceptualization of privacy and that societies' notions of its scope have evolved in response to changing political contexts and technological landscapes.[2] Privacy has often been linked to the interests of limiting access to the self and exercising control over one's personal information and actions.[3] In its diverse characterizations, privacy has been closely linked to human dignity.

The right to privacy is protected in the International Covenant on Civil and Political Rights (ICCPR), which had 168 state parties as of November 2016. Article 17 provides the following:

1. No one shall be subjected to arbitrary or unlawful interference with his privacy, family, home or correspondence, nor to unlawful attacks upon his honour and reputation.
2. Everyone has the right to the protection of the law against such interference or attacks.[4]

---

[2] *Report of the Special Rapporteur on the Right to Privacy, Joseph A. Cannataci*, ¶ 20, U.N. Doc. A/HRC/31/64 (March 8, 2016) ("Cannataci Report"); D. Solove, "Conceptualizing Privacy" (2002) 90 *California Law Review* 1088–89; D. Banisar and S. Davies, "Global Trends in Privacy Protection: An International Survey of Privacy, Data Protection, and Surveillance Laws and Developments" (1999) 18 *John Marshall Journal of Computer & Information Law* 1–113 at 6–8.
[3] See, e.g., H. Nissenbaum, *Privacy in Context: Technology, Policy, and the Integrity of Social Life* (Stanford, CA: Stanford University Press, 2010), pp. 69–70, 81–88; Solove, "Conceptualizing Privacy," at 1109–24. Solove has identified at least six different but interrelated conceptualizations of the essence of privacy: 1) the right to be let alone, 2) limited access to self, 3) secrecy, 4) control over personal information, 5) personhood (the protection of one's personality, individuality, and dignity), and 6) intimacy (control over one's intimate relations or aspects of life).
[4] International Covenant on Civil and Political Rights, in force March 23, 1976, GA res. 2200A (XXI), 21 UN GAOR Supp. (No. 16) at 52, 999 UNTS 171, art. 17.

Article 12 of the Universal Declaration of Human Rights contains a nearly identical formulation,[5] and the right is also protected in the European Convention for the Protection of Human Rights and Fundamental Freedoms (European Convention),[6] the Charter of Fundamental Rights of the European Union,[7] the American Convention on Human Rights,[8] and the Arab Charter on Human Rights.[9]

International and domestic tribunals have interpreted the right to privacy as protecting an individual's capacity to decide with whom she has intimate relationships,[10] when to have a family and who forms part of it,[11] and even when to end her own life.[12] Privacy in one's correspondence serves to limit the government's power to monitor its subjects, and it protects a sphere in which individuals can develop and express ideas, exchange confidences, and build relationships. When surveillance of communications occurs or is perceived to occur, individuals are inhibited from seeking and disseminating ideas, and self-censorship results.[13] In light of its relation

---

[5] Universal Declaration of Human Rights, GA res. 217A (III), U.N. Doc. A/810 at 71 (1948). Article 12 omits the two occurrences of the word "unlawful" from the first paragraph.

[6] European Convention for the Protection of Human Rights and Fundamental Freedoms, Rome, November 4, 1950, in force September 3, 1953, ETS 5; 213 UNTS 221. According to Article 8, "1. Everyone has the right to respect for his private and family life, his home and his correspondence. 2. There shall be no interference by a public authority with the exercise of this right except such as is in accordance with the law and is necessary in a democratic society in the interests of national security, public safety or the economic wellbeing of the country, for the prevention of disorder or crime, for the protection of health or morals, or for the protection of the rights and freedoms of others."

[7] Charter of Fundamental Rights of the European Union, October 26, 2012, in force December 1, 2009, 2010 O.J. (C83) 389 (March 30, 2010), art. 7 ("Everyone has the right to respect for his or her private and family life, home and communications.").

[8] American Convention on Human Rights, San Jose, November 22, 1969, in force July 18, 1978, OAS Treaty Series No. 36; 1144 UNTS 123. Article 11 establishes: "1. Everyone has the right to have his honor respected and his dignity recognized. 2. No one may be the object of arbitrary or abusive interference with his private life, his family, his home, or his correspondence, or of unlawful attacks on his honor or reputation. 3. Everyone has the right to the protection of the law against such interference or attacks."

[9] League of Arab States, *Arab Charter on Human Rights*, September 15, 1999. Article 17 establishes: "Private life is sacred, and violation of that sanctity is a crime. Private life includes family privacy, the sanctity of the home, and the secrecy of correspondence and other forms of private communication."

[10] *Dudgeon v. United Kingdom*, Eur. Ct. H.R., App. No. 7525/76 (October 22, 1981).

[11] *Atala Riffo and daughters v. Chile*, Judgement, Inter-Am. Ct. H.R. (ser. C) No. 239, ¶¶ 161–78 (February 24, 2012); *Artavia Murillo et al. ("In Vitro Fertilization") v. Costa Rica*, Judgement, Inter-Am. Ct. H.R. (ser. C) No. 257 (November 28, 2012); *Airey v. Ireland*, Eur. Ct. H.R., App. No. 6829/73 (October 9, 1979).

[12] *See, e.g., Cruzan v. Director, Missouri Department of Health*, 497 U.S. 261 (1990) (describing lower court judgments on individual decisions to terminate medical treatment that were framed in terms of privacy rights, but declining to address the case in those terms).

[13] *Report of the Special Rapporteur on the promotion and protection of the right to freedom of opinion and expression, Frank La Rue*, ¶ 24, U.N. Doc. A/HRC/23/40 (2013) ("La Rue Report 2013"); PEN America, *Global Chilling: The Impact of Mass Surveillance on International Writers* (2015); Human Rights Watch and the American Civil Liberties Union, *With Liberty*

to all of these interests and other human rights, the right to privacy has been called "an essential condition for the free development of the personality."[14]

In human rights law, the state's duty to respect and ensure rights entails negative and positive obligations. The state fulfills its negative obligation by not interfering with an individual's right unless it acts in accordance with the law, in pursuit of a legitimate interest, and in a manner that is necessary and proportionate to the fulfillment of that interest.[15] The positive obligation encompasses "the duty of the States Parties to organize the governmental apparatus and, in general, all the structures through which public power is exercised, so that they are capable of juridically ensuring the free and full enjoyment of human rights."[16] With respect to the right to privacy, the UN Human Rights Committee[17] has affirmed that states must establish privacy protections in law as part of their duty to ensure rights.[18]

European institutions have led the way in interpreting the scope of the right to privacy in communications, and particularly in balancing it with the state's interests in gathering information for law enforcement and national security purposes. In 1978, the European Court of Human Rights established that "[p]owers of secret surveillance of citizens, characterising as they do the police state, are tolerable under the Convention only in so far as strictly necessary for safeguarding the democratic institutions."[19] European leadership in this area stems from the region's experience during the Second World War, when census records facilitated the identification of the Jewish population and other groups targeted for persecution and extermination by Nazi and Nazi-influenced regimes.[20] Germany's particularly staunch defense of the right to privacy is also linked to the widespread use of surveillance by the Stasi secret police in East Germany and the elaborate files in which it detailed individuals' private lives.[21]

The European approach initially contrasted with the more stringent approach of the UN Human Rights Committee, whose 1988 General Comment on the right to privacy indicated that "[s]urveillance, whether electronic or otherwise, interceptions

---

to Monitor All: How Large-Scale U.S. Surveillance is Harming Journalism, Law, and American Democracy (July 2014).

[14] *In Vitro Fertilization,* ¶ 143; see also Cannataci Report, ¶ 8.

[15] See, e.g., *General Comment No. 31, Nature of the General Legal Obligation on States Parties to the Covenant,* U.N. Doc. CCPR/C/21/Rev.1/Add. 13 (2004), ¶6.

[16] *Velasquez Rodriguez v. Honduras,* Judgement, Inter-Am. Ct. H.R. (ser. C) No. 4, ¶ 166 (July 29, 1988); see also General Comment No. 31, ¶¶ 7, 13; *Airey v. Ireland,* ¶ 32.

[17] The Human Rights Committee is the UN body charged with receiving periodic reports from states parties to the ICCPR on implementation of the treaty, as well as interpreting the ICCPR through its general comments and, where a state has recognized its competence, through reports issued in response to communications. International Covenant on Civil and Political Rights, arts. 28, 40–41.

[18] *General Comment No. 31,* ¶ 8.

[19] *Klass and others v. Germany,* Eur. Ct. H.R., App. No. 5029/71 (September 6, 1978), ¶ 42.

[20] W. Seltzer, "Population Statistics, The Holocaust, and the Nuremberg Trials" (1998) 24 *Population and Development Review* 511–52.

[21] See, e.g., T. Coombes, "Lessons from the Stasi," *The European* (April 1, 2015).

of telephonic, telegraphic and other forms of communication, wire-tapping and recording of conversations should be prohibited."[22] This pronouncement appears strikingly categorical and out of step with state practice. It has historically been regarded as a legitimate state interest to gather foreign intelligence in order to prevent, detect, and prosecute crime and threats to national security.[23]

Over time, however, a more uniform set of global standards on the right to privacy in digital communications has formed, and other human rights institutions have looked to the European Court's extensive case law to inform interpretations of this right. Until the beginning of this century, interpretations of the right to privacy in communications by the European Court and UN mechanisms generally focused on articulating guidelines for conducting targeted surveillance. But advances in technology, coupled with rising national security concerns, have facilitated and incentivized the amassing of large quantities of data by governments. Revelations by Edward Snowden and others have demonstrated the areas in which Western states fall short of meeting existing human rights standards, as well as the areas in which these standards are poorly developed or absent.

B *The Impact of the Snowden Revelations on Privacy in the Digital Age*

Beginning in June 2013, the Snowden disclosures gave the public a wealth of detail about the scope and nature of government surveillance of communications in the digital age, primarily focusing on intelligence programs in the United States and the United Kingdom. The documents describe how the US government collected call detail records of millions of individuals from telecommunications companies on an ongoing basis, performed queries on the records in order to identify potential suspects of terrorism and other international crimes, and used "contact-chaining" to review the records of individuals within three levels of communication of the initial suspect to identify other potential suspects.[24] Through the PRISM program, the US government

---

[22] *General Comment No. 16: Article 17 (Right to Privacy)*, U.N. Doc. CCPR/GC/16 (1988), ¶ 8.
[23] *See, e.g.*, A. Deeks, "An International Legal Framework for Surveillance" (2015) **55** *Virginia Journal International Law* 291–368 at 300, 301–05, 313 ("Most scholars agree that international law either fails to regulate spying or affirmatively permits it."); R. J. Bettauer, "Questions Relating to the Seizure and Detention of Certain Documents and Data (*Timor-Leste* v. *Australia*). Provisional Measures Order" (2014) **108** *American Journal of International Law* 763–69. In its first case involving espionage issues, the International Court of Justice determined that "a State has a plausible right to the protection of its communications with counsel relating to an arbitration or to negotiations, in particular, to the protection of correspondence between them, as well as to the protection of confidentiality of any documents and data prepared by counsel to advise that State in such a context." *Questions Relating to the Seizure and Detention of Certain Documents and Data (Timor Leste v. Australia)*, International Court of Justice, Request for the Indication of Provisional Measures, Order of March 3, 2014, ¶ 27.
[24] G. Greenwald, "NSA collecting phone records of millions of Verizon customers daily," *The Guardian*, June 6, 2013; Privacy and Civil Liberties Oversight Board (United States), *Report on the Telephone Records Program Conducted under Section 215 of the USA PATRIOT Act and on*

compelled electronic communications service providers to provide the contents of online communications in response to requests that identified specific attributes of interest (i.e., "selectors"). Through the "upstream" method of surveillance, authorities gained access to the contents of telephone and Internet communications from the cables that transmit the communications internationally.[25]

The Snowden documents suggested that the United Kingdom had obtained the contents of communications in bulk by tapping undersea cables[26] and had intercepted and stored webcam images (including a large number of nude images) from nearly two million user accounts globally.[27] Agencies of both governments purportedly defeated encryption standards to access secure communications,[28] intercepted the communications of diplomatic missions and world leaders, including Angela Merkel and Dilma Rousseff,[29] and used listening stations in their foreign embassies to intercept communications traffic abroad.[30]

Although the Snowden revelations largely focused on the United States, the United Kingdom, and their English-speaking partners in Canada, Australia, and New Zealand (the Five Eyes Alliance), information has also been published suggesting that large-scale surveillance programs exist in France,[31] Sweden,[32] Russia,[33]

---

the Operations of the Foreign Intelligence Surveillance Court (January 23, 2014), pp. 21–31 ("PCLOB Report on Section 215"). Through these methods, it was estimated that the US government may have retained records related to more than 120 million telephone numbers.

[25] Privacy and Civil Liberties Oversight Board (United States), Report on the Surveillance Program Operated Pursuant to Section 702 of the Foreign Intelligence Surveillance Act (July 2, 2014), pp. 32–41; G. Greenwald and E. MacAskill, "NSA Prism program taps into user data of Apple, Google, and others," The Guardian, June 7, 2013. For a description of several US intelligence programs disclosed by Edward Snowden, see A. Toh, F. Patel, and E. Gotein, "Overseas Surveillance in an Interconnected World," Brennan Center for Justice, New York University School of Law (2016), pp. 5–10.

[26] E. MacAskill et al., "GCHQ taps fibre-optic cables for secret access to world's communications," The Guardian, June 21, 2013.

[27] S. Ackerman and J. Ball, "Optic Nerve: Millions of Yahoo webcam images intercepted by GCHQ," The Guardian, February 28, 2014.

[28] J. Ball, J. Borger, and G. Greenwald, "Revealed: How U.S. and U.K. spy agencies defeat internet privacy and security," The Guardian, September 6, 2013.

[29] E. MacAskill et al., "GCHQ intercepted foreign politicians' communications at G20 summits," The Guardian, June 17, 2013; E. MacAskill and J. Borger, "New NSA leaks show how US is bugging its European allies," The Guardian, June 30, 2013; J. Burke, "NSA spied on Indian embassy and UN mission, Edward Snowden files reveal," The Guardian, September 25, 2013; J. Ball, "NSA monitored calls of 35 world leaders after US official handed over contacts," The Guardian, October 25, 2013.

[30] "The NSA's Secret Spy Hub in Berlin," Der Spiegel, October 27, 2013.

[31] F. Johannes and J. Follorou, "In English: Revelations on the French Big Brother," Le Monde, July 4, 2013.

[32] J. Borger, "GCHQ and European spy agencies worked together on mass surveillance," The Guardian, November 1, 2013.

[33] I. Poetranto, "The Kremlin's new Internet surveillance plan goes live today," The Citizen Lab, November 1, 2012, https://citizenlab.ca/2012/11/the-kremlins-new-internet-surveillance-plan-goes-live-today/; S. Walker, "Russia to monitor 'all communications' at Winter Olympics in Sochi," The Guardian, October 6, 2013.

China,[34] Ethiopia,[35] and Colombia,[36] among other countries. Researchers and WikiLeaks have alleged that government authorities in the Middle East, Africa, and Latin America have obtained spyware that allows them to hack into communications devices remotely in order to monitor individuals.[37]

The Snowden revelations had a more direct impact on international law than prior reports because they also signaled that US and UK surveillance programs targeted powerful allies. Germany, Brazil, and other states brought their grievances to the United Nations, and in December 2013, the General Assembly called on states "[t]o review their procedures, practices and legislation regarding the surveillance of communications, their interception and the collection of personal data, including mass surveillance, interception and collection, with a view to upholding the right to privacy by ensuring the full and effective implementation of all their obligations under international human rights law."[38] The General Assembly requested the Office of the High Commissioner for Human Rights (OHCHR) to prepare a report on the right to privacy in the digital age, and the following year it encouraged the Human Rights Council to create a special mandate dedicated to the subject.[39] Joseph Cannataci was appointed as the first Special Rapporteur on the right to privacy in 2015, with a mandate to gather information and raise awareness regarding challenges facing the right to privacy, both generally and in the digital age.[40] Civil society organizations have also advocated for limitations on state surveillance at the international level, developing the Necessary and Proportionate Principles, which are based on the international human rights legal standards described below.[41]

The US government responded to the Snowden revelations by terminating its bulk collection of telephony metadata under one legal authority and committing to greater transparency regarding its communications surveillance programs.[42] Seven months

---

[34] OpenNet Initiative, *Internet Filtering in China* (2009), pp. 14–17; Human Rights Watch, *Freedom of Expression and the Internet in China* (2001).

[35] Human Rights Watch, *They Know Everything We Do: Telecom and Internet Surveillance in Ethiopia* (2014).

[36] Privacy International, *Shadow State: Surveillance, Law and Order in Colombia* (2015), pp. 27–31.

[37] See, e.g., B. Marczak et al., "Mapping Hacking Team's 'Untraceable' Spyware," The Citizen Lab, February 2014; W. R. Marczak, J. Scott-Railton, and M. Marquis-Boire, "When Governments Hack Opponents: A Look at Actors and Technology," Twenty-Third USENIX Security Symposium (August 2014); see also A. Hern, "Hacking Team hack casts spotlight on murky world of state surveillance," *The Guardian*, July 11, 2015; WikiLeaks, "The Hacking Team Archives," July 8, 2015, https://wikileaks.org/hackingteam/emails/.

[38] "The Right to Privacy in the Digital Age," U.N. Doc. A/RES/68/167, December 18, 2013.

[39] "The Right to Privacy in the Digital Age," U.N. Doc. A/RES/69/66, December 18, 2014.

[40] "The Right to Privacy in the Digital Age," U.N. Doc. A/HRC/RES/28/16, April 1, 2015;

[41] A. Alexander, "Digital surveillance 'worse than Orwell,' says new UN privacy chief," *The Guardian*, August 24, 2015.

[42] Uniting and Strengthening America by Fulfilling Rights and Ensuring Effective Discipline Over Monitoring Act of 2015 (USA FREEDOM Act), Public Law 114-23, 129 Stat. 268 (June 2, 2015).

after the revelations, its signals intelligence policy was updated to establish principles circumscribing the collection and use of signals intelligence.[43] The policy directive recognized the "legitimate privacy interests" of all persons, and it required that intelligence gathering "include appropriate safeguards for the personal information of all individuals" regardless of their nationality or location. These steps represent progress, but debate about the proportionality of surveillance programs operated by US authorities continues.

On the opposite side of the Atlantic, the United Kingdom, France, and Switzerland have recently passed new laws expanding their surveillance powers.[44] The UK Investigatory Powers Act establishes broad powers for the government to engage in bulk collection of communications data, obtain data located overseas from companies with a UK presence, require the decryption of communications, and perform "bulk equipment interference."[45] Some experts have praised the clarity of the bill and its oversight provisions; privacy experts and advocates have been highly critical of its sweeping powers.[46]

The next section discusses the well-developed body of international human rights law that applies to the surveillance programs revealed by Edward Snowden. While these standards are not well defined in a few areas, such as bulk collection of data, the tribunals and experts that interpret them are moving to fill these gaps.

III HUMAN RIGHTS LAW AND PRIVACY IN DIGITAL COMMUNICATIONS

The language of human rights treaties is general, and it falls to international tribunals, human rights mandate holders, expert bodies, and national courts to interpret the scope and meaning of a right. The European Court of Human Rights defines the obligations of the forty-seven contracting parties of the European Convention on Human Rights. Interpretations of the ICCPR, in turn, are generated by UN bodies including the International Court of Justice (ICJ), the

---

[43] The White House, Presidential Policy Directive/PPD-28, Signals Intelligence Activities, January 17, 2014.

[44] Loi No. 2015–912 of July 24, 2015 (France); "Swiss endorse new surveillance powers," BBC, September 25, 2016; "Switzerland votes in favour of greater surveillance," AFP, September 25, 2016; "Wet op de inlichtingen – en veiligheidsdiensten 20" (Netherlands), September 1, 2015; Y. Bahceli, "'Dutch intelligence-gathering reform bill sparks privacy concerns," Reuters, September 1, 2015.

[45] Investigatory Powers Act of 2016 (November 29, 2016), www.legislation.gov.uk/ukpga/2016/25/contents/enacted/data.htm.

[46] See, e.g., D. Anderson QC, "Oral Evidence Taken Before the Joint Committee for the Investigatory Powers Bill," December 2, 2015, Questions 61–75, www.parliament.uk/documents/joint-committees/draft-investigatory-powers-bill/oral-evidence-draft-investigatory-powers-committee.pdf. But see Cannataci Report, ¶ 39; E. MacAskill, "'Extreme surveillance' becomes UK law with barely a whimper," The Guardian, November 19, 2016; I. Ashok, "UK passes Investigatory Powers Bill that gives government sweeping powers to spy," International Business Times, November 18, 2016.

Human Rights Committee, special mandate holders, and the Office of the High Commissioner for Human Rights (but only the decisions of the ICJ are legally binding on parties). The Court of Justice of the European Union has also begun to interpret the rights to privacy and data protection as contained in the EU Charter of Fundamental Rights. The Inter-American Commission and Inter-American Court of Human Rights interpret the American Convention on Human Rights. Consistent with the principle that human rights are universal, these entities draw on one another's interpretations of rights and have thereby begun generating a fairly uniform body of international law on the right to privacy.

A *Legality, Necessity, and Proportionality*

Human rights law is implicated when a state interferes with the right to privacy, which occurs when the contents of communications or communications data are collected by state authorities, regardless of whether the data is examined.[47] Once authorities examine data that has been collected, a second interference takes place. Retaining data over time interferes with the right to privacy,[48] as does sharing communications data with other parties.[49] Restricting anonymity in digital communications is also considered to be an interference with the right to privacy, because anonymous and secure communications allow the free exchange of information and ideas, and anonymity "may be the only way in which many can explore basic aspects of identity, such as one's gender, religion, ethnicity, national origin or sexuality."[50]

In order to be consistent with international human rights law, an interference with a qualified right such as privacy must meet the tests of legality, necessity, and

---

[47] See, e.g., *Malone v. United Kingdom*, Eur. Ct. H.R., App. No. 8691/79, ¶ 84 (August 2, 1984); Office of the UN High Commissioner for Human Rights, "The right to privacy in the digital age" ("OHCHR Report"), U.N. Doc. A/HRC/27/37, ¶ 19; *Escher v. Brazil*, Judgement, Inter-Am. Ct. H.R. (ser. C) No. 200, ¶ 114 (July 6, 2009).

[48] See, e.g., *Amann v. Switzerland*, Eur. Ct. H.R., App. No. 27798/95, ¶ 69 (February 16, 2000); *Rotaru v. Romania*, Eur. Ct. H.R. App. No. 28341/95, ¶ 46 (Grand Chamber, May 5, 2000); *S. and Marper v. United Kingdom*, App. Nos. 30562/04 and 30566/04, ¶ 86 (Grand Chamber, December 4, 2008); *Digital Rights Ireland v. Minister of Communications*, Eur. Ct. H.R., App. Nos. 293/12 and 594/12, ¶¶ 34–35 (April 8, 2014).

[49] *Report of the Special Rapporteur on the promotion and protection of human rights and fundamental freedoms while countering terrorism*, Martin Scheinin, ¶¶ 26–28, U.N. Doc. A/HRC/14/46 (May 17, 2010) ("Scheinin Report 2010"); *Report of the Special Rapporteur on the promotion and protection of human rights and fundamental freedoms while countering terrorism*, Martin Scheinin, ¶¶ 35, 48, U.N. Doc. A/HRC/10/3 (Feb. 4, 2009) ("Scheinin Report I 2009"); *Tristan Donoso v. Panama*, Judgement, Inter-Am. Ct. H.R. (ser. C) No. 193, ¶ 83 (January 27, 2009).

[50] *Report of the Special Rapporteur for the promotion and protection of the right to freedom of opinion and expression*, David Kaye, ¶ 12, U.N. Doc. A/HRC/29/32 (May 22, 2015). As David Kaye has noted, "[e]ncryption and anonymity provide individuals and groups with a zone of privacy online to hold opinions and exercise freedom of expression without arbitrary and unlawful interference or attacks." Ibid., ¶ 16; *see also* La Rue Report 2013, ¶¶ 23, 47–49.

proportionality.[51] In terms of legality, the action constituting the interference (such as interception of communications) must be previously established in a law that is publicly accessible, clear, and precise, meaning that its consequences are foreseeable.[52] An interference must be in pursuit of a legitimate aim, and it must be a necessary and proportionate means of achieving that aim. For the European Court of Human Rights, the measure must be "necessary in a democratic society," meaning that it must answer a "pressing social need," and state authorities must provide "relevant and sufficient" justifications for the measure.[53]

The court has established that states have a margin of appreciation in determining whether a measure is necessary and proportionate, particularly when the protection of national security is concerned.[54] When a state engages in secret surveillance, the analysis focuses on whether the measures are "strictly necessary for safeguarding the democratic institutions" and whether "adequate and effective guarantees against abuse" are in place.[55] Because individual applicants can rarely prove that they have been the subject of such surveillance, the European Court has permitted challenges to intelligence laws *in abstracto* in certain circumstances, at times finding a violation of Article 8 where the legal framework did not meet the legality test,[56] and at other times looking at whether the law itself is necessary and proportionate.[57]

For the European Court, laws containing a great degree of specificity are more likely to be deemed consistent with the European Convention. The law should specify the nature of the offenses for which surveillance can be ordered,[58] which

---

[51] While Article 8(2) of the European Convention specifies this, human rights bodies, experts, and tribunals have interpreted the ICCPR and the American Convention to require this test as well. See, e.g., *Report of the Special Rapporteur on the promotion and protection of human rights and fundamental freedoms while countering terrorism, Martin Scheinin*, ¶¶ 16–19, U.N. Doc. A/HRC/13/37 (December 28, 2009) ("Scheinin Report II 2009"); OHCHR Report ¶ 23; *Escher v. Brazil*, ¶ 116; *Weber and Saravia v. Germany*, Eur. Ct. H.R., App. No. 54934/00, ¶ 80 (June 29, 2006).

[52] OHCHR Report, ¶ 23; *Escher v. Brazil*, ¶¶ 130–31; *Zakharov v. Russia*, Eur. Ct. H.R., App. No. 47143/06, ¶ 229 (Grand Chamber, December 4, 2015).

[53] See, e.g., *S. and Marper v. United Kingdom*, ¶ 101.

[54] *Leander v. Sweden*, Eur. Ct. H.R., App. No. 9248/41, ¶ 59 (March 26, 1987); *S. and Marper v. United Kingdom*, ¶ 102; *Weber and Saravia*, ¶ 106; *Zakharov v. Russia*, ¶ 232.

[55] *Klass v. Germany*, ¶ 42; *Weber and Saravia v. Germany*, ¶ 106; *Zakharov v. Russia*, ¶ 232; see also OHCHR Report, ¶ 25.

[56] See, e.g., *Liberty and others v. United Kingdom*, Eur. Ct. H.R., App. No. 58243/00, ¶¶ 64–70 (July 1, 2008); *Malone v. United Kingdom*, ¶¶ 80–82; *Rotaru v. Romania*, ¶ 62; *Amann v. Switzerland*, ¶ 63; *Association for European Integration and Human Rights and Ekimdzhiev v. Bulgaria*, Eur. Ct. H.R., App. No. 62540/00, ¶ 93 (June 28, 2007).

[57] See, e.g., *Kennedy v. United Kingdom*, Eur. Ct. H.R., App. No. 26839/05, ¶ 155 (March 18, 2010); see also *Klass v. Germany*; *Zakharov v. Russia*; *Szabo and Vissy v. Hungary*, Eur. Ct. H.R., App. No. 37138/14 (January 12, 2016).

[58] *Kennedy v. United Kingdom*. Although *Weber and Saravia* was an admissibility decision, the court deemed the German G-10 law to be *prima facie* consistent with the European Convention. The law provided for nontargeted communications surveillance in order to identify or prevent six specific offenses: "1) an armed attack on the Federal Republic of Germany; 2) the commission of international terrorist attacks in the Federal Republic of Germany; 3)

individuals' communications can be monitored,[59] and which authorities are empowered to request, order, and carry out surveillance, as well as the procedure to be followed.[60] It should provide for "a limit on the duration of telephone tapping; the procedure to be followed for examining, using and storing the data obtained; the precautions to be taken when communicating the data to other parties; and the circumstances in which recordings may or must be erased or destroyed."[61] Laws that restrict the right to privacy "must not render the essence of the right meaningless and must be consistent with other human rights, including the prohibition of discrimination."[62]

The European Court of Human Rights has determined on two occasions that the German G-10 Act of 1968 satisfied the rigorous standards for legality that a communications surveillance law must meet.[63] It has also approved provisions of the UK Regulation of Investigatory Powers Act on the interception of domestic communications.[64] In contrast, the court has found that other laws in the United Kingdom, as well as in Russia, Switzerland, Bulgaria, Romania, and Hungary, lacked the necessary specificity and gave the authorities overly broad discretion to conduct communications surveillance.[65]

B *The Necessity and Proportionality of Bulk Collection*

For years, human rights bodies have emphasized that although advances in communications technology require evolution in legal safeguards, the tests of legality, necessity, and proportionality continue to apply.[66] Yet many have questioned

---

international arms trafficking within the meaning of the Control of Weapons of War Act and prohibited external trade in goods, data-processing programmes and technologies in cases of considerable importance; 4) the illegal importation of drugs in substantial quantities into the territory of the Federal Republic of Germany; 5) the counterfeiting of money (Geldfälschung) committed abroad; 6) the laundering of money in the context of the acts listed under points 3 to 5." *Weber and Saravia*, ¶ 27.

[59] See, e.g., *Klass v. Germany*, ¶ 51.
[60] See, e.g., *Escher v. Brazil*, ¶ 131.
[61] *Szabo and Vissy v. Hungary*, ¶ 56; *Weber and Saravia v. Germany*, ¶ 95; see also *Escher v. Brazil*, ¶ 131; OHCHR Report, ¶ 28; La Rue Report 2013, ¶ 81.
[62] OHCHR Report, ¶ 23; *Klass v. Germany*, ¶ 51.
[63] *Klass v. Germany*; *Weber and Saravia v. Germany*. As mentioned above, *Weber and Saravia* was an admissibility decision rather than a judgment on the merits, but the court conducted a thorough examination of the G-10 law and determined that there were "adequate and effective guarantees against abuses of the State's strategic monitoring powers," making the applicants' claims under Article 8 "manifestly ill-founded." *Weber and Saravia*, ¶¶ 137–38.
[64] *Kennedy v. United Kingdom*.
[65] *Zakharov v. Russia*, ¶¶ 244–52; *Amann v. Switzerland*; *Malone v. United Kingdom*; *Association for European Integration and Human Rights and Ekimdzhiev v. Bulgaria*; *Rotaru v. Romania*; *Szabo and Vissy v. Hungary*.
[66] See, e.g., *Klass v. Germany*, ¶ 48; La Rue Report 2013, ¶ 50; *Szabo and Vissy*, ¶¶ 68–70; Report on Terrorism and Human Rights, I/A C.H.R., OEA/Ser.L/V/II.116 Doc. 5 rev. 1 corr. (2002) ¶ 371; *Escher v. Brazil*, ¶ 115.

whether programs that collect or retain data from millions of individuals who are not implicated in criminal activity or terrorism can ever be necessary and proportionate means of protecting the state and its people. For several UN Special Rapporteurs, the answer is no.[67] The OHCHR, the European Court of Human Rights, and the Court of Justice of the European Union have taken a more measured approach. While they have condemned indiscriminate or generalized surveillance measures, they have indicated that the principles that apply to targeted interception of communications and large-scale collection are generally the same.[68]

When analyzing bulk surveillance programs, the European Court employs a higher level of scrutiny, and it has found that programs that are clearly circumscribed by law and accompanied by robust oversight mechanisms can be consistent with the right to privacy.[69] In *Weber and Saravia v. Germany*, the court deemed "strategic monitoring" of communications to be consistent with the European Convention, because the law provided sufficient guarantees against abuses of state power.[70] The law permitted interception based on "catchwords" designed to identify communications linked to one or more of six specific crimes. The guarantees included clear rules governing every aspect of data collection and use, as well as oversight by the three branches of government and a civilian agency.[71]

In contrast, bulk surveillance programs that do not clearly circumscribe state power in law and in practice have been deemed inconsistent with Article 8 of the Convention. The court has ruled that the indefinite retention of biometric data of persons who were suspected (but not convicted) of committing criminal offenses was not necessary in a democratic society.[72] In *Liberty v. United Kingdom*, the bulk interception of external communications pursuant to a 1985 law was deemed to violate Article 8 because it gave the executive unfettered discretion as to which of the intercepted communications could be examined.[73] In the 2015 case *Zakharov v. Russia*, the court found the government's system of direct access to communications networks by state authorities (known as "SORM") inconsistent with the European Convention. The court noted that interception could take place for a broad range of offenses (including pickpocketing), and that judges had limited powers to order and oversee interception.[74] Because interception orders were not

---

[67] La Rue Report 2013, ¶ 62; Scheinin Report I 2009, ¶ 30.
[68] *Liberty v. United Kingdom*, ¶ 63; OHCHR Report, ¶ 20.
[69] OHCHR Report, ¶ 20.
[70] *Weber and Saravia v. Germany*, ¶¶ 117, 137–38. *But see* La Rue Report 2013, ¶ 59 (suggesting that the G-10 law's provisions on warrantless interception for national security purposes is overly broad).
[71] *Weber and Saravia*, ¶¶ 96–102, 115–22, 137.
[72] *S. and Marper v. United Kingdom*, Eur. Ct. H.R., App. No. 30562/04 (Grand Chamber, December 12, 2008) ¶ 125.
[73] *Liberty v. United Kingdom*, ¶ 64.
[74] *Zakharov v. Russia*, ¶¶ 244–52.

presented to communications service providers, the court questioned whether judicial control existed in practice.⁷⁵

Most recently, in *Szabo and Vissy* v. *Hungary*, the court determined that broadly drafted laws and weak oversight of surveillance (primarily by political officials of the same agency that conducted the surveillance) rendered bulk interception of communications inconsistent with the Convention. Deeming "strategic, large-scale interception" for national security purposes to be "a matter of serious concern," the court stated: "A measure of secret surveillance can be found as being in compliance with the Convention only if it is strictly necessary, as a general consideration, for the safeguarding [of] the democratic institutions and, moreover, if it is strictly necessary, as a particular consideration, for the obtaining of vital intelligence in an individual operation."⁷⁶ "An individual operation" might be one with a specific target⁷⁷; it might also be an effort to locate and apprehend a terrorist by collecting all communications in a certain area during a particular period. Both *Weber and Saravia* and the recent *Tele2 Sverige* judgment of the Court of Justice of the European Union support the latter position. The court will have more opportunities to determine whether bulk collection should be further circumscribed, as at least three cases challenging bulk surveillance programs in the United Kingdom are pending before it.⁷⁸

For their part, several UN human rights experts have concluded that the bulk surveillance of communications is inherently incompatible with the protection of Article 17 of the ICCPR. The former UN Special Rapporteur for counterterrorism and human rights, Martin Scheinin, has indicated that intelligence-gathering programs should be "case-specific interferences [with the right to privacy], on the basis of a warrant issued by a judge on showing of probable cause or reasonable grounds."⁷⁹ The current Special Rapporteur, Ben Emmerson, and the Special Rapporteur on the right to privacy, Joseph Cannataci, have made similar determinations.⁸⁰ While Scheinin and others have emphasized the need for strong oversight mechanisms and strict regulations on the use of data that is collected, he and the

---

⁷⁵ Ibid., ¶¶ 261–72.
⁷⁶ *Szabo and Vissy* v. *Hungary*, ¶¶ 69, 73.
⁷⁷ *See* S. St. Vincent, "Did the European Court of Human Rights Just Outlaw 'Massive Monitoring of Communications' in Europe?," Center for Democracy and Technology, January 13, 2016, https://cdt.org/blog/did-the-european-court-of-human-rights-just-outlaw-massive-monitoring-of-communications-in-europe/.
⁷⁸ *Big Brother Watch and others* v. *United Kingdom*, Eur. Ct. H.R., App. No. 58170/13 (September 4, 2013); *Bureau of Investigative Journalism and Alice Ross* v. *United Kingdom*, Eur. Ct. H.R., App. No. 62322/14 (September 11, 2014); *10 Human Rights Organizations and others* v. *United Kingdom*, Eur. Ct. H.R., App. No. 24960/15 (May 20, 2015).
⁷⁹ Scheinin Report I 2009, ¶ 30.
⁸⁰ *Report of the Special Rapporteur on the promotion and protection of human rights and fundamental freedoms while countering terrorism, Ben Emmerson*, U.N. Doc. A/HRC/25/59 (March 11, 2014) ¶¶ 52, 59; Cannataci Report, ¶ 39. *But see* La Rue Report 2013 (which does not state that bulk surveillance is per se incompatible with the ICCPR).

other experts suggest that these safeguards are insufficient to make bulk surveillance consistent with the right to privacy.[81]

It seems unlikely that the European Court will shift to the UN rapporteurs' more categorical condemnation of bulk collection, especially as the Court of Justice of the European Union has recently reaffirmed the standards of its case law to date. The European Court's position is logical: Communications surveillance is not prohibited by international law, and it is practiced by prominent European states. As a policy matter, however, it is problematic that human rights law should legitimize a practice that few states will conduct in a rights-respecting manner, and which leads to ever-increasing amounts of data being accessible to actors with a variety of motivations.

C *Effective Oversight of Communications Surveillance*

International human rights law generally provides that large-scale surveillance can be consistent with the right to privacy if it is accompanied by robust oversight mechanisms. Yet oversight of intelligence services and their covert operations has always proved challenging, even in societies where the rule of law is well established. Legislative committees conduct oversight of the intelligence services in the United States and the United Kingdom, but the Snowden revelations raised doubts as to whether these committees have access to the information necessary to perform their roles effectively.[82] In the United States, oversight of signals intelligence activities conducted by executive order is limited.[83] Additionally, while the US Foreign Intelligence Surveillance Court provides judicial authorization and oversight of several intelligence-gathering programs, for many years the confidential nature of its opinions obscured its surprisingly broad interpretation of a provision that permitted the collection of information "relevant to an authorized investigation."[84] That court's authority to examine the collection of foreign intelligence under the PRISM and upstream programs revealed by Snowden is also limited to assessing the government's targeting and minimization procedures.[85]

UN bodies and the European Court have recognized that *ex ante* authorization of communications surveillance by the judiciary provides a powerful safeguard against

---

[81] Scheinin Report I 2009, ¶¶ 37, 74.
[82] *See, e.g.*, D. Feinstein, "Feinstein Statement on Intelligence Collection of Foreign Leaders," (October 28, 2013); Z. Carpenter, "Can Congress Oversee the NSA?" *The Nation*, January 30, 2014; House of Commons Home Affairs Committee [United Kingdom], Seventeenth Report: Counterterrorism, Chapter 6 (April 30, 2014).
[83] "Overseas Surveillance in an Interconnected World," 32–34.
[84] E. Gotein and F. Patel, "What Went Wrong with the FISA Court," Brennan Center for Justice (2015), 22; PCLOB Report on Section 215, 59–60; *American Civil Liberties Union v. Clapper*, 785 F.3d 787, 811–19 (2d Cir. 2015).
[85] PCLOB Report on Section 215, 177; "What Went Wrong with the FISA Court," 27, 29.

abuse,[86] but they have declined to deem it a requirement of adequate surveillance laws, given the often limited powers of the judiciary to access relevant information or to assess the necessity and proportionality of surveillance.[87] Instead, they recommend that oversight be performed by all branches of government, including executive inspectors general or supervisory bodies, as well as civilian agencies.[88] For these authorities, oversight mechanisms must have sufficient resources and access to pertinent information in order to serve as an effective check on the power of law enforcement or security agencies.[89] There must also be a measure of public scrutiny; for example, anyone should be able to bring a claim before an oversight body, and its periodic reports and decisions about individual complaints should be publicly accessible.[90]

As the European Court recognized in *Zakharov*, communications service providers also have the potential to be a check on intelligence services and law enforcement agencies.[91] Communications service providers execute judicial orders for surveillance and can challenge those that are overly broad or illegal.[92] They can also increase transparency about how surveillance is conducted by disclosing the numbers of requests for interception and communications data that they receive.[93] Whistleblowers offer another potential check on the power of public authorities to conduct surveillance, and experts have emphasized the need for protections for those who act in good faith when disclosing information "to the media or the public at large if they are made as a last resort and pertain to matters of significant public concern."[94]

---

[86] La Rue Report 2013, ¶ 81; Office of the Special Rapporteur for Freedom of Expression, Inter-American Commission on Human Rights, *Freedom of Expression and the Internet*, OEA/Ser.L/V/II.CIDH/RELE/INF. 11/13 (December 31, 2013) ¶ 165.

[87] See, e.g., OHCHR Report, ¶ 37; *Association for European Integration and Human Rights and Ekimdzhiev v. Bulgaria*, ¶¶ 84, 87; *Rotaru v. Romania*, ¶ 59; *Zakharov v. Russia*, ¶¶ 258–63.

[88] See OHCHR Report, ¶ 37; Scheinin Report 2010, ¶ 8.

[89] See, e.g., Scheinin Report 2010, ¶ 9; *Zakharov v. Russia*, ¶¶ 274–81. The former UN Special Rapporteur for counterterrorism and human rights has praised the Norwegian parliamentary oversight mechanism, and the European Court has approved systems in Germany and the United Kingdom. Scheinin Report I 2009, ¶ 45; *Klass and others v. Germany*; *Weber and Saravia v. Germany*; *Kennedy v. United Kingdom*, ¶¶ 166–68; *Szabo and Vissy v. Hungary*, ¶¶ 82–83. *But see* La Rue Report 2013, ¶ 59 (expressing concern that the German G-10 law permits warrantless surveillance of communications by the intelligence services).

[90] See, e.g., *Szabo and Vissy v. Hungary*, ¶¶ 82–83.

[91] *Zakharov v. Russia*, ¶ 270.

[92] OHCHR Report, ¶ 38.

[93] La Rue Report 2013, ¶ 92; Inter-American Commission for Human Rights, *Freedom of Expression and the Internet*, ¶ 113; "Report of the Freedom Online Coalition Working Group Three, Privacy and Transparency Online," November 2015, pp. 43–45.

[94] Scheinin Report II 2009, ¶ 16; La Rue Report 2013, ¶¶ 52, 79, 84; UN Special Rapporteur on the promotion and protection of the right to freedom of opinion and expression and the Inter-American Commission on Human Rights Special Rapporteur for Freedom of Expression, "Joint Statement on WikiLeaks" (December 21, 2010).

D Access to Effective Remedy

Closely linked to oversight is the requirement that states ensure access to an effective remedy for anyone who claims that her rights have been violated.[95] The remedy may be through a judicial or nonjudicial mechanism that has the capacity to bring about the investigation, prosecution, and sanction of those responsible for violations (if applicable) and to provide an adequate remedy for the victim.[96] Any mechanism should be independent and have access to the evidence necessary to determine claims before it.[97]

The secret nature of communications surveillance can render access to justice more tenuous for those who claim a violation of their right to privacy. As a result, human rights tribunals and experts are increasingly recommending that authorities provide notice to targets of surveillance once the surveillance has ceased.[98] States, however, have generally resisted this practice as impractical or detrimental to surveillance operations and methods. If a state does not provide notice, it should have liberal rules on standing to bring claims that challenge covert surveillance regimes.[99] If an individual's right to privacy is found to have been violated, adequate remedies may include a declaratory judgment, damages, and injunctive relief against the orders that permit data to be intercepted or retained. Publication of decisions determining the rights of complainants also contributes to transparency and constitutes part of such a remedy.[100]

Although significant gaps between law and practice remain, a fairly comprehensive set of rules has emerged in the jurisprudence of the European Court of Human Rights. Surveillance programs are more likely to be consistent with international human rights law when they are strictly regulated by law, overseen by a number of independent and properly resourced bodies, capable of being challenged, and marked by the greatest degree of transparency possible. At the same time, human rights law itself has fallen short in two respects. First, its rules apply to states, rather than to the private actors who hold this personal data, and second, it has only recently begun to address the privacy protections that should apply to communications when they transit borders. The next section examines how European

---

[95] ICCPR art. 2(3); European Convention art. XX; American Convention art. 25.
[96] See, e.g., General Comment No. 31, ¶¶ 8, 15–19; Velasquez Rodriguez v. Honduras, ¶¶ 174 et seq.
[97] Scheinin Report II 2009, ¶¶ 10–12.
[98] See, e.g., La Rue Report 2013, ¶ 82; Weber and Saravia v. Germany, ¶ 135; Zakharov v. Russia, ¶ 287. But see OHCHR Report, ¶ 40 (determining that subsequent notice is not necessary but closely related to the question of effective remedy); Tele2 Sverige AB v Post-och telestyrelsen and Secretary of State for the Home Department v. Tom Watson and Others, Eur. Ct. H.R., App. Nos. 203/15 and 698/15, ¶ 121 (December 21, 2016).
[99] Zakharov v. Russia, ¶¶ 171, 298.
[100] See, e.g., Association for European Integration and Human Rights and Ekimdzhiev v. Bulgaria, ¶ 102; Kennedy v. United Kingdom, ¶ 167.

institutions seek to fill the first gap by interpreting EU data protection norms in light of the rights to privacy and data protection. The following section describes how both UN and European interpretations of the right to privacy are evolving to address the flow of digital communications across national borders.

IV DATA PROTECTION AND THE RIGHT TO PRIVACY

While human rights law sets out the obligations of states that are parties to human rights treaties, data protection laws and principles regulate practices of both state and private actors that can affect the right to privacy. The protection of personal information has historically been regarded as a component of the right to privacy,[101] yet with the adoption of the Charter of Fundamental Rights of the European Union in 2009, data protection became a distinct fundamental right in Europe.[102] UN Special Rapporteur Martin Scheinin has opined that a right to data protection is emerging at a global level as well.[103] While it is not recognized as such in human rights treaties outside of Europe, interpretations of data protection law that are closely tied to international human rights standards may convert this body of law into an effective tool for protecting rights at the domestic and international levels.

In terms of international law and guidelines, data protection principles are contained in the Council of Europe's Data Protection Convention,[104] the OECD Privacy Framework,[105] and the Asia Pacific Economic Cooperation Privacy Framework.[106] They are reflected in the newly adopted EU General Data Protection Regulation, which applies in the 28 EU member states, and in the proposed EU Regulation on Privacy and Electronic Communications.[107] They include the

---

[101] *See, e.g.*, Convention for the Protection of Individuals with regard to Automatic Processing of Personal Data, Strasbourg, January 8, 1981, C.E.T.S. No. 108, in force October 1, 1985, art. 1 ("Data Protection Convention"); *S. and Marper v. United Kingdom*, ¶ 103; *Van Hulst v. Netherlands*, Comm. No. 903/1999, U.N. Doc. CCPR/C/82/D/903/1999 (November 15, 2004) ¶ 7.9; Scheinin Report II 2009, ¶ 55; *General Comment No. 16*, ¶ 10; Inter-American Commission on Human Rights, *Freedom of Expression and the Internet*, ¶¶ 138–42; Solove, "Conceptualizing Privacy."

[102] Charter of Fundamental Rights of the European Union, art. 8(1); Treaty on the Functioning of the European Union, OJ C 326, 26.10.2012, pp. 47–390, art. 16(1).

[103] Scheinin Report II 2009, ¶ 12.

[104] Data Protection Convention, arts. 5–8.

[105] Organisation for Economic Co-operation and Development, Recommendation of the Council Concerning Guidelines governing the Protection of Privacy and Transborder Flows of Personal Data, C(80)58/FINAL, as amended on July 11, 2013 by C(2013)79 ("OECD Privacy Framework").

[106] APEC Privacy Framework, Publication APEC#205-SO-01.2 (December 2005).

[107] Regulation (EU) 2016/679 of the European Parliament and of the Council of April 27, 2016, on the protection of natural persons with regard to the processing of personal data and on the free movement of such data, and repealing Directive 95/46/EC (General Data Protection Regulation), OJ L 119, 4.5.2016, pp. 1–88; European Commission, Proposal for a Regulation of the European Parliament and of the Council concerning the respect for private life and the protection of personal data in electronic communications and repealing Directive 2002/58/

principles that the collection and use of personal data – including communications data – should be in accordance with the law, subject to limitations, and strictly for the fulfillment of purposes that are clearly articulated to the data subject. Data should be deleted when it is no longer necessary for the purposes that justified collection. The entity collecting personal data should only disclose that data to other parties by the authority of the law or if the data subject has consented. Individuals should have notice about, and a measure of control over, the ways in which their data is collected, used, and shared, as well as ways to hold states and private actors accountable for violations.[108] These principles echo the international human rights standards laid out in the previous section, and they form the basis of strong domestic data protection laws in states such as Canada, Argentina, Israel, and Japan.[109]

The Court of Justice of the European Union (CJEU) has interpreted EU data-protection law in light of the rights to privacy and data protection established in the EU Charter of Fundamental Rights, and its recent decisions have had sweeping impacts on public and private actors in Europe and beyond its borders. In 2014, the CJEU ruled that an EU law that allowed member states to mandate the storage of communications metadata for periods of between six months and two years was inconsistent with the rights to data protection and privacy.[110] According to the CJEU, telephony metadata "may allow very precise conclusions to be drawn concerning the private lives of the persons whose data has been retained." It determined that the retention of data of persons who were not linked to crimes was problematic, and the legal framework lacked clear rules as to how authorities should access and use that data.[111]

The CJEU reiterated its holding in *Tele2 Sverige*, indicating that "the general and indiscriminate retention of all traffic and location data" was not strictly necessary to achieve the aim of fighting serious crime and terrorism.[112] It added that member states' laws could permit the targeted retention of metadata for the purpose of fighting serious crime; they could also permit the retention of data from one or more geographical areas where "objective evidence" demonstrates a clear link "to fighting serious crime or to preventing a serious risk to public security."[113] These holdings are consistent with *Weber and Saravia, S. and Marper*, and other case law

---

EC (Regulation on Privacy and Electronic Communications), COM(2017) 10 (final), January 10, 2017.
[108] See Data Protection Convention, arts. 5–10; OECD Privacy Framework.
[109] See, e.g., Federal Law for the Protection of Personal Data in the Possession of Private Actors (Mexico) (July 5, 2010); Law on the Protection of Personal Data (Argentina), Law 25.326 (October 30, 2000); see also "Global Trends in Privacy Protection."
[110] *Digital Rights Ireland*.
[111] Ibid., ¶¶ 27, 58–68.
[112] *Tele2 Sverige*, ¶ 103. Where *Digital Rights Ireland* dealt with the EU Data Retention Directive, *Tele2 Sverige* addressed domestic data-retention laws in the United Kingdom and Sweden.
[113] Ibid., ¶¶ 106–111. The CJEU also held that authorities must notify individuals whose data has been retained once notification is unlikely to jeopardize the relevant investigations. Ibid., ¶ 121.

of the European Court of Human Rights,[114] but unlike the latter judgments, they could be implemented immediately by private actors, who were no longer subject to the retention mandate. As such, the judgments had the practical effect of limiting the amount of data accessible to state authorities for surveillance.

In the *Google Spain* case, the CJEU further demonstrated the capacity of data-protection law to regulate the privacy practices of non-state actors. The CJEU held that search engine providers must respond to requests from individuals to de-index their names from search results. Such requests must be honored when the information linked to their names is "inadequate, irrelevant or no longer relevant, or excessive in relation to the purposes of the processing at issue," unless the public interest in finding this information is determined to outweigh the individual's privacy rights.[115] Several civil society organizations have argued that the decision improperly placed private companies in the role of public authorities charged with balancing rights and interests. The counterpoint is that perhaps any actor that can impact an individual's fundamental rights, as defined in the EU Charter, should assume this level of responsibility.

By providing an explicit legal link between the practices of some of the largest multinational corporations and human rights, EU law creates more opportunities for individuals to challenge the practices of large entities. Similarly, it increases the power of European authorities to regulate these companies, both in Europe and abroad. The CJEU's decisions may also help to define the scope of companies' responsibility to respect users' privacy rights, a topic that is explored in greater depth in Chapter 11 of this volume.[116]

As human rights norms become a greater foundation for data protection law, EU authorities are also increasingly applying the latter to data that crosses international borders. The next section examines how the challenge of cross-border data flows is gradually being met by developments in both international human rights law and EU data protection law. It also notes the outstanding dilemmas to which neither body of law has definitively spoken yet.

---

[114] *S. and Marper v. United Kingdom*, ¶ 125. In this case, the European Court looked to the Data Protection Convention to interpret the scope of the right to privacy enshrined in Article 8 and held that the indefinite retention of biometric data of individuals suspected of committing criminal offenses was inconsistent with Article 8.

[115] *Google Spain SL and Google Inc. v. Agencia Española de Protección de Datos (AEPD) and Mario Costeja González*, CJEU, C-131/12, ECLI:EU:C:2014:317 (May 13, 2014), ¶¶ 94, 97.

[116] Civil society organizations and human rights experts have increasingly analyzed private companies' data-collection practices in light of their responsibilities per the UN Guiding Principles on Business and Human Rights. *See, e.g., Report of the Special Rapporteur on the promotion and protection of the right to freedom of opinion and expression, David Kaye,* A/HRC/32/38 (May 11, 2016); Cannataci Report ¶ 46(f); Inter-American Commission on Human Rights, *Freedom of Expression and the Internet,* ¶ 112; "Report of the Freedom Online Coalition Working Group Three, Privacy and Transparency Online"; Ranking Digital Rights, 2015 Corporate Accountability Index (November 2015), pp. 16–18.

V ENSURING THE RIGHT TO PRIVACY EXTRATERRITORIALLY

The privacy protections contained in human rights law have traditionally addressed states' conduct regarding their subjects' data within their own borders. But digital communications flow seamlessly across borders, challenging traditional paradigms of jurisdiction over individuals and information.[117] This means that privacy protections may be illusory when governments with sophisticated surveillance capabilities can access the communications data of people who are not subject to their jurisdiction.

A *The Extraterritorial Application of the Right to Privacy*

International human rights law provides little guidance as to the obligations of states vis-à-vis non-nationals located beyond their territories whose communications are targeted or simply swept up in bulk surveillance programs.[118] The ICCPR requires a state party "to respect and to ensure to all individuals within its territory and subject to its jurisdiction" the rights contained in the Convention without discrimination.[119] The Human Rights Committee and the ICJ have interpreted this language as a disjunctive, meaning that a state's duty extends to "anyone within the power or effective control of that State Party, even if not situated within the territory of the State Party."[120] A contrary interpretation would allow states to avoid their human rights obligations when exercising jurisdiction outside of their territories and be inconsistent with the object and purpose of the treaty.[121] The United States and Israel have disagreed with this position, and for many years the United States advocated a "strict territoriality" reading of Article 2 of the ICCPR, although its position seems to have softened in recent years.[122]

When the European Court of Human Rights has addressed the extraterritorial conduct of its Contracting Parties, it has found effective control to be present in two types of situations: when state agents "exerci[se] control and authority over an

---

[117] *See, e.g.*, J. Daskal, "The Un-territoriality of Data (2015) **125** *Yale Law Journal* 326–397.

[118] *See, e.g., Weber v. Saravia*, ¶72 (in which the European Court declined to determine whether a complaint filed by applicants located outside of Germany alleging violations of their privacy rights by the German state was admissible *ratione personae*); *see also* Submission of Privacy International et al., OHCHR consultation in connection with General Assembly Resolution 68/167, "The right to privacy in the digital age," April 1, 2014.

[119] ICCPR art. 2(1).

[120] *General Comment No. 31*, ¶ 10; *see also Legal Consequences of the Construction of a Wall in the Occupied Palestinian Territory*, Advisory Opinion, 2004 ICJ Rep. 136, ¶¶ 109–11; I/A C.H.R., Report No. 109/99, Case 10.951, *Coard et al.* (United States), September 29, 1999, ¶ 37.

[121] *Legal Consequences of the Construction of a Wall*, ¶ 109.

[122] *See, e.g.*, United States Department of State, Memorandum Opinion on the Geographic Scope of the International Covenant on Civil and Political Rights, October 19, 2010; UN Human Rights Committee, "Human Rights Committee considers report of the United States," March 14, 2014, www.ohchr.org/EN/NewsEvents/Pages/DisplayNews.aspx?NewsID=14783; *Legal Consequences of the Construction of a Wall in the Occupied Palestinian Territory*, ¶ 110.

individual" (the personal model of jurisdiction), or when a state occupies a foreign territory through military action and assumes responsibility for some or all of the public functions normally performed by the government in that territory (the spatial model).[123] Yet this analysis of the degree to which state agents exercise physical control over individuals is ill-suited to the nature of communications surveillance, where control over infrastructure and individuals is virtual.[124] Communications surveillance programs most often involve a state's collection and review of data from its own territory, even though the communications may originate and terminate in other states and the rights holders may be beyond the collecting state's jurisdiction.[125] Some types of collection more clearly involve extraterritorial action – e.g., a state's interception of communications traffic via equipment located in its embassies abroad – but the impact on rights occurs in a different manner from the exercise of "effective control" over persons or territory.

Noting the mismatch between the prevailing test for extraterritorial obligations and the facts surrounding communications surveillance, several human rights experts have maintained that when analyzing a state's exercise of jurisdiction, one should look at its control over *rights* rather than over individuals or territory. Therefore, in the context of communications surveillance, it is the assertion of authority in ways that affect the rights of individuals that triggers a state's human rights obligations, even with respect to a person with no connection to that state.[126] For Marko Milanovic, in most (if not all) of the situations described in the Snowden documents, the state's obligation to *respect* the human rights of impacted individuals outside of its territory should apply.[127] Consequently, the state's interference with an

---

[123] *Al-Skeini v. United Kingdom*, Eur. Ct. H.R., App. No. 55721/07, ¶¶ 133–40 (Grand Chamber, July 7, 2011). Notably, the two cases before the European Court dealing with extraterritorial communications surveillance involved applicants who were both nationals and non-nationals, and the court did not address the state's obligations to the latter. *Weber and Saravia*, ¶¶ 72; *Liberty v. United Kingdom*, Eur. Ct. H.R., App. No. 58243/00 (July 1, 2008).

[124] *See* J. Daskal, "Extraterritorial Surveillance under the ICCPR ... The Treaty Allows It!," *Just Security*, March 7, 2014, www.justsecurity.org/7966/extraterritorial-surveillance-iccpr-its-allowed/.

[125] *See, e.g.*, M. Milanovic, "Human Rights Treaties and Foreign Surveillance: Privacy in the Digital Age" (2015) **56**(1) *Harvard International Law Journal* 81–146.

[126] *See, e.g.*, Letter to the Editor from M. Nowak, "What does extraterritorial application of human rights treaties mean in practice?," *Just Security*, March 11, 2014, www.justsecurity.org/8087/letter-editor-manfred-nowak-extraterritorial-application-human-rights-practice/; Letter to the Editor from Former Member of the Human Rights Committee, M. Scheinin, *Just Security*, March 10, 2014, www.justsecurity.org/8049/letter-editor-martin-scheinin/; P. Margulies, "The NSA in Global Perspective: Surveillance, Human Rights, and International Counterterrorism" (2014) **82** *Fordham Law Review* 2137–2167 at 2148–52 (arguing that a state exercises "virtual control" over communications infrastructure when it conducts surveillance).

[127] *See, e.g.*, "The NSA in Global Perspective"; "Human Rights Treaties and Foreign Surveillance," 118–119; *see also* Memorandum Opinion on the Geographic Scope of the International Covenant on Civil and Political Rights, pp. 49–50, 55–56 (arguing that a state may have obligations based on a sliding scale, and proposing that "once a state exercises authority or effective control over an individual or context, it becomes obligated to respect Covenant rights to the extent of that exercise of authority").

individual's privacy rights must be in pursuit of a legitimate aim and be a necessary and proportionate means of achieving that aim. The state's positive obligation to ensure rights, however, would only apply to individuals located within its territory. Others would eschew the control test entirely and contend that laws that offer distinct protections based on the nationality or location of the subject of surveillance are difficult to justify under human rights law.[128]

In *The Right to Privacy in the Digital Age*, the OHCHR found several of the aforementioned arguments regarding a state's extraterritorial human rights obligations to be compelling at a high level, writing:

> [D]igital surveillance therefore may engage a State's human rights obligations if that surveillance involves the State's exercise of power or effective control in relation to digital communications infrastructure, wherever found, for example, through direct tapping or penetration of that infrastructure. Equally, where the State exercises regulatory jurisdiction over a third party that physically controls the data, that State also would have obligations under the Covenant. If a country seeks to assert jurisdiction over the data of private companies as a result of the incorporation of those companies in that country, then human rights protections must be extended to those whose privacy is being interfered with, whether in the country of incorporation or beyond.[129]

The report adds that, according to the principle of nondiscrimination contained in the ICCPR, states must respect the legality, necessity, and proportionality principles regardless of the nationality or location of the subject of communications surveillance.[130]

The OHCHR explicitly declined to limit the scope of the state's obligations to subjects of communications surveillance beyond its borders to that of merely respecting rights, in a manner similar to the statements of the ICJ and the Human Rights Committee. This leaves open the question of whether, under the ICCPR, a state may have a duty to *ensure* the rights of these individuals, even though the basis for jurisdiction may be a fleeting or virtual action. If this is the case, many of the obligations outlined above could flow to state action that has a definitive impact on the privacy rights of individuals beyond its territory. Extraterritorial surveillance would have to be based on laws that are consistent with international human rights standards and be subject to effective oversight. Any individual whose rights were

---

[128] A. Deeks, "An International Legal Framework for Surveillance" (2015) 55 *Virginia Journal of International Law* 251–367 at 310–11; see also Daskal, "Extraterritorial Surveillance" (arguing that the conduct of surveillance on foreign nationals abroad is not covered by the ICCPR).
[129] OHCHR report, ¶ 34.
[130] Ibid. ¶ 36; see also La Rue Report 2013, ¶¶ 64, 87; *Concluding observations on the fourth periodic report of the United States of America*, U.N. Doc. CCPR/C/USA/CO/4 (April 23, 2014), ¶ 22 (maintaining that the state's obligations in the realm of privacy rights do not differ depending on the nationality or location of the target of surveillance).

impacted must have access to an effective remedy, and regulation of non-state actors would extend to extraterritorial actions as well.

The United States' 2014 update to its signals intelligence policy, requiring that intelligence gathering "include appropriate safeguards for the personal information of all individuals" irrespective of their nationality or location,[131] is the most explicit action taken by a state to date to extend protections to those impacted by its extraterritorial surveillance. In light of the broad powers contained in the UK Investigatory Powers Act and other laws, more detailed interpretations of these obligations from UN mechanisms or from the European Court are needed to guide state action.

B EU Data Protection Law and Extraterritorial Privacy Protections

This chapter has argued that European authorities are interpreting data protection law in a way that fills the gaps in privacy protections left by international human rights law. As part of this effort, they are also increasingly applying EU data protection law extraterritorially, in an attempt to fill the void of uncertainty regarding the protections that adhere to individuals' communications data when it crosses borders. In doing so, EU authorities may ultimately elevate privacy protections for communications well beyond the European continent.

The new EU General Data Protection Regulation and the proposed Privacy and Electronic Communications Regulation specify that they are binding on companies located outside of the EU that offer services to data subjects within the EU or otherwise monitor their behavior.[132] Since 1995, EU law has restricted the transfer of personal data outside of Europe to states that are deemed to have an adequate level of legal protection for the privacy rights of individuals.[133] The CJEU has interpreted this provision to mean that a third country must offer "a level of protection of fundamental rights and freedoms that is *essentially equivalent* to that guaranteed within the European Union" in order for general transfers to that country to be approved.[134] A multinational company may also transfer data to a state that has not been deemed adequate if the company commits to providing adequate safeguards.[135] Furthermore, the recently adopted EU-US Umbrella Agreement establishes privacy protections for the personal data of Europeans (as well as persons from the United

---

[131] Presidential Policy Directive/PPD-28, Signals Intelligence Activities, The White House, January 17, 2014.
[132] Regulation (EU) 2016/679, art. 3; Proposal for a Regulation on Privacy and Electronic Communications, art. 3.
[133] Directive 95/46/EC of the European Parliament and of the Council of 24 October 1995 on the protection of individuals with regard to the processing of personal data and on the free movement of such data, Official Journal L 281, 23/11/1995 P. 0031–0050, arts. 25, 26(2).
[134] *Maximilian Schrems v. Data Protection Commissioner*, CJEU, C-362/14, ECLI:EU:C:2015:650, ¶ 73 (emphasis added).
[135] Regulation (EU) 2016/679, arts. 44–50.

States) in the context of criminal law enforcement cooperation.¹³⁶ With these instruments, EU authorities aim to achieve a baseline level of privacy protection for their subjects' communications and other personal data vis-à-vis foreign actors from the private and public sectors, regardless of where they are located or where they handle that data.

National authorities in the EU are also seeking to apply EU data protection law extraterritorially by requiring companies to comply on a worldwide basis, as opposed to only with reference to sites directly aimed at the specific jurisdiction in question. For example, following the CJEU's *Google Spain* decision, French data protection authorities ordered Google to de-index search results that fit the judgment's criteria on a global scale, in order to protect data subjects' privacy rights more effectively.¹³⁷ Google had previously ensured that no users located in the European Union could access de-indexed results, but French authorities seek to make de-indexing decisions applicable across the global Internet. If upheld on appeal, this judgment could extend the reach of certain European data protection norms internationally.¹³⁸

In addition to strengthening protections for the privacy rights of Europeans regardless of where their data flows, the European approach may also elevate privacy protections for individuals outside of the region. A handful of non-EU states have been designated as having adequate data protection standards by the European Commission, and this stable basis for data transfer is attractive for trading partners. In the wake of the Snowden revelations, the CJEU used this mechanism to push for changes in US surveillance law. In the *Schrems* case of 2015, the CJEU invalidated the European Commission's decision that the US legal regime offered an adequate level of protection for data subjects under the Safe Harbor Agreement reached between the US government and the European Commission. The CJEU determined that "legislation permitting the public authorities to have access on a generalised basis to the content of electronic communications" for national security purposes was inconsistent with the right to privacy.¹³⁹

The *Schrems* decision had the potential to halt a significant portion of the transatlantic flow of personal data, prompting US and EU authorities to negotiate the Privacy Shield agreement as a replacement.¹⁴⁰ US authorities have also supplemented the agreement with detailed explanations of US surveillance law and

---

¹³⁶ Council of the European Union, "Umbrella agreement: EU ready to conclude with the US," December 2, 2016, www.consilium.europa.eu/en/press/press-releases/2016/12/02-umbrella-agreement/.

¹³⁷ A. Hern, "Google says non to French demand to expand right to be forgotten worldwide," *The Guardian*, July 30, 2015.

¹³⁸ A. Hern, "Google takes right to be forgotten battle to France's highest court," *The Guardian*, May 19, 2016.

¹³⁹ *Schrems*, ¶ 93. The CJEU also found insufficient evidence that Europeans could obtain an effective remedy for violations of their privacy rights. Ibid. ¶ 95.

¹⁴⁰ *See* www.privacyshield.gov.

practice. Nevertheless, the adequacy of the US legal regime continues to be impugned.[141] States beyond Europe are also following the region's example when updating data protection laws, by limiting the legal bases for collecting personal data and restricting the flow of data to states that are deemed adequate.[142] Thus, the ultimate legacy of the *Schrems* case may be a gradual harmonization of data protection standards among key parts of the data economy, with EU rules serving as the foundation.

Despite the evolution of international human rights law and EU data protection law regarding privacy and cross-border data flows, clear rules have not yet emerged to address which state's privacy protections should apply to communications data when multiple governments assert jurisdiction over it.[143] In a case involving Microsoft in the United States, a federal appeals court ruled that the location of the data should determine which state may claim jurisdiction (and which privacy protections apply).[144] The UK Investigatory Powers Act allows the government to issue extraterritorial warrants for communications data if the data is held by a company that is subject to its regulatory jurisdiction.[145] For Jennifer Daskal, both approaches to jurisdiction are unsatisfactory, given the mobility of data, the incentives for companies and governments to decide who may access data based on where it is stored, and the conflict of laws which companies may face.[146] Instead, Daskal poses that the law should allow for multiple jurisdictional triggers to be evaluated, including the nationality and location of the data subject.[147] The absence of clear rules on jurisdiction and privacy protections in this scenario has led to calls for international law to fill the void through the negotiation of an international treaty[148] or smaller bilateral or multilateral agreements.[149] From a human rights perspective, the OHCHR's position should guide the development of any such framework: The

---

[141] *Digital Rights Ireland v. Commission*, T-670/16, Action brought September 16, 2016.
[142] *See, e.g.*, E. Kosinski and S. Asayama, "Transfer of Personal Data under Japan's Amended Personal Information Protection Act," White and Case Technology Newsflash, October 13, 2015; P. A. Palazzi, "New Draft of Argentine Data Protection Law Open for Comment," International Association of Privacy Professionals, February 8, 2017; "Brazil Releases Draft Data Protection Bill," Hunton and Williams Privacy and Information Security Law Blog, February 6, 2015.
[143] Daskal, "The Un-territoriality of Data" at 365–70, 373–78; J. Daskal, "Law Enforcement Access to Data Across Borders: The Evolving Security and Rights Issues" (2016) 8 *Journal of National Security Law & Policy* 473–501.
[144] *In re Warrant to Search a Certain Email Account Controlled and Maintained by Microsoft Corp.*, Case 14–2985, Document 286-1 (2d Cir. July 14, 2016).
[145] Investigatory Powers Act, §§ 41–43, 85, 52, 126–27, 149, 168–69, 190.
[146] "The Un-territoriality of Data," 389–95; "Law Enforcement Access to Data Across Borders," 487–91.
[147] "The Un-territoriality of Data," 395.
[148] B. Smith, "Time for an international convention on government access to data," Microsoft Corporate Blogs, January 20, 2014.
[149] Daskal, "Law Enforcement Access to Data Across Borders: The Evolving Security and Rights Issues" at 492–94.

privacy protections that attach to a person's communications when she transits borders or when jurisdiction is disputed should be those that are contained in international human rights law. Any state that impacts those rights – by accessing the data or sharing it with another state – should be required to ensure those protections. UN experts and the European Court of Human Rights can support efforts to establish robust and predictable privacy protections that transcend borders by continuing to develop standards on the universality of privacy rights in the digital age.

VI CONCLUSION

Developments in communications technology, coupled with revelations by Edward Snowden and others, have demonstrated that while human rights law has a well-developed body of standards on the right to privacy in communications, there are key areas where these standards fall short. The bulk collection of communications data seems generally permitted but circumscribed in human rights law, although few states appear to conduct such surveillance in accordance with these limits. Rules regarding the protections that apply to communications and other personal data when they are in the hands of private companies or when they transit borders are evolving, but at present are incomplete.

The most impactful recent development in this space may be the interpretation of EU data protection law in a way that incorporates or converges with the right to privacy. EU institutions are using data protection norms and enforcement mechanisms to give individuals stronger protections against the public and private actors that access their communications, regardless of location. This approach has the potential to contribute to stronger privacy protections beyond Europe, as its norms are increasingly replicated by other states seeking determinations of adequacy. Ideally, the European approach will also prompt UN mechanisms and governments to come together to devise more global solutions for the protection of privacy in the digital age, with international human rights law as their foundation.

11

Human Rights and Private Actors in the Online Domain

Rikke Frank Jørgensen

I INTRODUCTION

The UN Special Rapporteur on the Promotion and Protection of the Right to Freedom of Opinion and Expression recently stated that the Internet has become the central global public forum with profound value for human rights.[1] In this global public forum, control over infrastructure and services is largely in the hands of companies, with some of the most powerful being from the United States. To participate in the online sphere, individuals must engage with online platforms such as Google and Facebook and rely on them for exercising rights and freedoms such as freedom of expression, freedom of information, and freedom of assembly.

In this sense, these companies have increasing power to influence rights in the online domain. The power of the major platforms flow from their control over a wide range of resources crucial to information search and public participation in the online realm. In 2013, *The New York Times* had a print and digital circulation of nearly two million and claimed to be the most visited newspaper site, with nearly thirty-one million unique visitors every month. YouTube, in contrast, had one billion unique visitors a month in 2014, or as many in a day as *The New York Times* has in a month.[2] In terms of company valuations, as of April 2014 *The Times*'s market value was around 1 percent of the value of Facebook or Google.[3] By the end

---

[1] *Report of the Special Rapporteur on the Promotion and Protection of the Right to Freedom of Opinion and Expression, David Kaye*, ¶ 11, U.N. Doc. A/HRC/29/32 (May 22, 2015) ("2015 Kaye Report").

[2] M. Ammori, "The 'New' New York Times: Free Speech Lawyering in the Age of Google and Twitter" (2014) 127 *Harvard Law Review* 2259–94 at 2266.

[3] Ibid., at 2267.

of 2015, Facebook had more than 1.6 billion users a month[4] and Google more than one billion searches a month.[5]

Online platforms are used every day by billions of people to express themselves and to comment on, debate, critique, search, create, and share views and content. As such, the Internet's distributed architecture and the decrease in communications costs have fundamentally altered the capacity of individuals to be active participants in the public sphere. On a positive note, this networked public sphere facilitates new means for civic engagement, public participation, social change, and countering repressive governments.[6] On a more cautious note, scholars have warned that the new infrastructure for exercising freedom of expression carries with it new modalities of interference with fundamental rights, and that adequate legal responses have yet to be found.[7]

One area of concern – not least among legal scholars, data protection authorities, and groups set up to protect fundamental rights on the Internet[8] – is the privatized law enforcement and self-regulatory measures of these corporate platforms. The concern is particularly related to the platforms' means of "content regulation" and privacy practices; for example, their day-to-day decisions on which content to remove or leave up, and the extent to which they collect, process, and exchange personal data with third parties.[9] Several cases in the United States and Europe have addressed this concern, and new cases continue to appear.[10] Scholars have also warned of a governance gap, where private actors with strong human rights impacts

---

[4] "Number of monthly active Facebook users worldwide as of 2nd quarter 2016," Statista, www.statista.com/statistics/264810/number-of-monthly-active-facebook-users-worldwide/.

[5] C. Smith, "100 Google Search Statistics and Fun Facts," DMR, http://expandedramblings.com/index.php/by-the-numbers-a-gigantic-list-of-google-stats-and-facts/.

[6] Yochai Benkler argues that the Internet and the networked information economy provide us with distinct improvements in the structure of the public sphere over mass media. This is due to the information produced by and cultural activity of non-market actors, which the Internet enables, and which essentially allow a large number of actors to see themselves as potential contributors to public discourse and potential actors in political arenas. Y. Benkler, *Wealth of Networks: How Social Production Transforms Markets and Freedom* (New Haven, CT: Yale University Press, 2006), p. 220. "The network allows all citizens to change their relationship to the public sphere. They no longer need to be consumers and passive spectators. They can become creators and primary subjects. It is in this sense that the internet democratizes." Ibid., p. 272.

[7] J. M. Balkin, "Old-School/New-School Speech Regulation" (2014) 127 *Harvard Law Review* 2296–342.

[8] This includes groups and networks such as the Electronic Privacy Information Center (US), the Electronic Frontier Foundation (US), Privacy International (UK/global), European Digital Rights (European), Access Now(US), and the Association for Progressive Communications (global).

[9] A key issue in the human rights context may be that content with historical or legal value – e.g., information that may serve as evidence of a human rights violation or war crime – is taken down for violation of terms of service or community standards.

[10] In the United States, the Federal Trade Commission has focused on Internet platforms on several occasions. For example, since 2011, Facebook Inc. has been under a consent order by the FTC for deceiving consumers by telling them they could keep their information on

operate within the soft regime of guidelines and corporate social responsibility with no direct human rights obligations.[11] International human rights law is binding on states only, and despite an increasing take-up of human rights discourse within Internet companies, their commitment remains voluntary and nonbinding. In addition, limited information is available in the public domain concerning the corporate practices that affect freedom of expression and privacy.

Although a part of public discourse has always unfolded within private domains, from coffeehouses to mass media, the current situation is different in scope and character. In the online realm, *the vast majority* of social interactions, discussions, expressions, and controversies take place on platforms and services provided by private companies. As such, an increasing portion of our sociality is conducted in privately owned spaces. In addition, these practices are entangled in a business model in which the conversations and interactions that make up online life are directly linked to revenue. This arguably represents yet another stage of the trend of privatization. Prior examples include the dominance of corporate-owned media over the civic public sphere, the outsourcing of government functions to private contractors, and the reduction of public spaces to malls and privately owned town squares.[12] However, the increasing significance of online platforms for public life gives rise to a large number of unresolved questions related to the techno-social design, regulation, and human rights impact of these companies as "curators of public discourse."[13]

As several scholars have argued, these online platforms have an enormous impact on human rights globally through the policies they adopt for their users. Within "Facebookistan" and "Twitterland,"[14] these polices have just as much

> Facebook private and then repeatedly allowing it to be shared and made public. The order requires that Facebook obtain periodic assessments of its privacy practices by independent third-party auditors for the next twenty years. For more information on the case, please refer to "Facebook, Inc.," Federal Trade Commission, www.ftc.gov/enforcement/cases-proceedings/092-3184/facebook-inc. In Europe, the Dutch Data Protection Authority (DPA) imposed an incremental penalty payment on Google in 2013 based on practices introduced with Google's privacy policy in 2012. According to the DPA, Google combines the personal data collected by all kinds of different Google services without adequately informing users in advance and without asking for their consent. In July 2015, the DPA announced that Google had revised its privacy policy following the demands of the DPA, and that Google had until the end of December 2015 to obtain the unambiguous consent of all of its users at each step. For more information on the case, please refer to "Dutch DPA: privacy policy Google in breach of data protection law," *Autoriteit Persoongegevens*, https://cbpweb.nl/en/news/dutch-dpa-privacy-policy-google-breach-data-protection-law.

[11] E. B. Laidlaw, *Regulating Speech in Cyberspace: Gatekeepers, Human Rights and Corporate Responsibility* (Cambridge: Cambridge University Press, 2015).
[12] Z. Tufekci, "Facebook: The Privatization of our Privates and Life in the Company Town," Technosociology: our tools, ourselves, http://technosociology.org/?p=131.
[13] T. L. Gillespie, "The Politics of Platforms" (2010) 12 *New Media & Society* 347–64 at 347.
[14] In *Consent of the Networked: The World-Wide Struggle for Internet Freedom* (New York: Basic Books, 2012), p. 150, Rebecca MacKinnon refers to Facebook's "digital kingdom" as Facebookistan. In *Foreign Policy*, she further argues that "Facebook is not a physical country, but with

validity as traditional legal rules and standards.[15] Moreover, the companies have wide discretion in enforcing the policies, as they weigh potential precedents, norms, competing interests, and administrability in developing the rules of expression and privacy that effectively govern their users worldwide. Arguably, Google's lawyers and executives have as much power to determine who may speak and who may be heard around the world than does any president, king, or Supreme Court justice[16] – or, as expressed by Marvin Ammori, "Technology lawyers are among the most influential free expression lawyers practicing today."[17] At the same time, the core business of these companies is built around expression, and most of them talk about their business in the language of freedom of expression and freedom of information. Google's official mission is "to organize the world's information and make it universally accessible and useful."[18] Twitter stresses that its goal is "to instantly connect people everywhere to what is most meaningful to them. For this to happen, freedom of expression is essential."[19] Twitter also states that tweets must flow as a default principle. Facebook's vision is to "give people the power to share and make the world more open and connected."[20]

In relation to privacy, the online infrastructure of free expression is increasingly merging with the infrastructure of content regulation and surveillance. The technologies, institutions, and practices that people rely on to communicate with one another are the same technologies, institutions, and practices that public and private parties employ for surveillance.[21] The online infrastructure simultaneously facilitates and controls freedom of expression, surveillance, and data mining. As such, it has become a new target for governments and corporate interests alike.

Since 2009, several of the major Internet companies have upgraded and formalized their human rights commitment. Most notably this has been via industry initiatives, such as the Global Network Initiative, that focus on a company's compliance with international human rights standards on privacy and freedom of

---

900 million users, its 'population' comes third after China and India. It may not be able to tax or jail its inhabitants, but its executives, programmers, and engineers do exercise a form of governance over people's online activities and identities." R. MacKinnon, "Ruling Facebookistan," *Foreign Policy*, June 14, 2012, http://foreignpolicy.com/2012/06/14/ruling-facebookistan/; see also A. Chander, "Facebookistan" (2012) 90 *North Carolina Law Review* 1807–42.

[15] Ammori, "The 'New' New York Times" at 2263.
[16] J. Rosen, "The Deciders: The Future of Privacy and Free Speech in the Age of Facebook and Google" (2012) 80 *Fordham Law* Review 1525–38 at 1536.
[17] Ammori, "The 'New' New York Times" at 2265.
[18] Google, www.google.com/about/company/.
[19] B. Stone, "The Tweets Must Flow," Twitter, January 28, 2001, https://blog.twitter.com/2011/tweets-must-flow.
[20] Facebook, http://investor.fb.com/faq.cfm.
[21] *See* Balkin, "Old-School/New-School Speech Regulation" at 2296–342.

expression.[22] Also, as Lisl Brunner points out in Chapter 10, in the wake of the NSA contractor Edward Snowden's revelations of state surveillance, there has been increasing public focus on the exchange of personal data between Internet companies and government agencies. As a result, several companies have started to publish transparency reports to document (at an aggregated level) the numbers and types of content removal requests they receive and accommodate.[23] Despite these efforts, there is still limited public knowledge of companies' internal mechanisms of governance; e.g., how they decide cases with freedom of expression implications or how they harness user data.[24] As illustrated by a number of cases in the European Union as well as in Europe more broadly, a number of human rights related practices continue to cause concern among scholars and regulators alike.[25]

Using the example of Internet companies, this chapter will critically examine current challenges related to human rights protection in the online domain. This will include questions such as: How shall we understand the role of Internet companies vis-à-vis freedom of expression? What does human rights law – and soft law such as the UN Guiding Principles on Business and Human Rights – say about private actors and their human rights responsibilities? How have major Internet companies taken up these challenges in their discourse and practices? What are some of the dynamics that work for or against stronger human rights protection online? And are the frameworks that currently govern the activities of these Internet companies sufficient to provide the standards and mechanisms needed to protect and respect human rights online?

II THE ROLE OF INTERNET COMPANIES IN THE ONLINE DOMAIN

Over the past ten years, the Internet's potential positive and negative impacts on human rights have been iterated time and again by the UN World Summit on the

---

[22] Global Network Initiative, www.globalnetworkinitiative.org.
[23] A transparency report discloses statistics related to government requests for user data or content over a certain period of time. Google was the first online platform to publish a transparency report in 2010, with Twitter following in 2012.
[24] Ranking Digital Rights published its first annual Corporate Accountability Index in November 2015. The index ranks sixteen Internet and telecommunication companies according to thirty-one indicators, focused on corporate disclosure of policies and practices that affect users' freedom of expression and privacy. Ranking Digital Rights, https://rankingdigitalrights.org.
[25] Examples include the US Federal Trade Commission (FTC) investigation into Google's practices in connection with its YouTube Kids app (2015), the FTC Consent Order on Facebook (2011), the Dutch Data Protection Authority case against Google (2013), the Austrian class action privacy lawsuit against Facebook (rejected by the Austrian Court in July 2015 due to the lack of jurisdiction), the Google/Spain ruling of the European Court of Justice (2014), the Belgian Privacy Commissioners' recommendations to Facebook (2015), the Irish Data Protection Authority's audit of, and recommendations to, Facebook (2011), the European Union's antitrust case against Google (2015), and the Article 29 Working Party's examination of Google's Privacy Policy (2012).

Information Society,[26] the UN Human Rights Council,[27] and UN thematic rapporteurs.[28] The former UN Special Rapporteur on the Promotion and Protection of Freedom of Opinion and Expression, Frank La Rue, for example, has emphasized the unprecedented opportunity presented by the Internet to expand the possibilities for individuals to exercise a wide range of human rights, with freedom of opinion and of expression as prominent examples.[29] Special Rapporteur La Rue also expressed concerns about the multiple measures taken by states to prevent or restrict the flow of information online, and he highlighted the inadequate protection of the right to privacy on the Internet.[30] Of specific relevance to this chapter is his emphasis on the way private actors may contribute to violating human rights online, given that Internet services are run and maintained by companies.[31] In parallel to this, policy reports and scholarship have increasingly addressed the specific challenges related to human rights protection in the online domain.[32]

---

[26] At the first UN World Summit on the Information Society (WSIS), held in two phases in 2003 and 2005, it was confirmed that international human rights law serves as the baseline for information and communications technology (ICT)-related policy. Since WSIS, UN agencies such as the International Telecommunication Union, the United Nations Development Programme, and UNESCO have been responsible for follow-up action to ensure that the WSIS vision is implemented. This implementation process was reviewed in December 2015. World Summit on the Information Society, www.itu.int/wsis/review/2014.html.

[27] Human Rights Council, Res. 20/8, *The Promotion, Protection and Enjoyment of Human Rights on the Internet*, U.N. Doc. A/HRC/RES/20/8 (July 16, 2012); Human Rights Council Res. 26/13, *The Promotion, Protection and Enjoyment of Human Rights on the Internet*, U.N. Doc. A/HRC/RES/26/13 (July 14, 2014); "The right to privacy in the digital age," U.N. Doc. A/HRC/27/37 (June 30, 2014).

[28] *Report of the Special Rapporteur on the Promotion and Protection of the Right to Freedom of Opinion and Expression, Frank La Rue*, U.N. Doc. A/HRC/17/27 (May 16, 2011) ("2011 La Rue Report"); *Report of the Special Rapporteur on the Promotion and Protection of the Right to Freedom of Opinion and Expression, Frank La Rue*, U.N. Doc. A/HRC/23/40 (April 17, 2013) ("2013 La Rue Report"); 2015 Kaye Report.

[29] 2011 La Rue Report.

[30] 2013 La Rue Report.

[31] 2011 La Rue Report, ¶ 44.

[32] R. F. Jørgensen (ed.), *Human Rights in the Global Information Society* (Cambridge, MA: MIT Press, 2006); C. Garipidis and N. Akrivopoulou, *Human Rights and Risks in the Digital Era: Globalization and the Effects of Information Technologies* (Hershey, PA: Information Science Reference, 2012); W. Benedek and R. Madanmohan, "Human Rights and Information and Communication Technology – Background Paper," in *Proceedings of the 12th Informal Asia-Europe Meeting (ASEM) Seminar on Human Rights* (Singapore: Asia-Europe Foundation, 2013), Seoul, Republic of Korea, June 27–29, 2012, pp. 34–87; D. Korff, *The Rule of Law on the Internet and in the Wider Digital World – Issue Paper for the Council of Europe* (Strasbourg: Council of Europe, 2014); R. F. Jørgensen, *Framing the Net: The Internet and Human Rights* (Cheltenham: Edward Elgar Publishing, 2013); C. Padovani, F. Musiani, and E. Pavan, "Investigating Evolving Discourses on Human Rights in the Digital Age: Emerging Norms and Policy Challenges" (2010) 72 *International Communication Gazette*, 4–5, 359–78; L. Horner, D. Hawtin, and A. Puddephatt, Directorate-General for External Policies of the Union Study, "Information and Communication Technologies and Human Rights," EXPO/B/DROI/2009/24 (June 2010).

It is now widely recognized that access to the Internet and participation in discourse through the Internet have become integral parts of democratic life. What is less debated is the fact that facilitating this democratic potential critically relies on private actors. Access to the Internet takes place through Internet service providers, information search is facilitated by search engines, social life plays out via online platforms, and so on. Despite the increasing role that these private actors play in facilitating democratic experience online, the governance of this social infrastructure has largely been left to companies to address through corporate social responsibility frameworks, terms of service, and industry initiatives such as the Global Network Initiative.[33] Moreover, there is limited research critically assessing the frameworks that govern the activities of these Internet companies and questioning whether they are sufficient to provide the standards and compliance mechanisms needed to protect and respect human rights online.

The Internet's democratic potential is rooted in its ability to promote "a culture in which individuals have a fair opportunity to participate in the forms of meaning making that constitute them as individuals."[34] Democratic culture in this sense is more than political participation; it encompasses broad civic participation where anyone, in principle, may participate in the production and distribution of culture. This democratic potential is linked to the Internet's ability to provide its users with unprecedented access to information and to decentralized means of political and cultural participation.[35] By decentralizing the production of content, supplementing mass media with new means of self-expression, and enabling collective action across borders, the Internet has the potential to be a more participatory public sphere. This potential has been widely addressed in the body of literature that considers the Internet as a new or extended public sphere, yet with limited evidence of the actual democratic impact of these new modalities.[36] Moreover, the democratic implications of having private actors with no public interest mandate controlling the sphere is still not sufficiently clear, yet several challenges surface.

A *No Public Streets on the Internet*

In the United States, the protections of a speaker's right to speech vary based on the chosen forum. The Supreme Court distinguishes among three types of forums: traditional public forums, designated forums, and nonpublic forums.[37] The traditional public forum doctrine protects speech in public places such as streets,

---

[33] Laidlaw, *Regulating Speech in Cyberspace*, p. 59.
[34] J. M. Balkin, "Digital Speech and Democratic Culture: A Theory of Freedom of Expression for the Information Society" (2004) 79 *New York University Law Review* 1–55 at 40.
[35] Laidlaw, *Regulating Speech in Cyberspace*, p. 18.
[36] For an elaboration of the Internet as a new kind of public sphere, please refer to Jørgensen, *Framing the Net*, pp. 81–106.
[37] *Perry Educ. Ass'n v. Perry Educators' Ass'n*, 460 U.S. 37 (1983).

sidewalks, and parks, which are traditionally recognized as being held in common for the public good.[38] Expressive activity in these spaces can, in specific and narrowly defined cases, be subject to "time, place, and manner restrictions," but only in exceptional cases can such restrictions be based on the messages themselves.[39] In contrast, the owners of private property are relatively free in the restrictions they may place on the speech that takes place on their property.

When Internet users search for information, express opinions, debate, or assemble, they largely do so within privately owned forums. Accordingly, the company that provides the service is free to set the conditions for allowed expressions and actions on its platform. As Stacey Schesser explains, "Each private URL owner controls the traffic on his or her website, therefore limiting the application of the First Amendment to the site. Although a website author may choose not to censor postings on her blog or remove discussion threads on his bulletin board, each URL owner retains the right to do so as a private actor."[40] Legally speaking, the online sphere holds no public streets or parks, and social media platforms such as Facebook and Google Plus do not constitute public forums, but rather private property made open to the public. In line with this, there is no First Amendment protection of speech on these platforms. On the contrary, the communications that users provide as they tweet or contribute to Facebook, Google, or LinkedIn is largely private property, owned by the company that provides the service.[41]

Moreover, these companies have broad power to restrict speech that would otherwise be protected by the First Amendment. The highly praised liability regime for online Internet services in the United States, which immunizes intermediaries from liability for third-party content[42] as codified in Section 230 of the Communication Decency Act, effectively gives Internet companies the discretion to regulate content. Without Section 230, Internet companies could be secondarily responsible for the content posted on their platforms, including defamatory speech, if they took steps to censor this content to remove speech that might be offensive to other users. Section 230's so-called Good Samaritan provision protects Internet services from liability if they restrict access to material or give others the technical means to do so.[43]

---

[38] R. Moon, "Access to State-Owned Property," in *The Constitutional Protection of Freedom of Expression* (Toronto: University of Toronto Press, 2000), pp. 148–81.
[39] Ibid., p. 148.
[40] Stacey D. Schesser, "A New Domain for Public Speech: Opening Public Spaces Online" (2006) 94 *California Law Review* 1791–825 at 92.
[41] Arguably, public streets and parks today are less significant than online platforms as spaces for public discourse.
[42] See, e.g., "CDA 230: The Most Important Law Protecting Internet Speech," Electronic Frontier Foundation, www.eff.org/issues/cda230.
[43] Schesser, "A New Domain for Public Speech" at 99. At the other end of the spectrum are countries where the state imposes liability regimes on Internet intermediaries in order to control online content. For a global overview of such practices and their negative impact on online freedom of expression, see, for example, the global surveys presented by the OpenNet

B Online Gatekeepers

In an attempt to categorize the Internet companies in control of the online public sphere, Emily Laidlaw focuses on their democratic impact, identifying three different types of gatekeepers: micro gatekeepers, authority gatekeepers, and macro gatekeepers.[44] According to this typology, macro gatekeepers maintain significant information control due to their size, influence, or scope, and due to the fact that users must pass through them to use the Internet. Examples of companies in this category would be Internet service providers, mobile network providers, and major search engines. Authority gatekeepers control high amounts of information traffic and information flow, although users are not dependent on them to use the Internet. Examples include sites such as Wikipedia and Facebook. In contrast, micro gatekeepers are sites that play a less important role as sources of information, but still facilitate information and debates of democratic significance, such as certain news sites.[45] Laidlaw's framework suggests that the human rights obligations of Internet gatekeepers should increase when they have the power to influence democratic life in a way traditionally reserved for public bodies. The scale of responsibility is reflected not only in the reach of the gatekeeper, but also in the infiltration of that information, process, site, or tool in democratic culture[46].

C Expressions Are Products

The current communications environment is also unique because user expressions constitute the products on which the business models of Internet companies are built. The business models of most, if not all, of the major online services are based on targeted advertising, which means that when individuals participate online – for example, by engaging in conversation or searching for information – these actions are captured, retained, and used for advertising purposes and, as such, constitute products that feed into the online business model. This is essentially different from the predigital age, when individuals' conversations, social networks, preferences, and information searches were neither captured nor *the* core element of the intermediary's business model.

Initiative (ONI), https://opennet.net/. Please note that the ONI stopped collecting data as of December 2014.

[44] This model, which builds on Karine Barzilai-Nahon's network gatekeepers theory (K. Barzilai-Nahon, "Toward a Theory of Network Gatekeeping: A Framework for Exploring Information Control," [2008] 59 *Journal of the American Society for Information Science and Technology* 1493–512), is elaborated in Laidlaw, *Regulating Speech in Cyberspace*, pp. 44–46.

[45] Laidlaw, *Regulating Speech in Cyberspace*, p. 53.

[46] Ibid., p. 48.

Because expressions are products, the relationships that people have with Internet companies are fundamentally different from traditional company-customer relationships. As Bruce Schneier explains:

> Our relationship with many of the internet companies we rely on is not a traditional company-customer relationship. That's primarily because we're not customers. We're products those companies sell to their *real* customers. The relationship is more feudal than commercial. The companies are analogous to feudal lords, and we are their vassals, peasants, and – on a bad day – serfs. We are tenant farmers for these companies, working on their land by producing data that they in turn sell for profit.[47]

Although this feudal analogy may appear extreme, Schneier reminds us that what appear to be free products are not. The information and communications that users provide when using the services are essential elements in the online business model and, as such, represent the core source of income for the companies.

There should be nothing new or controversial about an Internet company seeking to optimize its revenue via advertising. The disturbing bit is that these platforms *de facto* control large chunks of the online public sphere and users have limited choice to opt out of the business scheme. There are no public streets on the Internet, and there are limited means of participating in political or cultural life outside the commercial realm. Moreover, contributing to the online economy via online expressions, habits, and preferences has become a premise for participation in the networked public sphere. Thus, according to Schneier: "It's not reasonable to tell people that if they don't like data collection, they shouldn't e-mail, shop online, use Facebook, or have a cell phone. . . . Opting out just isn't a viable choice for most of us, most of the time; it violates what have become very real norms of contemporary life."[48]

On an equally skeptical note, Shoshana Zuboff argues that the economic characteristics of the online business model are in the process of undermining long-established freedoms and represent a largely uncontested new expression of power.[49] Scholars such as Julie Cohen and Niva Elkin-Koren have cautioned that the digital era represents threats to fundamental freedoms whose ramifications we are yet to understand.[50] Elkin-Koren notes, "As information becomes crucial to every aspect of everyday life, control over information (or lack thereof) may affect our ability to participate in modern life as independent, autonomous human beings."[51]

---

[47] B. Schneier, *Data and Goliath: The Hidden Battles to Capture Your Data and Control Your World* (New York: W. W. Norton & Company, 2015), p. 58.
[48] Ibid., pp. 60–61.
[49] S. Zuboff, "Big Other: Surveillance Capitalism and the Prospects of an Information Civilization" (2015) 30 *Journal of Information Technology* 75–89.
[50] J. E. Cohen, *Configuring the Networked Self: Law, Code, and the Play of Everyday Practice* (New Haven, CT: Yale University Press, 2012).
[51] N. Elkin-Koren, "Affordances of Freedom: Theorizing the Rights of Users in the Digital Era" (2012) 6 *Jerusalem Review of Legal Studies* 96–109 at 97.

Thus, access to the Internet and participation in discourse through the Internet have become integral parts of modern life. The exercise of this public life, however, takes place almost exclusively via privately owned platforms. Moreover, it is entangled in a business model in which knowledge of individual behavior and preferences is closely linked to revenue. In effect, this means that private actors have unprecedented power to impact the way that billions of users are able to express themselves, search and share information, and protect their privacy. Yet as private actors, they remain largely outside the reach of human rights law.

In the following, I will examine some of the legal and extralegal dimensions of this challenge. First, what does human rights law say about the obligations of private actors? Second, how have the companies themselves responded to these challenges? And third, do these approaches suffice to protect human rights online?

III HUMAN RIGHTS LAW AND PRIVATE ACTORS

Human rights law is state-centric in nature in the sense that states – not individuals, not companies – are the primary duty bearers. Legally speaking, only the state can be brought before a human rights court, such as the European Court of Human Rights, and examined for alleged human rights violations. Part of this obligation, however, is a duty upon the state to ensure that private actors do not violate human rights, referred to as the horizontal effect of human rights law. National regulation related to labor rights or data protection, for example, serves as machinery for enforcing human rights standards in the realm of private parties.

Whereas human rights law is focused on the vertical relation (state obligations to the individual), it recognizes the horizontal effect that may arise in the sphere between private parties.[52] The horizontal effect implies a state duty to protect human rights in the realm of private parties, for example, via industry regulation. A large amount of the literature related to online freedoms has been occupied with new means of state interference with human rights, for example, through new means of restricting content, engaging in surveillance, or involving Internet companies in law enforcement. These new means of state interference have been explored in several comprehensive studies, for example, by the Open Net Initiative[53] and by scholars such as Jack Balkin, who have examined the characteristics of "old-school" (pre-Internet) versus "new school" speech regulation. In contrast, less attention has been paid to the implications that arise in the sphere of horizontal relations, such as when companies, on their own initiative, remove content because it violates their terms of service, or when they exchange personal data with third parties as part of their

---

[52] P. van Dijk et al. (eds.), *Theory and Practice of the European Convention on Human Rights* (Antwerp, Oxford: Intersentia, 2006), p. 6.
[53] R. Deibert et al. (eds.), *Access Controlled: The Shaping of Power, Rights, and Rule in Cyberspace* (Cambridge, MA: MIT Press, 2010); R. Deibert et al. (eds.), *Access Denied the Practice and Policy of Global Internet Filtering* (Cambridge, MA: MIT Press, 2008).

business model. In the analysis that follows, emphasis will be on horizontal relations and the human rights duties and responsibilities that may be invoked in this realm.

Over the past decade, the interface between human rights law and private actors has been the focus of considerable attention, resulting in the adoption of broad soft law standards[54] and the launch of many multistakeholder initiatives, including the UN Global Compact. The UN Global Compact represents one of the core platforms for promoting corporate social responsibility (CSR), a concept that refers to a company's efforts to integrate social and environmental concerns into its business operations and stakeholder interactions. According to the UN Global Compact's framing of corporate social responsibility, businesses are responsible for human rights within their sphere of influence. While the sphere of influence concept is not defined in detail by international human rights standards, it tends to include the individuals to whom a company has a certain political, contractual, economic, or geographic proximity.[55] Arguably, CSR has some normative base in the human rights discourse, but these rights have not been well integrated:

> On the whole, relatively few national CSR policies or guidelines explicitly refer to international human rights standards. They may highlight general principles or initiatives that include human rights elements, notably the OECD Guidelines and the Global Compact, but without further indicating what companies should do operationally. Other policies are vaguer still, merely asking companies to consider social and environmental "concerns," without explaining what that may entail in practice.[56]

Even where CSR pays attention to human rights, it primarily addresses social and economic rights, in particular as it relates to working conditions and environmental and community impact, with limited attention to civil and political rights.[57] The critique of the CSR framework that it was too limited in scope, with a focus on selected rights only, was one of the drivers of the work of John Ruggie, who served as the special representative to the secretary general on issues of human rights and transnational corporation from 2005 to 2011.

---

[54] See "OECD Guidelines for Multinational Enterprises," Organization for Economic Co-operation and Development, http://mneguidelines.oecd.org/text/; "ILO Declaration on Fundamental Principles and Rights at Work," International Labour Organization, www.ilo.org/declaration/lang–en/index.htm.

[55] Business Leaders Initiative on Human Rights, U.N. Global Compact, and Office of the High Commissioner for Human Rights, "A Guide for Integrating Human Rights into Business Management," (2007), p. 8, www.ohchr.org/Documents/Publications/GuideHRBusinessen.pdf. For literature on the normative grounding of CSR in the human rights discourse, see, e.g., T. Campbell, "The Normative Grounding of Corporate Social Responsibility: A Human Rights Approach," in D. McBarnet (ed.), *The New Corporate Accountability: Corporate Social Responsibility and the Law* (Cambridge: Cambridge University Press, 2007). According to Campbell, human rights offers primarily a discursive rather than legal framework for CSR.

[56] *Report of the Special Representative of the Secretary-General on the Issue of Human Rights and Transnational Corporations and Other Business Enterprises, John Ruggie,* ¶ 35, U.N. Doc. A/HRC/14/27 (April 9, 2010).

[57] United Nations Global Compact, www.unglobalcompact.org/.

In 2011, Ruggie's work culminated with an endorsement of the United Nations' Guiding Principles on Business and Human Rights (UNGP).[58] The UNGP provides a set of principles that states and businesses should apply to prevent, mitigate, and redress corporate-related human rights abuses. Contrary to the sphere of influence approach, the UNGP focuses on the potential and actual human rights impact of any business conduct.[59] The UNGP elaborates the distinction that exists between the state *duty to protect human rights* and the corporate *responsibility to respect human rights* based on three pillars, often called the "Protect, Respect, and Remedy" framework. The first pillar (Protect) focuses on the role of the state in protecting individuals' human rights against abuses committed by non-state actors; the second pillar (Respect) addresses the corporate responsibility to respect human rights; and the third pillar (Remedy) explores the roles of state and non-state actors in securing access to remedy. Ruggie's report to the Human Rights Council, which provided the basis for the UNGP, explains:

> Each pillar is an essential component in an inter-related and dynamic system of preventative and remedial measures: the State duty to protect because it lies at the very core of the international human rights regime; the corporate responsibility to respect because it is the basic expectation society has of business in relation to human rights; and access to remedy because even the most concerted efforts cannot prevent all abuse.[60]

The second pillar affords a central role for human rights due diligence by companies. Due diligence comprises four steps, taking the form of a continuous improvement cycle.[61] Companies must publish a policy commitment to respect human rights. As part of its due diligence process, a company must assess, using a human rights impact assessment, the actual and potential impacts of its business activities on human rights; remediate the findings of this assessment into company policies and practices; track how effective the company is in preventing adverse human rights impacts; and communicate publicly about the due diligence process and its results. Companies are expected to address all their impacts, though they may prioritize their actions. The UNGP recommends that companies first seek to prevent and mitigate their most severe impacts or those where a delay in response would make consequences irremediable.[62]

Since the corporate responsibility to respect human rights refers to all internationally recognized human rights, not just those in force in any one particular

---

[58] *Report of the Special Representative John Ruggie, Guiding Principles on Business and Human Rights: Implementing the United Nations 'Protect, Respect and Remedy' Framework*, U.N. Doc. A/HRC/17/31 (March 21, 2011) ("2011 Ruggie Report").

[59] See, for example, John Ruggie's annual reports from 2006 to 2011, http://business-humanrights.org/en/un-secretary-generals-special-representative-on-business-human-rights/reports-to-un-human-rights-council.

[60] 2011 Ruggie Report, p. 4.

[61] Ibid., pp. 17–20.

[62] Ibid., p. 24.

jurisdiction,⁶³ human rights due diligence should encompass, at minimum, all human rights enumerated in the International Bill of Human Rights.⁶⁴ The UNGP guidance on human rights impact assessments remains at a general level, without detailed descriptions of the process or orientation on how it should be adapted to particular industries. Various initiatives have since attempted to address this, which we will return to below.⁶⁵

Whereas pillars one and three combine existing state obligations under international human rights law with soft law recommendations, pillar two is soft law only, reflecting the lack of direct human rights obligations for companies under international law.⁶⁶ The debate on whether and how to create binding human rights obligations for companies has been ongoing for more than two decades, but there is little indication that companies will be bound by human rights law in the foreseeable future.⁶⁷

With regard to the state duties, the UNGP reiterates two existing human rights obligations. First, states must protect against human rights abuses within their territory and jurisdiction by third parties,⁶⁸ and second, states must provide individuals access to remedies for human rights abuses.⁶⁹ According to the first obligation, the state is required to take appropriate steps to prevent, investigate, punish, and redress private actors' human rights abuses that take place in its jurisdiction. Such steps include effective policies, legislation, and regulation; access to remedies; adjudication; and redress. The second obligation iterates that states must take appropriate steps to ensure that injured parties have access to effective remedies when business-related human rights abuses occur within the state's territory or jurisdiction. This includes remedies provided via judicial, administrative, legislative, or other appropriate means.

In line with this, the case law of the European Court of Human Rights (ECtHR) confirms that states have an obligation to protect individuals against violations by

---

[63] Ibid., p. 11.
[64] Ibid., p. 12. The International Bill of Human Rights consists of the Universal Declaration of Human Rights (1948), the International Covenant on Civil and Political Rights (1966), and the International Covenant on Economic, Social, and Cultural Rights (1966).
[65] For guidance on human rights impact assessment, see, for example, "Rights and Democracy, Getting it Right: Human Rights Impact Assessment Guide," International Centre for Human Rights and Democratic Development, http://hria.equalit.ie/en/; FIDH, "Community-based Human Rights Impact Assessments," www.fidh.org/en/issues/globalisation-human-rights/business-and-human-rights/community-based-human-rights-impact-assessments.
[66] For an elaboration of the argument see, for example, J. Knox "The Ruggie Rules: Applying Human Rights Law to Corporations," in R. Mares (ed.), *The UN Guiding Principles on Business and Human Rights: Foundations and Implementation* (Leiden, Boston: Martinus Nijhoff, 2012).
[67] For an account of this development, see J. Ruggie, "Business and Human Rights: The Evolving International Agenda" (2007) 101 *American Journal of International Law* 819–40; Mares, *UN Guiding Principles on Business and Human Rights*, pp. 1–49.
[68] 2011 Ruggie Report, p. 1.
[69] Ibid., p. 25.

business enterprises. This entails an obligation to protect individuals against violations by business enterprises as third parties as well as those acting as state agents. In the first case, the human rights violation is constituted by the state's failure to take reasonable measures to protect individuals against abuse by business enterprises; in the latter, the abusive act of the business enterprise is attributed to the state, so that the state is considered to directly interfere with the rights at stake.[70] The case law of the ECtHR on violations by business enterprises acting as state agents concerns both the case where the state owns or controls business enterprises and the case where private corporations exercise public functions through procurement contracts and privatization of public services.[71]

Ruggie's framework, which has been widely praised and endorsed by states as well as business enterprises, has also been criticized for its slow uptake, its ineffectiveness, and for not creating binding obligations on companies.[72] Yet, a hard-law punitive approach has also long had its skeptics, and numerous empirical studies have spoken to the significance of social factors, both internal and external, in affecting companies' behavior.[73]

The UNGP has resulted in several follow-up initiatives at both the global and regional level. At the global level, a UN working group on human rights and transnational corporations and other business enterprises was established in June 2011 to promote the effective and comprehensive dissemination and implementation of the UNGP.[74] After completing its initial three-year appointment in 2014, the group had its mandate extended for another three-year term.[75] The group has, among other things, produced a "Guidance" on the development of national action plans on business and human rights.

---

[70] S. Lagoutte, "The State Duty to Protect against Business-Related Human Rights Abuses: Unpacking Pillar 1 and 3 of the UN Guiding Principles on Human Rights and Business," Working Paper, *Human Rights' Research Papers*, No. 2014/1 (2014), p. 9.

[71] See, e.g., *Tatar v. Romania*, Eur. Ct. H.R., App. No. 67021/01 (January 27, 2009); *Fadeyeva v. Russia*, Eur. Ct. H.R., App. No. 55723/00 (June 9, 2005); *Öneryildiz v. Turkey*, Eur. Ct. H.R., App. No. 48939/99 (Grand Chamber, November 30, 2004); *Guerra & Others v. Italy*, Eur. Ct. H.R., App. No. 14967/89 (Grand Chamber, February 19, 1998); *López Ostra v. Spain*, Eur. Ct. H.R., App. No. 16798/90 (December 9, 1994).

[72] S. A. Aaronson and I. Higham, "'Re-Righting Business': John Ruggie and the Struggle to Develop International Human Rights Standards for Transnational Firms" (2013) 35 *Human Rights Quarterly* 333–64; D. Bilchitz, "A Chasm between 'Is' and 'Ought'?: A Critique of the Normative Foundations of the SRSG's Framework and the Guiding Principles," in S. Deva and D. Bilchitz (eds.), *Human Rights Obligations of Business: Beyond the Corporate Responsibility to Respect?* (Cambridge: Cambridge University Press, 2013), pp. 107–37.

[73] C. Methven O'Brien and S. Dhanarajan, "The Corporate Responsibility to Respect Human Rights: A Status Review," Working Paper, National University of Singapore, 2015/005 (2015), 4.

[74] See the presentation of the working group by the UN High Commissioner for Human Rights, www.ohchr.org/EN/Issues/Business/Pages/WGHRandtransnationalcorporationsandotherbusiness.aspx.

[75] The website of the working group is available at: www.ohchr.org/EN/Issues/Business/Pages/WGHRandtransnationalcorporationsandotherbusiness.aspx.

At the European level, the European Commission has produced sector-specific guides on UNGP implementation in relation to three business sectors, including the information and communication technology (ICT) sector.[76] The guide is not a legally binding document, but translates the expectations of the UNGP to the specifics of the business sector at a rather generic level. In relation to the ICT sector, the guide stresses that the right to privacy and to freedom of expression can be particularly impacted by companies in the ICT sector.[77] The guide focuses on the state pressure that companies may be subjected to when they operate in contexts where the national legal framework does not comply with international human rights standards (i.e., a vertical conflict). In contrast, the negative human rights impact that may flow from the company's governance of content or tracking of user behavior is not addressed, and, as such, the guide provides limited guidance on horizontal conflict (i.e., relations between private actors). This focus on the vertical conflict is also dominant in the Global Network Initiative (addressed below) and indicates that the human rights discourse by Internet companies tends to highlight push-back strategies against illegitimate government requests, with less attention being paid to the human rights impact of the company's own actions.

This points to an unanswered question: What would human rights law and supplementary guidelines such as the UNGP say about the responsibility of private actors that potentially affects the rights of billions of individuals worldwide?

As stated above, states are obligated to prevent human rights violations by private actors, and private actors have a moral obligation to respect human rights. States cannot delegate their human rights obligations to a private party, and they are obligated to ensure that appropriate regulations result in human rights–compliant business practices. Moreover, each company has a responsibility to assess its actual human rights impact, i.e., the way that its operational practices, services, and products impact on its users' human rights.

The state obligation to ensure human rights entails both a positive and negative element. It requires the state to refrain from certain conduct, but also to take positive steps to ensure the enjoyment of the right in question. Freedom of expression, for example, requires that the state refrain from engaging in censorship, but also that it – via national regulation – enables freedom of the press.[78] The measures and behavior required of businesses to fulfill their responsibility to respect human rights should be

---

[76] The guide is available at: https://ec.europa.eu/anti-trafficking/sites/antitrafficking/files/information_and_communication_technology_0.pdf.

[77] Institute for Human Rights and Business and SHIFT for the European Commission, "ICT Sector Guide on Implementing the UN Guiding Principles on Business and Human Rights," 2012, Section 2.

[78] In the ruling *Editorial Board of Pravoye Delo and Shtekel* v. *Ukraine*, Eur. Ct. H.R., App. No. 33014/05 (May 5, 2011), the European Court of Human Rights for the first time acknowledged that Article 10 imposes on states a positive obligation to create an appropriate regulatory framework to ensure effective protection of journalists' freedom of expression *on the Internet*.

provided for by each state's respective national laws and policies in all the various areas in which these laws and policies touch on business activities.[79]

Arguably, in many specific cases, such regulation exists and businesses do, to a large extent, respect human rights standards by complying with legal rules. It would be too optimistic, however, to assume that governments and subordinate public authorities always have the ability and the will to regulate business conduct in line with human rights requirements,[80] not least in relatively new policy areas such as online freedom of expression. Moreover, in the case of online service providers, there is an additional layer of responsibility. Not only does the company have responsibilities in relation to its employers and community, it also directly or indirectly affects its users, who in practice might be billions of people.

Historically, controversial cases have involved privacy and freedom of expression in particular, yet with some legal subtleties that distinguish the two rights in question. As Lisl Brunner notes in Chapter 10, the right to privacy has some protection in national legislation (in particular in Europe) in the form of data-protection laws that stipulate principles, procedures, and safeguards that public and private actors must adhere to when collecting and processing personal data.[81] In the EU context, for example, Google is subject to the European Data Protection Directive, which imposes conditions and safeguards for data collection, processing, and exchange on public institutions and private companies alike. When Google, as in the *Google Spain* case, violates a user's right to privacy, the company is the direct duty bearer under Spanish data protection legislation.[82]

In contrast, Internet platforms are rarely subject to regulation concerning the negative impact they may have on freedom of expression. When content is filtered, blocked, or taken down by Twitter because it allegedly violates the community standards, there is limited guidance in international human rights law, and rarely is there national legislation that applies. In these situations, the company is acting in a judicial capacity, deciding whether to allow content to stay up or to remove it according to internal governance practices and standards, but without the human rights requirements that would apply if Twitter were a state body rather than a private company. For example, if Twitter removes posts for violating its community standards, this does not trigger international human rights law. In contrast, if a state-owned Twitter were to remove content from the public domain, this practice would have to follow the three-part test governing limits on freedom of expression. According to the three-part test, any limitation on the right to freedom of expression must be provided by law that is clear and accessible to everyone; it must pursue one

---

[79] Methven O'Brien and Dhanarajan, "The Corporate Responsibility to Respect Human Rights," 5.
[80] Ibid.
[81] It should be noted that informational privacy covers only one aspect of privacy.
[82] *Google Spain SL, Google Inc. v. Agencia Española de Protección de Datos, Mario Costeja González*, CJEU, Case C-131/12 (May 13, 2014).

of the purposes set out in Article 19, paragraph 3 of the ICCPR; and it must be proven as necessary and the least restrictive means required to achieve the purported aim.[83]

A related challenge concerns the cases where content is taken down because it allegedly violates national law in the country of operation. As mentioned, Internet services hosted in the United States are insulated from liability under Section 230 of the Communications Decency Act. Concerning copyright infringements, however, Section 512 of the Digital Millennium Copyright Act codifies limited liability, which means that Internet services are required to remove alleged illegal content when notified of its presence on their service, or they will face liability for that content.[84] This is similar to the European approach, which imposes limited liability on online services through the Electronic Commerce Directives. Both regimes have been criticized for encouraging businesses to privately regulate their affairs, with freedom of expression implications.[85] When Facebook, for example, acts upon an alleged copyright violation by removing the content, it is making decisions with freedom of expression implications, yet as a private actor it is not obligated to follow the three-part test prescribed by human rights law. The regimes that insulate online platforms from liability for the third-party content they carry also effectively insulate them from liability when they take down protected content out of fear of liability (e.g., alleged copyright infringement).

In sum, the practices of online platforms (especially macro or authority gatekeepers) have effects on freedom of expression and privacy far beyond their roles as employers and members of a community. Do the power, influence, and capacity to affect democratic life qualify for a special class of public interest companies that invite additional corporate responsibilities beyond the duty to respect human rights?[86] Does it accentuate the positive obligation on the state to legislate the obligations of these companies? Although Ruggie briefly touched upon these issues (with prisons as an example), there is limited guidance in his work as to the answers. In addition, although these companies' negative impact on privacy is regulated in some regions of the world, their potential negative impact on freedom of expression is not. Neither the United States nor Europe has regulations to protect against the potential negative impact that the content-regulation practices of a major Internet company could have on freedom of expression. Moreover, the human rights

---

[83] 2011 La Rue Report.
[84] See US Copyright Office, "The Digital Millennium Copyright Act of 1998: U.S. Copyright Office Summary," www.copyright.gov/legislation/dmca.pdf.
[85] See Schesser, "A New Domain for Public Speech"; I. Brown, "Internet Self-Regulation and Fundamental Rights," *Index on Censorship* (2010), https://papers.ssrn.com/sol3/papers.cfm?abstract_id=1539942; B. Frydman and I. Rorive, "Regulating Internet Content through Intermediaries in Europe and the USA," (2002) 23(1) *Zeitschrift für Rechtssoziologie* 41–59.
[86] *Business and Human Rights: Towards Operationalizing the "Protect, Respect and Remedy" Framework*, U.N. Doc. A/HRC/11/13 (April 22, 2009), at 17.

responsibilities of Internet companies are largely discussed in relation to illegitimate government requests, as exemplified by the Global Network Initiative, addressed below.

The above challenges are rooted in the gray zone where human rights law ends and corporate social responsibility begins, and it is in this zone that online platforms operate. Their practices may affect human rights immensely, yet they are not regulated with a view to their impact on freedom of expression, freedom of information, or privacy, except in some specific cases. Moreover, even when Internet companies are subjected to regulation, such as on data protection, the past ten years have illustrated the tremendous challenge of holding those companies accountable to these standards. As such, there is a lacuna in checks and balances concerning these private actors. This is paradoxical, since online, these private actors are at the center of the Internet's democratizing force and exercise significant human rights impacts on their users.

IV THE UPTAKE OF HUMAN RIGHTS WITHIN INTERNET COMPANIES

As previously mentioned, most human rights cases related to the ICT sector have concerned freedom of expression and the right to privacy. In December 2008, this led to the launch of the first industry initiative concerned with the human rights compliance of Internet companies, the Global Network Initiative, addressed below. First, however, it should be noted that most major global platforms emphasize freedom of expression as a core element of their business. Facebook's mission, for example, is framed in the language of freedom of expression and association by its founder, Mark Zuckerberg:

> There is a huge need and a huge opportunity to get everyone in the world connected, to give everyone a voice and to help transform society for the future... . By giving people the power to share, we are starting to see people make their voices heard on a different scale from what has historically been possible. These voices will increase in number and volume. They cannot be ignored. Over time, we expect governments will become more responsive to issues and concerns raised directly by all their people rather than through intermediaries controlled by a select few.[87]

At Twitter, the company vision is closely linked to freedom of expression and the new digital means of realizing this right: "Our legal team's conceptualization of speech policies and practices emanate[s] straight from the idealism of our founders – that this would be a platform for free expression, a way for people to disseminate their ideas in the modern age. We're here in some sense to implement that vision."[88]

---

[87] S. Ard, "Mark Zuckerberg's IPO Letter: Why Facebook Exists," Yahoo! Finance, February 1, 2012, http://finance.yahoo.com/news/mark-zuckerberg's-ipo-letter–why-facebook-exists.html.

[88] Ammori, "The 'New' New York Times" at 70.

Google stresses that the company "[has] a bias in favor of people's right to free expression in everything we do."[89]

The Global Network Initiative (GNI) has, since 2008, been the common venue for some of the major Internet companies' discourse on human rights norms related to freedom of expression and privacy.[90] The GNI is a multistakeholder group of companies, members of civil society, investors, and academics that was launched in the United States. Formation of the GNI took place against the backdrop of two particular incidents. One was Yahoo's handover of user information to Chinese authorities, thereby exposing the identity of a Chinese journalist, leading to his arrest and imprisonment. The second was Google's launch of a censored search engine in China.[91]

The goal of the GNI is to "protect and advance freedom of expression and privacy in the ICT sector."[92] At the time of writing, Google, Yahoo, Facebook, Microsoft, and LinkedIn were the Internet company members, whereas seven of the big telecommunication companies – united in the parallel initiative Telecommunications Industry Dialogue – were admitted as members in 2017.[93] The baseline for GNI's work consists of four core documents, developed in broad collaboration among the participants: the "Principles," the "Implementation Guidelines," the "Accountability, Policy and Learning Framework," and the "Governance Charter." The Implementation Guidelines operationalize the overall principles in detailed guidance to companies, whereas the Governance Charter describes how the GNI is governed in order to ensure integrity, accountability, relevance, effectiveness, sustainability, and impact. The Accountability, Policy, and Learning Framework supplements the Governance Charter with more detail on how the work of the GNI is carried out.[94]

Since its inception, the GNI has been criticized for lack of participation (including by smaller and non-US companies), for not being independent enough in the

---

[89] C. Cain Miller, "Google Has No Plans to Rethink Video Status," *The New York Times*, September 14, 2012, http://perma.cc/LX2F-DKE9. Commentators have argued that Google's philosophy likely impacts the thinking at companies across Silicon Valley, since its alumni have been shaped by shared experiences and an ongoing informal network, which shares experiences on difficult questions. Ammori, "The 'New' New York Times" at 69.

[90] The website is available at: www.globalnetworkinitiative.org.

[91] See C. M. Maclay, "An Improbable Coalition: How Businesses, Non-Governmental Organizations, Investors and Academics Formed the Global Network Initiative to Promote Privacy and Free Expression Online," PhD thesis, Northeastern University (2014) (providing a detailed account of the formation of the GNI).

[92] The GNI Principles are available at: http://globalnetworkinitiative.org/principles/index.php.

[93] "The Global Network Initiative and the Telecommunications Industry Dialogue join forces to advance freedom of expression and privacy," Global Network Initiative, www.telecomindustrydialogue.org and http://globalnetworkinitiative.org/news/global-network-initiative-and-telecommunications-industry-dialogue-join-forces-advance-freedom.

[94] "Core Commitments," Global Network Initiative, https://globalnetworkinitiative.org/corecommitments/index.php.

assessment process,[95] for the lack of a remedy mechanism, for insufficient focus on privacy by design, and for a lack of accountability.[96] These criticisms speak to the inherent challenge of having an industry define its own standards and procedures for respecting users' rights to privacy and freedom of expression. Moreover, it has been argued that the protection of users' rights runs contrary to business interests.[97] In relation to the latter challenge, it is important to note some fundamental differences between the rights in question and the challenges they pose.

Both privacy and freedom of expression protect individual freedoms by setting limits on state (and private actor) intrusion. With privacy, these limits are formulated as principles that guide how and when personal information may be collected, processed, and exchanged with a third party. In relation to data protection, it seems paradoxical to expect that the boundaries for data collection and use will be most effectively protected by companies whose business model is built around harnessing personal data as part of their revenue model. Whereas companies may push back against illegitimate government requests for user data, they are less likely to be a sufficiently critical judge of their own business practices, not least when these are closely linked to their business model.

With freedom of expression, the issue is slightly different. Here, the potential conflict between human rights standards and business practices stems from several factors, more indirectly linked to the revenue model. These factors include: unclear liability regimes that might incentivize the company to remove alleged illegal content without sufficient due process safeguards and that position the company as the final authority regarding which content to remove; pressure from governments to block, filter, or remove content; and internally defined standards regarding content moderation and enforcement of the standards.

As reflected in its baseline documents, the GNI is strongly anchored in the initial narrative of providing guidance to Internet companies in countries where local laws conflict with international human rights standards, rather than the systematic human rights impact assessment suggested by the UNGP. The GNI Principles state:

> The right to freedom of expression should not be restricted by governments, except in narrowly defined circumstances based on internationally recognized laws or

---

[95] In June 2014, the GNI board consolidated the assessment process into a two-stage model: first, self-reporting from the companies to GNI after one year of membership; second, assessment of each company member every two years. The assessment is carried out by a list of GNI-approved assessors and examines the policies, systems, and procedures put in place by the company to comply with the GNI Principles.

[96] MacKinnon, *Consent of the Networked*, pp. 179–82. For news coverage on this, see, for example, L. Downes, "Why no one will join the Global Network Initiative," *Forbes*, March 30, 2011, https://www.forbes.com/sites/larrydownes/2011/03/30/why-no-one-will-join-the-global-network-initiative/#275f5878d782.

[97] D. Doane, *The Myth of CSR: The Problem with Assuming That Companies Can Do Well While Also Doing Good Is That Markets Don't Really Work That Way* (Stanford, CA: Stanford Graduate School of Business, 2005), pp. 22–29.

standards.... Participating companies will respect and protect the freedom of expression rights of their users when confronted with government demands, laws and regulations to suppress freedom of expression, remove content or otherwise limit access to information and ideas in a manner inconsistent with internationally recognized laws and standards.[98]

Similarly, the Implementation Guidelines for Freedom of Expression discuss company practices in relation to "Government Demands, Laws and Regulations"[99] rather than human rights impacts. These principles illustrate that for the GNI, threats to freedom of expression are framed as illegitimate government behavior, and its role is to assist companies with human rights–compliant conduct when confronted with, for example, an overly broad request for filtering or blocking of content.

While industry push-back against illegitimate government requests undoubtedly addresses a relevant human rights problem, it is not sufficient to comply with the responsibilities set out in the UNGP. Those responsibilities require companies to know their actual and potential human rights impacts, to prevent and mitigate abuses, and to address adverse impacts they are involved in. In other words, companies must carry out human rights due diligence across all operations and products. The process of identifying and addressing the human rights impact must include an assessment of all internal procedures and systems, as well as engagement with the users potentially affected by the company practices. It follows that for GNI members such as Yahoo, Facebook, and Google, it is not sufficient to focus on government requests and human rights–compliant practices in this realm. Rather, assessment is needed on the freedom of expression impacts that may flow from all company practices, including, for example, when the company enforces community standards or takes down content based on alleged copyright infringement.

Internet platforms such as Facebook and YouTube influence the boundaries of what users can say and view online via their terms of service. Enforcement of these terms of service must work effectively at a scale of millions of users, including in high-profile controversies such as the "Innocence of Muslims" video,[100] as well as in more routine cases where users report objectionable content. In practice, the terms

---

[98] "GNI Principles: Section on Freedom of Expression," Global Network Initiative, http://globalnetworkinitiative.org/principles/index.php#18.

[99] "GNI Implementation Guidelines: Section on Freedom of Expression," Global Initiative Network, http://globalnetworkinitiative.org/implementationguidelines/index.php#29.

[100] In 2006, the "Innocence of Muslims" video sparked outrage in countries throughout the Middle East for its perceived criticism of Islam. While YouTube allowed the video to remain online in the United States, stating that the video did not break US law, it was removed in countries where it violated local laws, as well as in Libya and Egypt, where it did not violate local laws. Commentators have argued that the case is illustrative of the way private companies carry out worldwide speech "regulation" – sometimes in response to government demands, sometimes to enforce their own terms of service. S. Benesch and R. MacKinnon, "The Innocence of YouTube," *Foreign Policy*, October 5, 2012, http://foreignpolicy.com/2012/10/05/the-innocence-of-youtube/.

are translated into specific definitions and guidelines that are operationalized by employees and contractors around the world, who "implement the speech jurisprudence"[101] by making decisions on which content to leave up or remove.[102] According to Google, for example, deciding on the limits of freedom of expression for a billion users is "a challenge we face many times every day."[103] Yet, an intermediary's terms of service and the means of enforcing those terms are not part of the GNI norms and standards.

A similar challenge is found in relation to privacy. The GNI Principles iterate that

> the right to privacy should not be restricted by governments, except in narrowly defined circumstances based on internationally recognized laws and standards.... Participating companies will respect and protect the privacy rights of users when confronted with government demands, laws or regulations that compromise privacy in a manner inconsistent with internationally recognized laws and standards.[104]

The corresponding section in the Implementation Guidelines addresses "Government Demands, Laws and Regulations" as well as "Data Collection." The latter is concerned with risk analysis of the specific national jurisdiction in which the company operates.[105] In line with its counterpart on freedom of expression, the GNI Principles and the attached Implementation Guidelines focus merely on the negative human rights impact caused by external pressure from governments, whereas internal mechanisms related to data processing and exchange remain unchallenged.

This is unfortunate, given that the business model of online platforms, which is based on targeted advertising, is increasingly accused of promoting privacy violations. On Facebook, for example, advertisements are targeted to individual users' interests, age, gender, location, and profile. This enables advertisers to select specific groups and target advertisements either on the Facebook website or on other websites using Facebook's advertising services. This business model has caused a number of privacy-related controversies. Most recently, in 2015, a Belgian research study criticized Facebook's data-processing practices and concluded, in relation to Facebook's social media plug-ins, that it processes the personal data of its users as well as the data of all Internet users who come into contact with Facebook, without the necessary consent for "tracking and tracing" or consent for the use of cookies.[106]

---

[101] Ammori, "The 'New' New York Times" at 76.
[102] The main channel for identifying objectionable content is user reporting enabled by technical features in the platform.
[103] Cain Miller, "Google Has No Plans to Rethink Video Status."
[104] "GNI Principles: Section on Privacy," Global Network Initiative, http://globalnetworkinitiative.org/principles/index.php#19.
[105] "GNI Implementation Guidelines, Section on Privacy," Global Network Initiative, http://globalnetworkinitiative.org/implementationguidelines/index.php#28.
[106] "KU Leuven Centre For IT & IP Law and Iminds-Smit Advise Belgian Privacy Commission in Facebook Investigation," KU Leuven, www.law.kuleuven.be/icri/en/news/item/icri-cir-advises-belgian-privacy-commission-in-facebook-investigation.

As a follow-up to the study, the Belgian Privacy Commissioner issued a set of recommendations to Facebook.[107] This is just one example of how Internet platforms can impact the privacy of their users due to their online business model rather than government pressure. Yet, these aspects of company practice in relation to privacy are not included in the GNI norms and standards.

In sum, several of the major Internet companies frame their core mission in terms of freedom of expression and engage in industry networks such as the GNI that are dedicated to protecting human rights norms and standards in the online domain. Yet, the effectiveness of the GNI to protect human rights is challenged by several factors. First, it is based on a voluntary commitment, with no binding obligations on companies. Second, it is largely occupied with limiting and safeguarding against undue government pressure on companies, whereas content regulation and user tracking and profiling are not covered, despite their potential human rights impact.

V CHALLENGES TO HUMAN RIGHTS PROTECTION IN A PRIVATIZED ONLINE DOMAIN

In this final section, I will discuss whether the frameworks that currently govern the activities of online platforms are sufficient to provide the standards and mechanisms needed to protect and respect human rights online, drawing on the challenges outlined in the previous section.

A first challenge relates to the circumstance that core civil and political rights (privacy, freedom to search for information, freedom to express opinion) are exercised within a commercial domain, with companies holding unprecedented power over the boundaries and conditions for exercising those rights. Arguably, some of the most widely used platforms and services may affect public and private life in a way traditionally reserved for public authorities, yet they are largely free from binding standards to protect freedom of expression and privacy. Whereas this governance gap may have a positive impact on rights and freedoms in a state-repressive context, it does not take away the challenges that this raises within democratic societies. Companies that have a substantial impact on the environment are increasingly subjected to national regulations for business conduct, yet similar attention has not been paid to online platforms. Scholarship is only now beginning to address the broader societal implications of private ownership of the online infrastructure of search, expression, and debate that results in the double logic of user empowerment and commodification of online activity.[108]

---

[107] Commission for the Protection of Privacy, Recommendation No. 04/2015 (May 13, 2015), www.privacycommission.be/sites/privacycommission/files/documents/recommendation_04_2015_0.pdf.

[108] J. Van Dijck and T. Poll, "Understanding Social Media Logic" (2013) 1 *Media and Communication* 2–14.

Human rights law is state-centric in nature and holds no direct human rights obligations for private actors. The governance gap accompanying globalization was a core driver for the development of the UNGP, and therefore for asserting the corporate responsibility to respect human rights as a freestanding, universally applicable minimum standard of business conduct – one driven by global social expectation while at the same time based on international law.[109] Nonetheless, the soft law framework of the UNGP, however widely endorsed, remains voluntary by nature, as do industry initiatives such as the GNI.

Further, even these soft law frameworks have significant gaps. In 2016, the five GNI member companies were positively assessed by GNI-appointed assessors for compliance with GNI norms and standards.[110] There are, however, several shortcomings to this assessment process. First, it does not entail a comprehensive human rights impact assessment of all business practices as prescribed by the UNGP, but instead focuses more narrowly on the issues that the GNI members have chosen to include in their development of norms and standards. This means that push-back strategies against illegitimate government requests are the focus of assessment, whereas the impact of business processes concerned with taking down content that does not adhere to internally defined business standards is not considered. Second, the terms and conditions of the assessment process (including the selection of assessors) are carried out within the circuit of the GNI, providing the companies subject to review with influence on the baseline for this review.

Another human rights weakness in these soft law frameworks concerns the limited access to remedy mechanisms. As emphasized by the third pillar of the UNGP, states must take appropriate steps to ensure access to an effective remedy when business-related human rights abuses occur within their jurisdiction. Despite the impact that online platforms have on users' rights of expression and privacy, limited channels exist for users to address potential or actual infringements of such rights.[111] In sum, given the impact that these companies potentially have on human rights in terms of scope and volume, the voluntary approach seems insufficient to provide the billions of Internet users with the level of protection they are entitled to according to international human rights law.

This brings us to the second challenge, namely, whether the state has a positive obligation to legislate the obligations of these companies. Does the character of major online platforms call upon states to provide human rights guidance and

---

[109] C. Methven O'Brien and S. Dhanarajan, "The Corporate Responsibility to Respect Human Rights" at 5.

[110] The assessment report from July 7, 2016, is available at: http://globalnetworkinitiative.org/content/public-report-201516-independent-company-assessments-0.

[111] The importance of access to remedies in an online context is stressed in the Council of Europe's guide to human rights for Internet users. Council of Europe, "Recommendation of the Committee of Ministers to Member States on a Guide on Human Rights for Internet Users," MSI-DUI (2013) 07Rev7 (April 16, 2014).

possible regulation of these actors? Until now, neither the United States nor Europe has taken up this challenge. In April 2016, the European Union concluded a four-year-long comprehensive data protection reform, including, among other things, increased focus on the practices of online platforms.[112] Yet while online platforms' negative impact on privacy has received some attention, their impact on freedom of expression has not. As such, there is no national regulation to protect against the potential negative impact that a major Internet platform may have on freedom of expression. As previously mentioned, in the United States, the First Amendment and the public forum doctrine protect expressions in the public domain, but on the Internet, private companies in control of communicative platforms are free to decide the types of speech they support. This includes taking down or blocking and filtering expression that would otherwise be protected by the First Amendment. In consequence, expression is less protected in the online domain, despite the wide opportunities online platforms provide for new means of realizing freedom of expression. Likewise, in the United States, there is no general data protection regulation covering these private actors, and thus no clear boundaries for the companies' handling of personal data.

However urgent, several factors indicate that a solution will not be forthcoming in this area any time soon. The transnational nature of online platforms makes it difficult for states to address their impact on freedom of expression or privacy domestically. Moreover, up till now, the United States and European states have been unable to agree on the scope of freedom of expression, for example concerning protected speech, and they have lacked a common standard for data protection. Whereas the European approach is geared toward both negative and positive state obligations in the area of freedom of expression and privacy (e.g., imposing regulations on private actors), the US approach has focused on the negative state obligation to avoid interference. While the issues raised have received some scholarly attention, they have not surfaced as prominent policy issues in either Washington or Brussels. As such, it is not realistic to expect common US/EU policy for the major online platforms in the foreseeable future.

If European states were willing to invoke their positive state obligation in order to protect freedom of expression online, they would have to apply national standards for protected speech to the online domain. In consequence, Internet platforms would have to comply with a number of different standards for protected speech, depending on the location of their users. Although this would most likely cause controversy and resistance from the companies, it is in principle no different from the current situation, in which platforms adhere to different regimes for unlawful

---

[112] The General Data Protection Regulation (EU 2016/679) has been highly controversial and its implications widely addressed by scholars and activists alike. *See, e.g.*, A. Dix "EU Data Protection Reform Opportunities and Concerns" (2013) 48 *Intereconomics* 268–86; D. Naranjo, "General Data Protection Regulation: Moving forward, slowly," *European Digital Rights*, June 3, 2015, https://edri.org/author/diego/page/5/.

content depending on the national context in which they operate. In other words, while both Facebook and Google have processes for dealing with alleged unlawful content in a specific national jurisdiction, they might also have processes for ensuring that no content is taken down unless it satisfies the criteria set out in human rights law. Such a mechanism would ensure that the companies' commitment to freedom of expression is operationalized not only in relation to government pressure, but also in relation to the day-to-day practices that govern their communities of users.

In conclusion, divergence in the US and European approaches to privacy and freedom of expression, as well as the complexity of defining legal responsibilities in the face of conflicting local laws, means that a concerted state effort in this field is unlikely. Yet authoritative human rights guidance for the major online platforms is urgently needed in order to clarify the scope of their responsibilities and, more importantly, to ensure that their impact on billions of users' rights is mitigated and potential violations are remedied.

12

Technology, Self-Inflicted Vulnerability, and Human Rights

G. Alex Sinha

I INTRODUCTION

Since 2013, perhaps no human rights issue has received as much sustained attention as the right to privacy. That was the year the first Snowden revelations reached the public, detailing sophisticated, large-scale US government surveillance programs designed to capture or analyze incredibly large volumes of digital data. In the weeks and months that followed, media reports confirmed that the US government had, at various recent points, run programs designed to scan and harvest data contained in e-mails, track Internet browsing activity, collect data from cell phones, collect digital contact lists (including e-mail and instant messaging contacts), and collect photographs of Internet users all over the world.[1] Other governments have been revealed to engage in similar practices.[2]

Targeting the use of digital technologies is an obviously fruitful approach for state surveillance programs. By the middle of 2016, one estimate placed worldwide Internet use at more than 3.5 billion people.[3] Cell phone use is even more widespread, with recent reports suggesting the world is approaching five billion

---

[1] For a brief summary of key Snowden revelations, see Human Rights Watch and American Civil Liberties Union, *With Liberty to Monitor All* (2014), pp. 8–11, www.hrw.org/sites/default/files/reports/usnsa0714_ForUPload_0.pdf. In the interest of full disclosure, note that I was the researcher and author of that report. *See also* J. Risen and L. Poitras, "N.S.A. Collecting Millions of Faces from Web Images," *The New York Times*, May 31, 2014, www.nytimes.com/2014/06/01/us/nsa-collecting-millions-of-faces-from-web-images.html (describing the collection of photographs for facial recognition purposes).

[2] *See, e.g.*, R. Gallagher, "U.K.'s Mass Surveillance Databases were Unlawful for 17 Years, Court Rules," *The Intercept*, October 17, 2016, https://theintercept.com/2016/10/17/gchq-mi5-investigatory-powers-tribunal-bulk-datasets/.

[3] *See* "World Internet Usage and Population Statistics," Miniwatts Marketing Group, www.internetworldstats.com/stats.htm.

mobile users.⁴ Significant numbers of people also use other technologies conducive to tracking, such as E-ZPass or Global Positioning System (GPS) devices. Those numbers are likely to increase in the coming years, which is particularly significant because of the way in which digital technologies lend themselves to insecure use and mass surveillance.

The ongoing global conversation about the legality of surveillance practices has focused on a number of dimensions of the human right to privacy, but there has been little serious discussion of a major factor in the expansion of the insecure use of digital technologies: the user. A significant portion of the information collected by surveillance (or otherwise made vulnerable to unintended recipients) is exposed voluntarily, sometimes deliberately or knowingly, or with unjustified ignorance of the risks of transmitting it in a particular manner. This chapter argues that, as human rights bodies, governments, and advocacy groups seek to understand the protections provided by the human right to privacy, it is also essential to clarify the conditions (if any) under which a person may waive those protections. The purpose of this chapter is therefore to help launch a conversation about waiving the human right to privacy and the role of the state in fostering the ability of individuals to make better choices in protecting their privacy.⁵

II INDIVIDUAL CHOICE AND THE HUMAN RIGHT TO PRIVACY

The Snowden revelations have triggered increased engagement on the right to privacy among civil society organizations,⁶ multiple votes within the United Nations General Assembly,⁷ research by the United Nations High Commissioner for Human Rights,⁸ and the establishment of a new special rapporteur on privacy by the Human Rights Council.⁹ Pressure is also building on the Human Rights Committee to

---

[4] "Number of mobile phone users worldwide from 2013 to 2019 (in billions)," Statista, www.statista.com/statistics/274774/forecast-of-mobile-phone-users-worldwide/.

[5] "Waiver" here is defined narrowly; it refers to discrete choices or events that remove specific, otherwise-protected information from under the umbrella of the human right to privacy (instead of a broad, blanket alienation of the right to privacy for all of one's protected matters).

[6] See, e.g., *With Liberty to Monitor All*; "Chilling Effects: NSA Surveillance Drives U.S. Writers to Self-Censor," PEN American Center, https://pen.org/chilling-effects; "Surveillance Self-Defense," Electronic Frontier Foundation, https://ssd.eff.org/en; A. Toh, F. Patel, and E. Goitein, "Overseas Surveillance in an Interconnected World," Brennan Center for Justice, www.brennancenter.org/publication/overseas-surveillance-interconnected-world.

[7] See G.A. Res. 68/167, *The right to privacy in the digital age*, U.N. Doc. A/Res/68/167 (December 18, 2013); G.A. Res. 69/166, *The right to privacy in the digital age*, U.N. Doc. A/Res/69/166 (December 18, 2014).

[8] Office of the United Nations High Commissioner for Human Rights, *The right to privacy in the digital age*, U.N. Doc. A/HRC/27/37 (June 30, 2014).

[9] See "Human Rights Council creates mandate of Special Rapporteur on the right to privacy," Office of the United Nations High Commissioner for Human Rights, March 26, 2015, www.ohchr.org/EN/NewsEvents/Pages/DisplayNews.aspx?NewsID=15763&LangID=E.

update its interpretation of the right to privacy under the International Covenant on Civil and Political Rights (ICCPR), primarily to account for changes in circumstances and technological developments that render its previous interpretation from the 1980s practically obsolete.[10]

This flurry of activity has largely left aside the relevance of individual users and the choices they make in the storage and transmission of their private information. Consider the state of the debate about US human rights obligations related to privacy – obligations that have occupied center stage since media outlets began publishing the Snowden revelations. Although a number of international agreements address the right to privacy,[11] a primary source of human rights obligations for the United States is the ICCPR, to which the United States has been a party since 1992. Article 17 of the ICCPR stipulates that "[n]o one shall be subjected to arbitrary or unlawful interference with his privacy, family, home or correspondence ... [and e]veryone has the right to the protection of the law against such interference."[12] Additional articles in the covenant inform the scope and rigidity of individual rights, such as by addressing the geographic range of state duties under the covenant or the conditions under which a right might be limited or outweighed by other considerations (such as protecting national security).

The ICCPR does not explicitly address the role of individual choice in connection with the right to privacy,[13] which means choice has not factored much into a debate that has largely followed the text of the covenant. For example, a significant dispute has arisen about the meaning of the covenant's ban on "arbitrary" and "unlawful" interference with protected privacy interests.[14] Multiple UN rights experts have recently concluded that non-arbitrariness requires states, *inter alia*, to ensure the *necessity* of interferences with the right to privacy and the *proportionality* of their invasive practices.[15] Such requirements are intended to ensure the

---

[10] See, e.g., American Civil Liberties Union, *Information Privacy in the Digital Age* (New York: ACLU Foundation, 2015), www.aclu.org/sites/default/files/field_document/informational_priv acy_in_the_digital_age_final.pdf; G. Alex Sinha, "NSA Surveillance Since 9/11 and the Human Right to Privacy" (2014) **59** *Loyola Law Review* 861–946.

[11] See, e.g., G.A. Res. 217 (III) A *Universal Declaration of Human Rights*, art. 12 (December 10, 1948); Inter-Am. Comm'n H.R. 9th Conf., *American Declaration of the Rights and Duties of Man*, art. V (May 2, 1948); Council of Europe, *European Convention on Human Rights*, art. 8.

[12] International Covenant on Civil and Political Rights ("ICCPR"), art. 17.

[13] Two ICCPR Articles, 18 and 19, do refer to individual choice (to one's right to choose a belief system and to choose preferred media, respectively). See ICCPR, arts. 18, 19. But those choices are essential to the rights themselves rather than related to the waiver of a covenant right. Ibid.

[14] ICCPR, art. 17.

[15] See, e.g., "Privacy in the Digital Age," ¶¶ 22–23; *Report of the Special Rapporteur on the promotion and protection of human rights and fundamental freedoms while countering terrorism, Martin Scheinin*, ¶ 17, U.N. Doc. A/HRC/13/37 (December 28, 2009); *Report of the Special Rapporteur on the promotion and protection of the right to freedom of opinion and expression, Frank La Rue*, ¶ 29, U.N. Doc. A/HRC/23/4 (April 17, 2003); *Report of the Special Rapporteur on the promotion and protection of the right to freedom of opinion and expression, David Kaye*, ¶ 29, U.N. Doc. A/HRC/29/32 (May 22, 2015).

continued relevance of the human right to privacy in the digital era and would, in theory, provide a check on large-scale surveillance of the sort revealed by Snowden.[16] Thus far, however, the United States has rejected requirements like necessity and proportionality, arguing that those standards do not necessarily follow from a ban on arbitrary and unlawful interference.[17] Instead, the United States insists that its programs need only be (and consistently are) "reasonable" because they are authorized by law and not arbitrary.[18]

Another dispute concerns the geographic scope of state duties under the covenant. Article 2(1) provides that "[e]ach State Party to the present Covenant undertakes to respect and to ensure to all individuals within its territory and subject to its jurisdiction the rights recognized in the present Covenant."[19] The United States typically interprets the phrase "within its territory and subject to its jurisdiction" conjunctively, such that it only accepts duties under the covenant toward people of whom both modifiers are true.[20] Key human rights bodies, including the Human Rights Committee (the UN body tasked with interpreting the ICCPR), have rejected that reading and interpret the phrase disjunctively.[21] This disagreement has garnered increased attention as a result of US surveillance

---

[16] In light of the consensus coalescing among those human rights bodies, advocacy groups like the American Civil Liberties Union have attempted to map out in substantial detail the nature of state obligations under Article 17. See, e.g., "Informational Privacy in the Digital Age."

[17] See, e.g., K. L. Razzouk, "Explanation of Position on draft resolution L.26/ Rev. 1 The Right to Privacy in the Digital Age," http://usun.state.gov/remarks/6259 (the link was active at the time of this writing).

[18] See ibid.; ICCPR art. 17(1).

[19] ICCPR art. 2(1).

[20] While he was a legal advisor at the US State Department, Harold Koh advocated for relaxing that standard and accepting that some ICCPR obligations might attach to US conduct outside of its own territory. See Harold Hongju Koh, "Memorandum Opinion on the Geographic Scope of the International Covenant on Civil and Political Rights," www.justsecurity.org/wp-content/uploads/2014/03/state-department-iccpr-memo.pdf. As recently as its 2015 submission to the Human Rights Committee, which monitors state compliance with the ICCPR, the United States has nevertheless continued to assert the original, narrower view. See Permanent Mission of the United States of America to the Office of the United Nations, "One-Year Follow-up Response of the United States of America to Priority Recommendations of the Human Rights Committee on its Fourth Periodic Report on Implementation of the International Covenant on Civil and Political Rights," ¶ 33, www.state.gov/documents/organization/242228.pdf. By contrast, the United States has softened its position on the extraterritorial obligations under another human rights convention, the Convention Against Torture. White House Office of the Press Secretary, "Statement by NSC Spokesperson Bernadette Meehan on the U.S. Presentation to the Committee Against Torture," www.whitehouse.gov/the-press-office/2014/11/12/statement-nsc-spokesperson-bernadette-meehan-us-presentation-committee-a.

[21] See, e.g., Human Rights Committee, "Concluding observations on the fourth report of the United States of America," ¶ 4, www.justsecurity.org/wp-content/uploads/2014/03/UN-ICCPR-Concluding-Observations-USA.pdf. See also Ryan Goodman, "UN Human Rights Committee Says ICCPR Applies to Extraterritorial Surveillance: But is that so novel?," www.justsecurity.org/8620/human-rights-committee-iccpr-applies-extraterritorial-surveillance-novel/.

revelations, because the narrower position turns would-be beneficiaries of human rights protection into unprotected targets of surveillance.

Yet another dimension of the ongoing debate about the human right to privacy concerns the limitations clauses built into the ICCPR. The United States has emphasized that it takes a broad reading of those limitations – especially the national security limitation articulated in, among other places, Articles 19, 21, and 22.[22] The US government routinely cites national security considerations to justify practices of concern to human rights bodies, including surveillance.[23] The implications of that approach are far-reaching in light of the ongoing War on Terror, which lacks any obvious or imminent endpoint.

Overall, the contours of this debate are unsurprising; the language of the covenant is a natural focal point for states and human rights bodies alike, and their disagreements, in turn, frame the contributions of interested advocacy groups. But the issue of personal choice casts a shadow over all such textual analysis. It is surely uncontroversial that one can, at least sometimes, waive privacy protection for particular pieces of information by exposing them to collection.[24] It should also be uncontroversial that some digital information, even if it can be obtained by intelligence agencies or hackers, remains protected by the human right to privacy. Yet through the insecure use of digital technologies, people increasingly expose enormous swaths of protected information with unclear levels of intentionality and culpability – even as the covenant remains silent on waiver generally and the ongoing debates about the human right to privacy fail to provide much clarity. The conditions under which one waives legal privacy protections are therefore both extremely important and extremely unclear.

Vulnerability can be chosen, such as when people share private information in public fora (like public websites). It can be recklessly or negligently assumed, such as when people undertake to store or transmit personal information in insecure ways (whether knowingly or because they lack a reasonable appreciation for the risk). It can arise through no fault of a user when it results from justifiable ignorance on the user's part. And it can be imposed by circumstance in spite of a user's best efforts, such as by the mere fact that surveillance authorities and hackers around the world typically have more power to harvest information than even the most committed individuals have to protect it. Any reasonable understanding of the waiver of the right to privacy must account for different notches on this spectrum.[25]

---

[22] See ICCPR, arts. 19, 21, 22.

[23] See, e.g., US Department of State, "Report of the United States of America Submitted to the U.N. High Commissioner for Human Rights in Conjunction with the Universal Periodic Review," ¶ 83, www.state.gov/j/drl/upr/2015/237250.htm; "One-Year Follow-up Response," ¶ 29.

[24] Truly public information – especially information one has *chosen* to make public – can likely be collected without interfering with a person's privacy, family, home, or correspondence. It is therefore difficult to see how it could fall within the scope of Article 17.

[25] In the context of US constitutional law, one is protected from searches and seizures by government agents when one has a reasonable expectation of privacy. Whether the proper

For simplicity, we might assume that posting private information – say, political leanings and hobbies – on a public Facebook page constitutes some sort of waiver of privacy protection for that information, meaning that such information would fall outside the scope of the protections contained in Article 17. But what about intermediate cases that expose us to broader-than-intended intrusions, such as posting the same information on a semipublic Facebook page that is set to restrict access only to approved "friends"? Or sending sensitive information via unencrypted e-mail to a single recipient? Or running a sensitive Google search from a personal computer alone in one's home? Or carrying a cell phone that could just as well connect to a stingray as a cell phone tower? Or attempting to send an encrypted message but accidentally sending the message unencrypted?

These examples underscore a complicating underlying factor: Even when we want to protect our privacy, we are often fundamentally incapable of employing digital technology securely, whether due to ignorance, lack of skill, or inherent limitations in the technology itself. Yet we constantly use such technology anyway. Consider the example of the United States, which in some ways is a particularly risky place for such a casual approach to technology. Not only does the United States aggressively gather as much data as it can through perhaps the most powerful surveillance apparatus in the world, but it also features some problematic legal precedents for privacy under domestic law.

A string of Supreme Court cases has established the legal principle that voluntary disclosure of information to third parties eliminates one's expectation of privacy for that information, thereby defeating constitutional privacy protections that would require law enforcement to get a warrant for it.[26] In several cases decided between 1952 and 1971, the court consistently held that the Fourth Amendment prohibition on unreasonable search and seizure does not apply to the contents of a person's utterances that are voluntarily communicated to a government agent or informant.[27] A second series of cases, decided between 1973 and 1980, extended that rule to business records that are provided to a third party. For example, in *United States v. Miller*, the Supreme Court ruled that there is no reasonable expectation of privacy in checks and deposit slips provided to banks, as those are "negotiable instruments" rather than "confidential communications" and the data they contain are

---

international law analysis deploys a similar concept, the spectrum laid out here is a useful starting point for identifying potentially relevant subjective and objective markers (such as intentions, expectations, reasonableness, and so forth). As discussed below, however, the proper approach to waiver under Article 17 is unlikely to mirror (closely, at any rate) the approach under US domestic law.

[26] For particularly helpful background on these cases, see Orin S. Kerr, "The Case for the Third-Party Doctrine" (2009) 107 *Michigan Law Review* 561, 567–70.

[27] See *On Lee v. United States*, 343 U.S. 747 (1952); *Lopez v. United States*, 373 U.S. 427 (1963); *Lewis v. United States*, 385 U.S. 206 (1966); *Hoffa v. United States*, 385 U.S. 293 (1966); *United States v. White*, 401 U.S. 745 (1971).

"voluntarily conveyed" to the banks.[28] In *Smith v. Maryland*, the Court reinforced its earlier holdings as applied to records of phone calls placed by a criminal suspect. The Court held that, because dialing numbers from one's phone involves providing those numbers to the phone company, the police can collect records of those calls, without a warrant, through the use of a pen register installed on the telephone company's (rather than the suspect's) property.[29]

In one sense, each of these rulings was quite narrow. The first cluster addressed a criminal defendant's communications with a government agent or informant, *Miller* concerned checks and deposit slips provided to a bank, and *Smith* addressed the right of a criminal suspect to assert Fourth Amendment protection for the numbers he dials from his home phone. Yet in all of these cases, the Court held that constitutional privacy protections under the Fourth Amendment to the US Constitution simply did not apply, at least in part because the parties asserting their rights had voluntarily disclosed the information in question to a third party. The cases are thus suggestive of a rule that extends to many other contexts.

As many have noted,[30] the underlying rule – sometimes referred to as the "third-party doctrine"[31] – has sweeping implications in the current era. Most people now turn over a significant and growing proportion of their private information to third-party service providers. E-mail providers like Google rather notoriously scan the text of our messages for key words so they can tailor their advertising to matters of interest to specific users. The specific websites we visit are recorded by Internet service providers (ISPs).[32] And, just like in Mr. Smith's case, significant information about our phone activity – now including text messages as well as calls – passes through the hands of our phone service providers.

In fairness, there is a question as to whether the third-party doctrine would extend to the content of communications (rather than metadata) passing through the hands of a service provider.[33] The phone numbers in *Smith* are considered metadata, and it is debatable whether the monetary values on checks and deposit slips from *Miller*

---

[28] 425 U.S. 435 (1976).
[29] *See Smith v. Maryland*, 442 U.S. 735 (1979). Other cases in this second series include *Couch v. United States*, 409 U.S. 322 (1973) and *United States v. Payner*, 447 U.S. 727 (1980).
[30] *See, e.g.*, J. Villasenor, "What You Need to Know about the Third-Party Doctrine," *The Atlantic*, www.theatlantic.com/technology/archive/2013/12/what-you-need-to-know-about-the-third-party-doctrine/282721/; C. Cohn and P. Higgins, "Rating Obama's NSA Reform Plan: EFF Scorecard Explained," Electronic Frontier Foundation, www.eff.org/deeplinks/2014/01/rating-obamas-nsa-reform-plan-eff-scorecard-explained.
[31] *See, e.g.*, Kerr, "The Case for the Third-Party Doctrine."
[32] Google records every search conducted on its search engine. B. Caddy, "Google tracks everything you do. Here's how to delete it," *Wired*, www.wired.co.uk/article/google-history-search-tracking-data-how-to-delete. That has driven some users to use search engines that claim they do not, such as DuckDuckGo. *See* DuckDuckGo, "Why You Should Care – Search History," https://duckduckgo.com/privacy#s2.
[33] *See, e.g., United States v. Warshak*, 631 F.3d 266 (6th Cir. 2010).

count as content.[34] The first series of cases discussed above concerned content, but that content was conveyed (sometimes unknowingly) *to* a government agent or informant rather than *through* a third-party service provider. One might attempt to distinguish these cases, as Orin Kerr has done, carving out Fourth Amendment protection for content but not metadata.[35] Others, like Greg Nojeim, argue that metadata can be sensitive enough to warrant Fourth Amendment protection on its own, even if precedent does not necessarily support that view.[36] There is no clear consensus on the matter in US courts, although the holding in *Smith* could arguably reach content, because the court does not explicitly distinguish between content and metadata: "[T]his Court consistently has held that a person has no legitimate expectation of privacy in information he voluntarily turns over to third parties."[37] And although Justice Sonia Sotomayor has questioned the third-party doctrine precisely for its implications at a time when so much of our daily activity involves third parties,[38] it remains unclear how US courts will continue to apply the doctrine.

In light of what appear to be tangible legal risks, not to mention the practical likelihood that various state intelligence agencies and others are likely to obtain nontrivial proportions of our digital data, it is therefore worth inquiring how deliberately and culpably people appear to store and transmit so much sensitive information insecurely.

III THE VOLUNTARINESS OF INSECURE USE OF TECHNOLOGY

It is impossible to deny that cell phones and the Internet are convenient for managing private matters, and many of us sometimes elect to use them when we know (or should know) that it is insecure to do so. But in light of the possible implications for waiving the human right to privacy, it is important to recognize that the use of digital technologies is often less voluntary than might appear at first glance. While there is obviously some element of choice in when and how people adopt technologies, there is a large measure of compulsion as well. Many employers assign e-mail addresses to employees, the use of which is more or less mandatory. Certain employers also issue smartphones to remain in better contact with their

---

[34] Some simply reject the distinction between metadata and content, such as the Office of the UN High Commissioner of Human Rights. *See* "The right to privacy in the digital age," ¶ 19.
[35] O. Kerr and G. Nojeim, "The Data Question: Should the Third-Party Doctrine Be Revisited?," *ABA Journal*, www.abajournal.com/magazine/article/the_data_question_should_the_third-party_records_doctrine_be_revisited/.
[36] Ibid.
[37] *Smith*, 442 U.S. at 743–44.
[38] "More fundamentally, it may be necessary to reconsider the premise that an individual has no reasonable expectation of privacy in information voluntarily disclosed to third parties. This approach is ill suited to the digital age, in which people reveal a great deal of information about themselves to third parties in the course of carrying out mundane tasks." *United States v. Jones*, 132 S. Ct. 945, 957, 181 L. Ed. 2d 911 (2012) (Sotomayor J. concurring) (internal citations omitted).

employees, or to enable employees to remain in better contact with clients. Universities often require students to use online systems to access course materials and other important information. Indeed, online research has become all but unavoidable for a number of jobs and scholarly endeavors. For many people, unplugging is not even a meaningful possibility.

There are also broader practical concerns that raise questions about the voluntariness of much Internet and phone use. The obvious utility of cell phones can make it prohibitive to eschew them. For example, cell phones are unparalleled as a parenting tool for remaining in contact with one's children. Even for those who can do without a cell phone, the choice to do so can impose substantial costs. Moreover, increasing cell phone use has correlated with a dramatic decline in the installation and maintenance of pay phones, making it ever more impractical to hold out for privacy reasons.[39]

Similarly, refusing to use the Internet may ultimately foreclose access to a range of substantive goods.[40] Remaining in touch with others is just one example. For those of limited means, long-distance communication without the Internet is especially difficult, as phone calls can be expensive and postal mail is slow. But instant messaging and voice-over-IP services (like Skype) permit free communication with other users all over the world. Similarly, staying up to date on current events, managing one's finances, and planning travel – not to mention applying for food assistance or other government benefits – are all increasingly difficult without the Internet, as companies and governments have scaled back support for non-Internet-based interaction. Resisting new technologies can be so costly that it hardly resembles a genuine choice.[41]

Moreover, there is good reason to conclude that many users of digital technologies simply fail to appreciate their vulnerability, rather than knowingly assuming that vulnerability.[42] People routinely surrender sensitive or damaging personal information to third parties that cannot safeguard it properly, as evidenced (for example) by recent hacks of Sony, Target, Yahoo, and the Ashley Madison website.[43] The Ashley

---

[39] D. Andreatta, "As pay phones vanish, so does lifeline for many," *USA Today*, www.usatoday.com/story/news/nation/2013/12/17/pay-phone-decline/4049599/.

[40] See Frank La Rue, "Report of the Special Rapporteur on the promotion and protection of the right to freedom of opinion and expression," ¶¶ 78, 85, www2.ohchr.org/english/bodies/hrcouncil/docs/17session/A.HRC.17.27_en.pdf. Some have interpreted this report as suggesting that access to the Internet is itself a human right. See, e.g., D. Kravets, "U.N. Report Declares Internet Access a Human Right," *Wired*, www.wired.com/2011/06/internet-a-human-right/.

[41] Richard Stallman has discussed a similar issue in the specific context of software selection. See R. Stallman, "National Institute of Technology – Trichy – India – 17 February 2004," Free Software Foundation, www.gnu.org/philosophy/nit-india.en.html.

[42] Whether that ignorance is justifiable may depend on a case-by-case analysis that considers the sophistication of the user and the nature of the technology at issue.

[43] See FBI National Press Office, "Update on Sony Investigation," www.fbi.gov/news/pressrel/press-releases/update-on-sony-investigation; M. Riley et. al., "Missed Alarms and 40 Million Stolen Credit Card Numbers: How Target Blew It," *Bloomberg*, www.bloomberg.com/news/

Madison hack, for instance, exposed the identities of many people who had used the site to set up extramarital affairs. Some of those people were seriously harmed by the revelation, including employees of the US federal government who used their official work e-mail addresses to set up accounts.[44] It is implausible that most of these people were indifferent to publication of their private information, and much more likely that they did not fully appreciate their vulnerability.

Further, many of those who recognize the privacy threats posed to electronic communication have limited resources and expertise to address the problem. For those who are not adroit with technology, it can be intimidating to pick among the options and get comfortable using the proper tools as a matter of course. Unsurprisingly, there are also conflicting opinions about the efficacy of various measures, and those who lack a technical background are not well placed to make a confident choice.

Security can also be both expensive and slow; at minimum, as one particularly tech-savvy individual put it, proper security measures can impose a significant tax on one's time.[45] Moreover, even those measures that are inexpensive and simple to use may be ineffective without buy-in from one's correspondents. For example, free, easy-to-use software is available for encrypting chats, text messages, and phone calls,[46] but encryption is pointless unless all parties to a communication are willing to use it. And the typical user also has no power whatsoever to improve the security of some of the systems many of us are obligated to use, such as e-mail provided by an employer.

Lacking the time, knowledge, power, and sometimes money necessary to invest heavily in data security, the average person finds himself trapped between two imperfect options: risk the insecurity that comes with the use of ubiquitous and convenient technologies, or forego some of the most efficient tools available for conducting personal or professional business. And, often enough, the world – whether through our employers, our schools, or the demands of our personal lives – picks the first option for us.

Even parties with significant resources strain to maintain security – not necessarily because they are careless or sloppy, but rather because digital data can be vulnerable to collection by any number of actors, and because it can be exceedingly difficult to

---

articles/2014-03-13/target-missed-warnings-in-epic-hack-of-credit-card-data; S. Thielman, "Yahoo hack: 1bn accounts compromised by biggest data breach in history," *The Guardian*, www.theguardian.com/technology/2016/dec/14/yahoo-hack-security-of-one-billion-accounts-breached; K. Zetter, "Hackers Finally Post Stolen Ashley Madison Data," *Wired*, www.wired.com/2015/08/happened-hackers-posted-stolen-ashley-madison-data/.

[44] E. Fink and L. Segall, "Government workers cope with fallout from Ashley Madison hack," CNN, http://money.cnn.com/2015/08/22/technology/ashley-madison-hack-government-workers/.
[45] See *With Liberty to Monitor All*, p. 34.
[46] Free software is also available for encrypting e-mails, but e-mail encryption remains notoriously clunky.

understand the nature of those vulnerabilities and to institute adequate protections. The US intelligence community, for instance, has failed repeatedly at protecting highly classified information from both whistleblowers and hackers.[47] It is not surprising, therefore, that sophisticated individuals struggle as well. I previously did research on surveillance for Human Rights Watch (HRW) and the American Civil Liberties Union (ACLU), for which I interviewed (among other people) nearly fifty journalists covering intelligence, national security, and law enforcement.[48] Those journalists have an overriding interest in protecting both the identities of their sources and the contents of their conversations. First and foremost, that interest arises from a general feeling of obligation to shield the identities of those who provide them with information. As a practical matter, the journalists also recognize that failure to protect sources effectively could compromise their ability to develop other sources in the future.

Many of the people I spoke with worked for major outlets, like *The New York Times*, *The Wall Street Journal*, *The Washington Post*, NPR, and ABC News.[49] Most also had years, even decades, of experience reporting on sensitive subjects, and a number had already won Pulitzer Prizes. As journalists go, therefore, the group I interviewed was elite in both their level of skill and their access to institutional support for the proper tools of the trade. All of that notwithstanding, these journalists consistently and vividly relayed to me significant ongoing challenges in using digital technologies securely.[50] Nearly all of them told me that the most secure method of doing their work involved avoiding technology as much as possible – meeting sources face-to-face while leaving cell phones at the office, or saving notes in hard copy rather than electronically.

Yet "going dark" by avoiding electronic devices significantly impeded their work, could still draw scrutiny, and was sometimes impossible.[51] To the extent that use of digital technologies is unavoidable, many of the journalists reported upgrading their

---

[47] Leaving aside the information disseminated by Snowden, the NSA also recently had exploits stolen by hackers. E. Nakashima, "Powerful NSA hacking tools have been revealed online," *The Washington Post*, www.washingtonpost.com/world/national-security/powerful-nsa-hacking-tools-have-been-revealed-online/2016/08/16/bce4f974-63c7-11e6-96c0-37533479f3f5_story.html. Separately, CIA Director John Brennan apparently had his private e-mail hacked by a teenager. K. Zetter, "Teen Who Hacked CIA Director's Email Tells How He Did It," *Wired*, www.wired.com/2015/10/hacker-who-broke-into-cia-director-john-brennan-email-tells-how-he-did-it/.
[48] *With Liberty to Monitor All*, p. 7.
[49] Ibid.
[50] The report details some of their challenges in greater detail. *See* ibid., pp. 22–48.
[51] One major challenge is making initial contact with a source without using e-mail or a phone. First there is a practical problem: you must find the source's precise physical location, which can be prohibitive if he or she is not nearby. Second, many sources do not take kindly to being accosted by a journalist they may not know seeking to develop a relationship they may have reservations about. On the other hand, using e-mail or phone, even with security measures in place, will nearly always leave a link between the journalist and the source that can be discovered later. *See* ibid.

security. For example, a number described learning how to use Tor, how to encrypt e-mails, chats, and texts, and how to purchase and set up air-gapped computers.[52] Some benefitted from data security training run by their outlets. Others with less institutional support described improvising new security measures, sometimes attempting to hide digital trails using methods that were in fact ineffective. One described sharing an e-mail account with a source and exchanging messages via saved, but unsent, drafts. That was the same technique David Petraeus reportedly used, unsuccessfully, in attempting to communicate secretly with his mistress.[53] Whether the journalists benefitted from professional security training or not, they were uniformly skeptical that it was even possible to be entirely secure in one's use of digital technologies. Interviewees commonly expressed the feeling that, when facing off against an adversary as powerful as the National Security Agency, the best one could hope to do is "raise the cost of surveillance."[54]

I found similar results from speaking to attorneys,[55] whose interests in digital security stem from, among other things, their professional responsibility to safeguard confidential client information.[56] I interviewed more than forty attorneys, most of them working on matters of potential interest to the US government, such as criminal defense in the national security context. Just as the journalists expressed significant concerns about managing their relationships with sources, the attorneys largely worried about managing their relationships with clients. Some found that merely warning their clients against using phones and e-mail made the clients increasingly mistrustful and hampered their ability to develop a relationship. And, like the journalists, a number of the lawyers expressed the belief that speaking face-to-face is now essential, notwithstanding the costs and time constraints associated with doing so.

It is noteworthy that these journalists and attorneys – educated, fairly well-resourced people with an unusually high interest in protecting the privacy of some of their interactions – have to wrestle so mightily with the issue of data security. Indeed, merely planning this research required a surprising amount of thought within HRW and the ACLU. I needed a way to reach out to my subjects electronically concerning sensitive matters, and a way to convey to them my competence in

---

[52] Air-gapped computers are computers that never connect to any insecure network (including the Internet), often configured to sit in a secure room. One journalist equated them to electronic typewriters. See *With Liberty to Monitor All*, p. 32.
[53] See D. Leinwand Leger and Y. Alcindor, "Petraeus and Broadwell used common e-mail trick," *USA Today*, www.usatoday.com/story/tech/2012/11/13/petraeus-broadwell-email/1702057/.
[54] *With Liberty to Monitor All*, p. 39.
[55] Ibid., pp. 49–65.
[56] See American Bar Association, "Model Rules of Professional Conduct: Rule 1.6: Confidentiality of Information," www.americanbar.org/groups/professional_responsibility/publications/model_rules_of_professional_conduct/rule_1_6_confidentiality_of_information.html. Some legal experts anticipate that failure to use secure technologies to store or communicate confidential information may soon become grounds for sanctions. *With Liberty to Monitor All*, pp. 58–59.

protecting any data they might provide. That was a difficult goal to achieve, especially because so much of security turns on the measures taken by one's correspondents. Like many of my research subjects, I therefore had to work with the institutions backing me to develop security protocols that were affordable and practical under the circumstances. Notwithstanding the assistance I had, the process involved some measure of trial and at least one embarrassing security error on my part. In short, true digital security is incredibly difficult or perhaps even impossible to attain – a point that cannot be lost in attempting to understand the relationship between the use of digital technologies and the waiver of the human right to privacy.

IV DRAWING SOME CONCLUSIONS

A modern understanding of the human right to privacy must contend with these questions, and the purpose of this chapter is to highlight the need for an extended conversation about the issue of waiver, especially among human rights authorities.[57] This section offers some preliminary conclusions about the conditions under which a technology user's actions might constitute a waiver of his or her privacy protections under Article 17 of the ICCPR, and the duties of the governments that are parties to the ICCPR to support the choices of individuals who take measures to secure their digital data. Per the Vienna Convention on the Law of Treaties, "[a] treaty shall be interpreted in good faith in accordance with the ordinary meaning to be given to the terms of the treaty in their context and in the light of its object and purpose."[58] When that approach alone "[l]eaves the [treaty's] meaning ambiguous or obscure," or "[l]eads to a result which is manifestly absurd or unreasonable," then "[r]ecourse may be had to supplementary means of interpretation, including the preparatory work of the treaty and the circumstances of its conclusion."[59] Application of these principles often permits and may at times require that a state's treaty obligations evolve with changing circumstances. As Eirik Bjorge has recently put it, "The wording [of a treaty] is important because it may lead us to ascertaining the intention of the parties, not because it is somehow an end in and of itself."[60] An evolutionary reading of a treaty may therefore "be required by good faith."[61] For the reasons laid

---

[57] It is noteworthy, for example, that the ten reports the first UN Special Rapporteur on privacy is scheduled to produce between 2017 and 2021 do not appear designed to address this subject at all. *See* Office of the High Commissioner for Human Rights, "Planned Thematic Reports and call for consultations," www.ohchr.org/EN/Issues/Privacy/SR/Pages/ThematicReports.aspx.

[58] Vienna Convention on the Law of Treaties, art. 31.

[59] Ibid., art. 32. To the extent that this chapter has focused on the United States, it is significant that the US accepts the Vienna Convention as informing interpretation of its treaty obligations. *See* E. Criddle, "The Vienna Convention on the Law of Treaties in U.S. Treaty Interpretation" (2004) 44 *Virginia Journal of International Law* 431, 443.

[60] E. Bjorge, *The Evolutionary Interpretation of Treaties* (Oxford: Oxford University Press, 2014), p. 189.

[61] Ibid., p. 190.

out below, the right to privacy reveals that the ICCPR is precisely the sort of treaty that requires an evolutionary reading.

A Waiver Should Be Understood Narrowly under the ICCPR

The text of Article 17 is the starting point in determining the conditions for waiver. As noted above, the covenant prohibits "arbitrary or unlawful interference" with four distinct but potentially overlapping items: privacy, family, home, and correspondence. Challenging questions arise immediately simply from the relationships among these categories. For example, privacy is distinctively diffuse relative to the other items; the concept of privacy neither encompasses everything about one's correspondence, family, and home, nor is it fully exhausted by those three domains. The language of Article 17 therefore immediately invites the possibility that certain information will warrant protection under more than one category. That possibility is even more complex than it appears at first glance because the protections under the different categories may not be symmetrical.

For example, whether state interference with a particular piece of information intrudes upon my privacy would seem to depend, at least in part, on my attitude toward that information or my previous choices about whether to publicize it. In other words, if I publicize certain information widely enough, then interference with that information is not, by definition, an interference with my privacy. By contrast, e-mail, text messages, and instant messages are almost certainly "correspondence" under the covenant, irrespective of whether I intend them to be private, and thus interference with these items could trigger Article 17 even if I had shared those e-mails, texts, or messages with many correspondents. By listing "correspondence" as its own protected category – separately from "privacy" – interference with correspondence could require an assessment of non-arbitrariness and lawfulness by default.[62]

This asymmetry may lead to irregular outcomes when assessing whether an individual has waived the protections of Article 17. Suppose a state takes the position that the interception of an unencrypted e-mail sent by me does not count as an interference with my privacy because the lack of encryption renders the e-mail non-private. Even so, the e-mail could still be protected under Article 17 as correspondence. By contrast, an analogous transaction that does not fall under the secondary protection of the "correspondence," "home," or "family" categories – say, storing certain work information unencrypted in the cloud – might not qualify for Article 17 protection at all. Those results may seem counterintuitive, raising further questions about whether "privacy" as a category should be understood as a catchall designed only to extend the protections of Article 17 to sensitive domains other than

---

[62] The fact that someone has made correspondence public may, of course, bear on the question of whether interference with that correspondence is arbitrary or unlawful.

one's correspondence, family, and home, rather than offering a separate layer of support that overlaps with those enumerated categories. In any event, how to untangle the relationships among these categories is exactly the sort of question this chapter argues is in need of an authoritative answer.

Notwithstanding such questions, however, and based on the covenant's object and purpose, it appears that covenant protections should generally be rounded up rather than down. The main objective of the covenant is described in broad terms in the preamble, which offers a sweeping account of the value of the rights it protects. Under the terms of the covenant, the enumerated rights "derive from the inherent dignity of the human person"; people must be able to enjoy those rights to achieve "the ideal of free human beings ... [living] in freedom from fear and want."[63] Although it can be tempting to link the spread of digital technologies and the rise of social media with a devaluation of privacy and conclude that attitudes toward privacy have shifted with technological advancement, that is neither obviously true nor especially relevant.[64] The object and purpose of the covenant would be undermined by permitting the casual or unintentional waiver of a core right simply because many people use digital technologies insecurely. Indeed, when only a minuscule, elite subset of the population is actually capable of safely maneuvering around technological vulnerabilities – and, even then, imperfectly – the appropriate conclusion is not that everyone else has chosen insecurity, but rather that security is too difficult to attain. Conditioning enjoyment of the right to privacy under the ICCPR on the secure use of digital technologies would render the right meaningless.

Among the implications of this conclusion is that strict application of the US third-party doctrine is likely incompatible with the ICCPR. Digital technologies nearly always involve the provision of information (whether content or metadata, for those who accept the distinction) to a third party. Were any disclosure to a third party enough to eliminate a user's privacy interest, users would lose privacy protections for nearly all of the information they store or transmit in digital form. Even if the doctrine only applied to metadata, it would be unacceptably broad under the

---

[63] ICCPR, preamble.
[64] Numerous opinion polls have been taken since the Snowden revelations, which appear to show that a majority of Americans continue to value privacy in a variety of contexts, even as they are willing to permit certain intrusions for the sake of protecting national security. See, e.g., M. Madden and L. Rainie, "Americans' Attitudes About Privacy, Security and Surveillance," Pew Research Center, www.pewinternet.org/2015/05/20/americans-attitudes-about-privacy-security-and-surveillance/; University of Southern California Annenberg School for Communication and Journalism, "Is online privacy over?," http://annenberg.usc.edu/news/around-usc-annenberg/online-privacy-over-findings-usc-annenberg-center-digital-future-show;
L. Cassani Davis, "How Do Americans Weigh Privacy Versus National Security?," *The Atlantic*, www.theatlantic.com/technology/archive/2016/02/heartland-monitor-privacy-security/459657/; L. Rainie and M. Duggan, "Privacy and Information Sharing," Pew Research Center, www.pewinternet.org/2016/01/14/privacy-and-information-sharing/.

covenant, especially because metadata can be as revealing as content, but is more poorly understood by the public at large (and therefore may be less voluntarily shared).

In the recent Supreme Court case where Justice Sotomayor questioned the wisdom of the third-party doctrine, Justice Samuel Alito contemplated the possible effects of new technology on privacy rights. He wrote:

> Dramatic technological change may lead to periods in which popular expectations are in flux and may ultimately produce significant changes in popular attitudes. New technology may provide increased convenience or security at the expense of privacy, and many people may find the tradeoff worthwhile. And even if the public does not welcome the diminution of privacy that new technology entails, they may eventually reconcile themselves to this development as inevitable.[65]

Depressing as these comments may be for privacy advocates, they are at least comprehensible in the context of a system that accepts the third-party doctrine. But that doctrine is questionable in the digital age, and it may never have taken root under the present circumstances. Many people *do* maintain some sort of subjective expectation of privacy in information they share with a third party. One might properly comprehend that an e-mail provider has access to one's e-mail content without also believing one's e-mails could just as well have been sent to *other* third parties, such as foreign governments searching for intelligence. The same is true for digital banking activity or text messages or any other manner of nonpublic transaction that is only possible with the assistance of a third party. In the abstract, and in the digital age, that expectation is not objectively unreasonable either – at least not obviously so.

Moreover, the third-party doctrine plays out differently under the covenant as compared to the US Constitution. Even beyond the wrinkle identified above with respect to the four types of interests protected by Article 17, there are a number of relevant differences between the rights to privacy guaranteed, respectively, by the covenant and the Fourth Amendment. For one, Article 17 makes no explicit reference to reasonableness or subjective expectations, unlike the Fourth Amendment to the Constitution, which bans, *inter alia*, "unreasonable searches and seizures." Article 17 also applies more broadly than the Fourth Amendment; whereas the Fourth Amendment specifically regulates US government action, Article 17 bans a variety of privacy interferences by government actors and also obliges governments to protect rights holders against interferences from private actors. Incorporating recognition of these points into an applicable waiver standard is essential to ensuring that the protections of Article 17 keep pace with technological change, thereby staying true to the object and purpose of the covenant.

---

[65] *United States v. Jones*, 132 S. Ct. 945, 962 (2012) (Alito J., concurring).

B States Should Support – and Certainly Should Not Interfere with – Active Steps Taken by Individuals to Protect Their Data

Under the ICCPR, states are bound both "to respect and [to] ensure" the rights in the covenant.[66] It has become common to describe those obligations by reference to the "respect, protect, and fulfill" framework, which in the context of privacy essentially captures the idea that states must avoid actively infringing on privacy rights, protect those rights from infringement by others, and take positive steps to support individual realization of privacy rights.[67] Recent discussions of privacy tend to focus on the negative obligations of states – the obligation not to violate privacy by engaging in improper surveillance, for example. But the positive obligations of states must remain part of the conversation as well.

By analogy, consider the human right to health, which is protected by the International Covenant on Economic, Social and Cultural Rights (ICESCR).[68] The ICESCR guarantees "the right of everyone to the enjoyment of the highest attainable standard of physical and mental health."[69] Properly understood, the right is not an entitlement *to be healthy*, but rather a right to have the state take adequate steps to facilitate the health of its population.[70] A state's full compliance with its obligations to ensure the right to health would not prevent members of its population from making poor health choices. Nevertheless, parties to the ICESCR are obligated to undertake measures to promote the health of their people.[71] Those measures might include reasonable efforts by the state to guarantee access to key resources for promoting health, such as by providing meaningful access to adequate medical care and nutrition.[72] They could also include the provision of information that facilitates informed health choices, such as nutrition labels on food packaging or disclosures about the side effects of various medical treatments.[73]

---

[66] See ICCPR, art. 2(1).
[67] See, e.g., U.N. Office of the High Commissioner for Human Rights, "International Human Rights Law," www.ohchr.org/EN/ProfessionalInterest/Pages/InternationalLaw.aspx.
[68] ICESCR, art. 12. The right is also suggested in the Universal Declaration of Human Rights, although somewhat less directly. See UDHR, art. 25.
[69] Note that the United States has not ratified the ICESCR, and therefore the right to health does not exert the same binding force on the US as the right to privacy does.
[70] UN Office of the High Commissioner for Human Rights and World Health Organization, "The Right to Health: Fact Sheet No. 31," at 5, www.ohchr.org/Documents/Publications/Factsheet31.pdf.
[71] States would also have negative obligations with respect to the right to health, such as to refrain from directly undermining the health of their populations (for example, by stripping large segments of the population of health insurance or polluting the drinking water).
[72] See ibid., at 3.
[73] See ibid.

Similarly, a state can properly ensure the right to privacy even as some of its citizens compromise their rights through poor choices about how to handle their own data.[74] The human right to privacy entitles one to protect certain information from invasion not just by one's own government, but by foreign governments, corporations, hackers, identity thieves, and most anyone else. As noted above, governments that have ratified the ICCPR are obligated to protect individuals from such other actors who might intrude on the right, and to facilitate the efforts of individuals who seek out tools to secure their own information.

This obligation to ensure the right to privacy has several implications. It means governments should work with tech companies to repair known software flaws rather than secretly hoarding those exploits and allowing their populations to be rendered vulnerable. It means governments should seek to prevent or discourage the hacking of personal information rather than tolerating or encouraging those hacks. It means governments should offer resources to educate the public on good technological hygiene – for example, managing passwords for their online accounts or securing their mobile devices – rather than making it easier for companies to collect and sell their digital information, such as their web browsing histories. And it means governments should support the development and use of technologies that make it genuinely possible for individuals to secure their information, such as end-to-end encryption for messaging services. At the very least, it certainly means states may not actively prevent people from accessing and using reasonable security and encryption tools.

States seeking to clear the way for aggressive surveillance might prefer simply to push their obligations aside. Consider once more the example of the United States, which has publicly opposed the proposals of technology companies to provide encryption as standard for various services.[75] In particular, law enforcement officials like former FBI Director James Comey have strenuously resisted "end-to-end" encryption as a default for various forms of communication, pressuring companies that offer such encryption to alter their "business model."[76] Similarly, former Director of National Intelligence James Clapper complained that the Snowden revelations accelerated personal use of encryption, which he claimed was "not a good thing."[77] Even President Barack Obama publicly stated opposition to

---

[74] There is something of an asymmetry between the right to privacy and the right to health, in that it may be less likely that a state would seek to limit the latter to advance an alternate interest, like national security.

[75] See J. Comey, "Encryption, Public Safety, and 'Going Dark'," Lawfare, www.lawfareblog.com/encryption-public-safety-and-going-dark; T. Schleifer, "FBI director: We can't yet restrain ISIS on social media," CNN, www.cnn.com/2015/06/18/politics/fbi-social-media-attacks/.

[76] D. Froomkin and J. McLaughlin, "Comey Calls on Tech Companies Offering End-to-End Encryption to Reconsider 'Their Business Model'," The Intercept, https://theintercept.com/2015/12/09/comey-calls-on-tech-companies-offering-end-to-end-encryption-to-reconsider-their-business-model/.

[77] J. McLaughlin, "Spy Chief Complains That Edward Snowden Sped Up Spread of Encryption by 7 Years," The Intercept, https://theintercept.com/2016/04/25/spy-chief-complains-that-edward-snowden-sped-up-spread-of-encryption-by-7-years/.

unbreakable encryption.[78] Moreover, as of 2011, the NSA had a policy of treating encrypted communications collected under one of its surveillance authorities as worthy of special scrutiny,[79] a point of concern for various privacy advocacy groups.[80] Although there were reports of serious debates within the Obama administration on encryption policy,[81] the public criticisms of encryption from prominent officials and the evidence that encrypted communications are viewed with suspicion by the government are discouraging. They are also legally relevant to the state's duty to ensure the right to privacy, for they aim to limit the use, availability, or utility of tools that help individuals secure that right for themselves.

V CONCLUSION

This chapter advocates for a serious conversation about waiving the human right to privacy. Many other elements of the right are appropriately on the table already, being dissected, debated, reinterpreted, and applied to novel circumstances. All essential questions about the right to privacy should be folded into that discussion. Privacy issues will only grow more significant as digital technologies and the surveillance programs that track them become more sophisticated and ubiquitous. It is important that we get these issues right, and now is the time to ensure that we do.

It bears emphasizing that a universally accepted standard for waiving the human right to privacy may prove to be elusive, especially given that states like the United States may contribute to the debate by drawing on excessively broad standards from their own domestic legal precedents. Moreover, an acceptable standard may prove to be challenging to implement in any event, and its creation would in no way diminish the importance of questions already being addressed, such as the definitions of the terms in Article 17 or the proper limitations on the right. Nevertheless, waiver has broad implications for the legality of common state practices; the persistence of questions about its application only sows doubt where clarity is essential.

---

[78] J. McLaughlin, "Obama Wants Nonexistent Middle Ground on Encryption, Warns Against 'Fetishizing Our Phones'," *The Intercept*, https://theintercept.com/2016/03/11/obama-wants-non existent-middle-ground-on-encryption-warns-against-fetishizing-our-phones/.
[79] See "Exhibit B: Minimization Procedures Used by the National Security Agency in Connection with Acquisitions of Foreign Intelligence Information Pursuant to Section 702 of the Foreign Intelligence Surveillance Act of 1978, as Amended," at 9, www.aclu.org/files/assets/ minimization_procedures_used_by_nsa_in_connection_with_fisa_sect_702.pdf.
[80] See, e.g., K. Opsahl and T. Timm, "In Depth Review: New NSA Documents Expose How Americans Can Be Spied on Without a Warrant," Electronic Frontier Foundation, www.eff .org/deeplinks/2013/06/depth-review-new-nsa-documents-expose-how-americans-can-be-spied-without-warrant.
[81] See S. Sorcher and J. Eaton, "What the US government really thinks about encryption," *The Christian Science Monitor*, www.csmonitor.com/World/Passcode/2016/0525/What-the-US-gov ernment-really-thinks-about-encryption.

13

The Future of Human Rights Technology

A Practitioner's View

Enrique Piracés

I INTRODUCTION

Technology has been extraordinarily effective in reducing distances between people and places, but it has created an increasing distance between the present and the future. The rates of new product introduction and adoption are speeding up. It took forty-six years for electricity to reach 25 percent of the US population. The same milestone took thirty-five years for the telephone and only seven for the Internet. For most of us, it is increasingly difficult to understand or anticipate long-term technological trends. It is common, especially in the context of human rights practice, that such inability stokes fears of a dystopian future in which ordinary people, especially those already marginalized or disenfranchised, become subjugated by technology rather than benefiting from it. This chapter is both an attempt to help practitioners cope with new technologies and a proposal to incorporate solidarity as the driving force for technology transfer.

It has become cliché to say that technology and its impact on society advance at a rapid pace. It is also commonplace to say that societies and legal frameworks have a hard time adapting to technology's pace and the behavioral changes it demands. But adaptation is a valuable goal, because there is no livable future without it. The human rights movement has taken note and, both systematically and spontaneously, looked for ways to adapt to the transformative era of the information society. Today, human rights campaigns rely heavily on social media and e-mail. The presentation of research results in courts, political offices, and public spaces commonly incorporates data visualization. Fact-finding practices often include the use of remote sensing and open source intelligence. Further, human rights research increasingly relies on computational analysis. Encrypted communications, and the tools and services that provide them, are now considered fundamental to the safety of human rights practitioners and their partners in the community. These are signs that, as the contributors to this volume remind us, the future of human rights will be intertwined with the advancement of technology.

The pace of technological change is unlikely to slow, and its relevance for human rights practice is unlikely to diminish. There is a valuable body of work, created over the past few decades, that focuses attention on the impact of technology on human rights. The lessons that we can extract from that literature will enrich our design for the future as well as our ability to evaluate the present.[1] Yet, as Molly Land and Jay Aronson point out in Chapter 1, the field of human rights technology is significantly undertheorized. I would add that the relationship between practice and theory has garnered even less attention. The contributors to this volume have gone a long way to redressing the first issue, especially with respect to human rights law. If we are to solve the second challenge, however, practitioners must help frame the debate in this interdisciplinary field. Doing so is essential to the advancement of effective human rights practice.

II WHERE DOES THE FUTURE BEGIN?

Over the past ten years, the notion of human rights technology as an area of practice has garnered attention across disciplines. The growing use of the term "human rights technology" signals the interest of technical, scientific, and practitioner communities in advancing it as a field of practice. An important example, and one of the likely origins of this multidisciplinary interest, occurred in 2009, when the Human Rights Center at the University of California, Berkeley called for "leading thinkers, civil society members, activists, programmers, and entrepreneurs to imagine, discover, share, solve, connect, and act together." This invitation materialized as an international conference, "The Soul of the New Machine: Human Rights, Technology & New Media,"[2] held in May 2009, and a follow-up conference, "Advancing the New Machine: A Conference on Human Rights and Technology,"[3] held in 2011, both in Berkeley. A diverse mix of academics, practitioners, and technologists attended those events, which launched a constructive debate about the uses of technology for human rights practice.

Since then, a growing number of efforts to create dialogue, promote debate, and engage technologists with rights defenders have emerged across the globe. Strategic donors to the human rights movement, like the MacArthur Foundation, the Ford Foundation, the Oak Foundation, Humanity United, and the Open Society Foundations, amplified these efforts. These foundations adapted their portfolios to help create the human rights technology field. Governments have also played a role,

---

[1] C. Weeramantry, *The Impact of Technology on Human Rights: Global Case-Studies* (Tokyo: United Nations University Press, 1993); J. Metzl, "Information Technology and Human Rights" (1996) 18(4) *Human Rights Quarterly* 705–46; R. Jørgensen et al., "ICT and Human Rights" (FRAME Deliverable No. 2.3, 2015).

[2] "Soul of the New Machine," UC Berkeley School of Law, www.law.berkeley.edu/research/human-rights-center/past-projects/technology-projects/soul-of-the-new-machine/.

[3] "Advancing the New Machine: A Conference on Human Rights and Technology," UC Berkeley School of Law, www.law.berkeley.edu/research/human-rights-center/past-projects/technology-projects/advancing-the-new-machine-a-conference-on-human-rights-and-technology/.

as can be seen in the programming of the Bureau of Democracy, Human Rights, and Labor at the US State Department,[4] the Open Technology Fund[5] of Radio Free Asia (an initiative of the US Broadcasting Board of Governors), and the Swedish International Development Agency.[6]

By now, there are dozens of international, regional, and national conferences and workshops each year that include debates on the use of technology for human rights.[7] Many organizations, like Benetech, HURIDOCS, and eQualit.ie, have carved a niche providing specialized technology and support to human rights practitioners. The growing interest can also be seen in the appearance of specialized and globally distributed communities of practice around issues of technology and human rights, such as the Internet Freedom Festival,[8] held yearly in Valencia, Spain, since 2015. This interest in technology has also reached traditional international actors like Amnesty International and Human Rights Watch, which have pioneered specialized programs within their organizations to address their remote sensing, data analysis, and digital security needs.[9] These examples are evidence of the growing and vibrant ecosystem interested in applying technology to solve human rights problems.

In order to frame how we think about the future of this field, it is essential to be aware of our own geopolitical and cultural positions. Human rights technology has not escaped some of the persistent problems that have faced the broader human rights movement. The most obvious, perhaps, has been the tendency to consolidate power in the economic capitals of the twenty-first century, geographically removed from most human rights crises. This can be acutely felt in the realm of technology, where investment in infrastructure can be too costly for grassroots organizations in the Global South. Current models of technology transfer reflect a unidirectional relationship, where technology is largely decided, designed, and created far away from the majority of people who need it. As Dalindyebo Shabalala reminds us in Chapter 3, funding and enforcement mechanisms for providing access to technology remain a challenge for effective technology transfer in international cooperation for adaptation to climate change.

---

[4] "Internet Freedom Funding Opportunity: State Department's Bureau of Democracy, Human Rights, and Labor (DRL)," Open Technology Fund, www.opentech.fund/article/internet-freedom-funding-opportunity-state-departments-bureau-democracy-human-rights-and.
[5] "About the program," Open Technology Fund, www.opentech.fund/about/program.
[6] "The Access Grants Program – an emerging initiative," Access Now, June 25, 2015, www.accessnow.org/the-access-grants-program-an-emerging-initiative/.
[7] "RightsCon Summit Series, www.rightscon.org/about-and-contact/; Y. Ulman, *Report on the International Conference on "Emerging Technologies and Human Rights"* Council of Europe Bioethics Committee, DH-BIO, Strasbourg, 4–5 May 2015 (December 2015).
[8] "History, Goals and Guiding Principles," Internet Freedom Festival, https://internetfreedomfestival.org/history/.
[9] "Remote Sensing for Human Rights," Amnesty International USA, www.amnestyusa.org/research/science-for-human-rights/remote-sensing-for-human-rights.

For human rights practice – understood as fact-finding, advocacy, and litigation toward accountability, transparency, and justice – the fundamental problems with technology transfer are not limited to funding, but also include decision-making and design. Most technology is designed in places like the United States and the United Kingdom for practitioners and activists in the Global South, but generally without their involvement or input. A concerning example of this can be seen in Google's Jigsaw project. Previously known as Google Ideas, it was re-launched in 2016 with the goal of "investing in and building technology to expand access to information for the world's most vulnerable populations."[10] Although this project may have been created in part out of genuine and bona fide good intentions, it is in reality an example of the kind of power-consolidating technology transfer that could harm the development of a sustainable and fair human rights technology ecosystem. As the technology law and policy scholar Julia Powles argues, human development and human rights are too complex and too culturally diverse to be addressed by profit-driven companies acting on their own initiative.[11] More to the point, as Rikke Frank Jørgensen points out in Chapter 11, the debate on binding human rights obligations upon companies has been ongoing for more than two decades, and the private sector has continued to be largely resistant to human rights frameworks.

The effect of this type of model – in which technology is designed for, but not with, practitioners – is twofold. First, it makes it more likely that a given technological "solution" will address a false dilemma, because there is little consideration of the context in which a particular technology will be deployed, what it may be displacing, and what social or cultural practices it may be enhancing or altering. Understanding the cultural impact of technology transfer is paramount, as technology is by nature disruptive. It would be naive, and potentially detrimental to the advancement of human rights, to think that the effects can be controlled and isolated to a particular issue. Designing technology without the stakeholders at the table could also mean a lost opportunity to learn from other approaches to problem solving, thus limiting the types of solutions that can be imagined.

Second, this model can lead to investments that are unsustainable on the ground. The yearly budget for a software developer in the Global North may be equivalent, for example, to the annual budget of a small organization that provides direct support to hundreds of migrants at the border between Mexico and Guatemala. Should we create expensive technology in their name from our comfortable seats in London, New York, or Palo Alto? Or should we bring them to the table to design a sustainable solution that recognizes their agency and goals? Should we even rely on for-profit companies to tackle complex

---

[10] E. Schmidt, "Google Ideas Becomes Jigsaw," Jigsaw, February 16, 2016, https://medium.com/jigsaw/google-ideas-becomes-jigsaw-bcb5bd08c423.

[11] J. Powles, "Google's Jigsaw project has new ideas, but an old imperial mindset," *The Guardian*, February 18, 2016, www.theguardian.com/technology/2016/feb/18/google-alphabet-jigsaw-geopolitical-games-technology.

geopolitical and cultural issues of global significance? Or should we create an open and distributed ecosystem that acts in the public interest?

When we think of the future, we must keep the sustainability of the human rights movement front and center. We need to guard against technology transfer creating dependence, exporting inequalities, or promoting a paternalistic relation between technology providers and human rights practitioners. The current approach to technology is instead largely based on the model of international cooperation for development, which Shabalala shows in Chapter 3 to be deficient on many levels. While his analysis focuses on new frameworks for organizing technology transfer at the government level, I wish to focus on efforts within the human rights community itself. In human rights practice, we can create better conditions for technology to effectively advance accountability, transparency, and justice if we move away from a technocratic approach and embrace the idea of transnational solidarity. International aid, like charity, is based on an asymmetrical relationship between a party in need and another party with resources or knowledge to share.[12] Relationships of that nature are prone to creating clientelism, dependency, and unidirectional knowledge transfer. A core motivation of this chapter is to suggest a solidarity-based framework as an alternative approach to technology transfer. A first step in that direction is for practitioners to educate themselves about the technology that will be the subject of that transfer.

III WHAT IS HUMAN RIGHTS TECHNOLOGY?

Human rights practitioners frequently work in under-resourced, high-pressure environments. They tend to use opportunistic and adaptive approaches to problem solving. Because of the financial constraints that most human rights practitioners face, few technologies have been developed specifically for human rights practice. Instead, practitioners have adapted the majority of tools they use in the field from existing technologies. There are a small number of exceptions, composed largely of software projects around information management or communications. This includes projects like Martus[13] and OpenEvsys,[14] which were created specifically for human rights documentation, and privacy-enhancing mobile apps like those created by the Guardian Project. It also includes projects like PGP encryption and the Tor Internet browser, which were created by forward-thinking individuals who understood very early on in the information era that privacy and anonymity were instrumental to human rights.

Beyond these examples, the vast majority of technologies used in human rights practice are based on creative or opportunistic adaptations of general-purpose

---

[12] B. Prainsack and A. Buyx, "Thinking Ethical and Regulatory Frameworks in Medicine from the Perspective of Solidarity on Both Sides of the Atlantic" (2016) 37(6) *Theoretical Medicine and Bioethics* 489–501.
[13] "Overview," Martus, https://martus.org/overview.html.
[14] "About OpenEvsys," OpenEvsys, http://openevsys.org/about-openevsys/.

technologies. Today, practitioners rely on WhatsApp and Telegram to communicate with their peers or the subjects of their work; WordPress or Drupal to promote their ideas; Dropbox or Google Drive to manage their files; Google Apps or G Suite to collaborate on documents; and Skype to engage in meetings and interviews.

A significant difference between the few examples of purpose-built human rights technology and the general-purpose technology adopted and adapted by practitioners is the nature of the software behind them. Those solutions that have been created for human rights-specific purposes are largely open source. This means that the developers made the code they used to build the technology publicly available for anyone to review and tinker with. The only requirement for those who make changes or additions to open source software is that they, in turn, allow others to freely use and modify their contributions.

The foundations and donors that support the human rights movement acted as positive agents of change in promoting the use of open source software. Nearly a decade ago, they began to request that the technology created with their support be designed as open and available to others. This is key for sustainability and replication, and quite likely allows donors to maximize the impact of their portfolios. This openness, especially if expanded beyond software, will be pivotal for the inclusion of Global South and grassroots organizations in the design, adoption, and evaluation of solutions that are tailored for them. Open source software is not necessarily cheaper to develop, but it is often available with few licensing and use restrictions. It also reduces dependency and promotes collaboration among distributed and culturally diverse communities.

An important consideration when thinking about technology is the fact that the same type of adaptation that human rights practitioners can make to advance accountability, transparency, and justice could be made by other actors – from governments and corporations to organized criminals and non-state actors. In that sense, most technologies could have dual or multiple uses, including for abuse and repression of human rights. For that reason, and as Lea Shaver concludes in Chapter 2, it is critical that human rights practitioners find avenues to exercise scrutiny and oversight over technological developments in order to minimize harm.

Finally, we must consider what type of technology we should be prepared to confront in the future. What most practitioners assume fits under "human rights technology" lies within the realm of information and communication technologies, or ICTs. But the uses of technology in the human rights context already go beyond this domain. Contemporary examples of this include the use of remote sensing by international organizations to find incidents of violence[15] or cultural heritage destruction,[16] the growing interest in unmanned aerial vehicles (UAVs, or drones)

---

[15] A. Marx and S. Goward, "Remote Sensing in Human Rights and International Humanitarian Law Monitoring: Concepts and Methods," (2013) 103(1) *Geographical Review* 100–11.
[16] "Case Against M. Al Mahdi," International Criminal Court, http://icc-mali.situplatform.com/.

to access unreachable areas,[17] and the use of DNA technology by forensic anthropologists to uncover evidence of mass atrocities.[18]

IV WHAT TECHNOLOGICAL TRENDS COULD SHAPE THE FUTURE OF HUMAN RIGHTS PRACTICE?

Popular culture plays an important role in shaping the way that human rights practitioners think about technology. We tend to be very generic when discussing the effects of technology in society. For example, it is common to see contemporary issues framed as "the impact of social media" on relationships or "the effect of mobile technology" on the economy, rather than on how companies, governments, communities, and individuals have integrated technology into our lives and societies. Thinking of technology as an entity divorced from human action is an inadequate starting point for discussing the future of human rights technology. If we were to follow that line of abstraction, we would risk ending up with a teleological framing of technology that authors like Kevin Kelly have proposed.[19] For Kelly, there is a super-organism of technology, a "technium," in the global interconnected system of technology that is "partly indigenous to the physics of technology itself." To think of the future of human rights technology, we need to avoid that path. Humans have created technology, and humans have used technology to alter society. We should avoid giving agency to technology and remind ourselves constantly that technology is created by people and organizations with agendas. These are agendas that will impact us, and we should aim to influence them.

To effectively shape these agendas, practitioners need a better and more specific understanding of the trends that will shape the future of human rights technology. In digital security, for example, we can expect an expanded use of technology, including end-to-end encryption, a system of communication in which encryption ensures that only the intended recipient can read the message; multifactor authentication, a method of computer access control in which a user is granted access only after successfully presenting several separate pieces of evidence to an authentication mechanism; and zero-knowledge encryption, a process that prevents a service provider from knowing anything about the user data that it is storing or transmitting.

In issues related to research and fact-finding, we can expect an increased use of UAVs, or drones, resulting in an increased availability of aerial images for

---

[17] D. Whetham, "Drones to Protect," (2015) 19(2) *The International Journal of Human Rights* 199–210.
[18] M. Doretti and C. Snow, "Forensic Anthropology and Human Rights," in D. Steadman (ed.), *Hard Evidence: Case Studies in Forensic Anthropology* (Upper Saddle River, NJ: Prentice Hall, 2003) pp. 290–310; S. Wagner, *To Know Where He Lies: DNA Technology and the Search for Srebrenica's Missing* (Oakland: University of California Press, 2008); A. Rosenblatt, *Digging for the Disappeared* (Redwood City, CA: Stanford University Press, 2015), p. 1.
[19] K. Kelly, *What Technology Wants* (New York: Penguin, 2010).

documentation of human rights and humanitarian situations[20]; an expanded use of remote sensing and satellite imagery, which has become less expensive and more available as more firms enter the market and satellite technology improves[21]; and an increased use of open source intelligence, knowledge produced from publicly available information that is collected, exploited, and disseminated in a timely manner to an appropriate audience for the purpose of addressing a specific investigative requirement.[22]

In the case of advocacy, we are likely to see an expanded use of complex visualization to support the narrative of human rights accountability efforts. The work of SITU Research, an organization working in design, visualization, and spatial analysis to facilitate the analysis and presentation of evidence documenting the destruction of sites of cultural heritage in Timbuktu, Mali, is an excellent example. Created in collaboration with the International Criminal Court's Office of the Prosecutor, SITU Research built a platform that combines geospatial information, historical satellite imagery, photographs, open source videos, and other forms of site documentation. The Office of the Prosecutor used SITU's tool successfully at the trial proceedings at the International Criminal Court in 2016.[23] This work is part an emergent field called forensic architecture, first developed at Goldsmiths College, University of London.[24] It refers to "the practice of treating common elements of our built environment as entry points through which to interrogate the present."[25]

The continued development of areas and projects like these will also be accompanied by new efforts in areas where technological trends are moving rapidly. While not exclusive, concepts like artificial intelligence, blockchain, sensors, open source hardware, and the Internet of Things reflect areas that are likely to offer fertile ground for the development of human rights technology and applications.

A *Artificial Intelligence*

Perhaps nothing embodies our fascination with and fear of technology better than artificial intelligence (AI). There are countless images in popular culture that

---

[20] K. Kakaes et al., *Drones and Aerial Observation: New Technologies for Property Rights, Human Rights, and Global Development: A Primer* (Washington, DC: New America, 2015).

[21] J. Kumagai, "9 Earth-Imaging Start-Ups to Watch," *IEEE Spectrum*, March 28, 2014, http://spectrum.ieee.org/aerospace/satellites/9-earthimaging-startups-to-watch.

[22] E. Higgins, "A New Age of Open Source Investigation: International Examples," in B. Akhgar et al. (eds.), *Open Source Intelligence Investigation* (New York: Springer International Publishing, 2016) pp. 189–196.

[23] See Kelly, *What Technology Wants*.

[24] E. Weizman, "Forensic Architecture: Violence at the Threshold of Detectability" (2015) 54(4) *E-flux Journal* 1–17.

[25] Y. Bois et al., "On Forensic Architecture: A Conversation with Eyal Weizman" (2016) 156 October 115–40.

evidence this, and while the reality is different than the anthropomorphic version of what we see on the big screen, AI is no less fascinating in reality.

AI is premised on the notion that "every aspect of learning or any other feature of intelligence can in principle be so precisely described that a machine can be made to simulate it."[26] Scientists have been working to make this dream a reality for several decades, but a critical milestone, the equivalent of the "man-on-the-moon moment," happened in late 2015 when AlphaGo, a computer program developed by Deep Mind, a UK company recently acquired by Google, was able to defeat the best human player in the world at the ancient game of Go.[27] The game of Go, which was invented in China thousands of years ago, has a number of possible legal moves larger than the number of atoms in the observable universe. It is this complexity that made it a sizable test for artificial intelligence.

Generally speaking, AI is divided into weak AI and strong AI. Most artificial intelligence applications so far are considered either an expert system (ES) or a knowledge-based system (KBS), which means that they rely on an existing model or corpus of knowledge. This is, in a way, the application of existing knowledge to assess the best answer to a question or problem. This form of AI is generally referred to as "weak AI" because it requires *a priori* knowledge to arrive at the answer to a question. "Strong AI," on the other hand, generally refers to the ability of a machine to perform "general intelligent action," which is why it is also referred to as artificial general intelligence. In the case of the AlphaGo scenario, this meant that instead of evaluating all possible moves to calculate all possible outcomes like an ES or KBS would do, AlphaGo thought and made decisions like a human. The extraordinary achievement of AlphaGo is that it is not an expert system, but rather relies on artificial general intelligence. In other words, it *learned* to play Go rather than being fed many possibilities and choosing the one that best fit a particular scenario. This is generally accepted as evidence that AI has reached a tipping point much sooner that most scientists thought it would.

How can all this be of use for human rights practice? Can a machine teach itself to solve human rights problems? Will this be an opportunity or a challenge for human rights practice? In thinking of the future, I would argue that it is more likely that human rights practice will first benefit from advances in specific areas of AI research like machine learning, computer vision, and natural language processing, not in automated decision-making. These advances will improve the ability of human rights researchers to discover, translate, and analyze relevant information.

To get a sense of what may be possible, we can look at some recent experimental uses of AI for human rights issues. Researchers at the University of Sheffield and the

---

[26] John McCarthy et al. "A Proposal for the Dartmouth Summer Research Project on Artificial Intelligence, August 31, 1955" (2006) 27(4) *AI Magazine* 12.
[27] C. Moyer, "How Google's AlphaGo Beat a Go World Champion," *The Atlantic*, March 28, 2016.

University of Pennsylvania have used AI to develop a method for accurately predicting the results of judicial decisions of the European Court of Human Rights. The research team identified 584 cases relating to three articles of the European Convention on Human Rights: Article 3, concerning torture and inhuman and degrading treatment; Article 6, which protects the right to a fair trial; and Article 8, on the right to respect for a private and family life. After running their machine learning algorithm against this dataset to find patterns in the text, the team was able to predict the verdicts at an accuracy of 79 percent. What this suggests is that AI could be used to build predictive models to discover patterns in judicial decisions. This approach could help increase the success and effectiveness of litigation in defense of human rights by assisting advocates and lawyers in planning their litigation strategy.

Another example of the potential use of AI to advance human rights practice can be found in the work of the Center for Human Rights Science (CHRS) at Carnegie Mellon University.[28] After hearing of the challenges that human rights organizations were facing in analyzing and verifying the large volume of online videos regarding human rights abuses, researchers at the CHRS began to experiment with AI applications to solve these problems. With the goal of creating efficient and manageable workflows for human rights practitioners, they have created computer vision and machine learning methods to rapidly process and analyze large amounts of video. Their tools help human rights practitioners detect audio like explosions, gunshots, or screaming in video collections; detect and count the number of people in a given frame of a video; aid in geolocation of a video; and synchronize multiple videos taken by different sources at the same time and place to create a composite view of an incident.

But perhaps the most sophisticated use of AI applied to human rights that we can find is in the center's Event Labeling through Analytic Media Processing (E-LAMP) system.[29] E-LAMP is a machine learning and computer vision–based video analysis system that is able to detect objects, sounds, speech, text, and event types (say, a news broadcast or a protest) in a video collection. In practice, this allows users to run semantic queries within video collections. If the system is properly trained, a user could ask it, for example, to find images of individuals performing a specific action or objects of a particular kind in a collection of thousands of videos. This means that practitioners can use a system that can search thousands or even millions of videos to answer questions like: How many videos show helicopters dropping things (e.g., barrel bombs or bodies)? How many videos may be communiques from a faction within a conflict? What are the commonalities among a group of videos? These search efforts can be done in a fraction of the time that it would take for a human

---

[28] The author is program manager and co-founder of the Technology Program at the Center for Human Rights Science, Carnegie Mellon University.

[29] Jay D. Aronson, Shicheng Xu, and Alex Hauptmann, "Video analytics for conflict monitoring and human rights documentation" (2015).

analyst to perform the same task. AI projects like E-LAMP will make practitioners more effective by allowing small teams to quickly examine and analyze large amounts of evidence. While systems like this could become valuable automated research assistants that aid in the process of knowledge discovery, they will remain instruments for human domain experts. E-LAMP cannot yet find all actions that are relevant for a case, for example, torture or physical abuse, but it is able to find potential markers for those actions that could then be reviewed by a practitioner.

The big opportunity for human rights practice lies in the extraordinary potential that artificial intelligence has to support problem solving, pattern detection, and knowledge discovery. But this kind of capability will not simply materialize from thin air. There is a time-bound opportunity for practitioners to influence artificial intelligence before it completely leaves its infancy. Legal experts could provide important guidance as to how, ethically, AI's findings could be verified in courts, how AI may shape the definition of legal personhood, and how data being analyzed in the cloud can be protected from exposure to nefarious actors. For this, human rights practitioners need to engage early and often with the technologists and organizations that are driving the technological future of AI.

B *Blockchain*

In 2008, a person or group of persons under the pseudonym Satoshi Nakamoto published a paper proposing Bitcoin, a peer-to-peer electronic currency aimed at supporting transactions without a central financial institution.[30] Since then, Bitcoin has drawn attention from a wide variety of actors and entities, ranging from banks and regulators to organized criminals and futurists. Looking back, it is not hard to see why it is considered a potential disrupter of national, regional, and international financial systems. It took only two years from its formal launch in 2009 for this revolutionary virtual currency to achieve parity with the US dollar.[31] And it took only a few additional years to reach an all-time-high $1,216.73 exchange rate.[32] Surprisingly, all of this happened with a decentralized, public, and open infrastructure.

But beyond its disruptive capacity and its direct challenge to institutions that reproduce and maintain inequalities, like banks and international financial regulators, there are other aspects of Bitcoin that could advance the future of transparency and accountability. Its potentially transformative power for human rights practice is anchored in the innovative design of the technology underneath the currency that facilitates public trust without the need for a third party controlling the currency. This technology is commonly referred to as "blockchain."

---

[30] S. Nakamoto, "Bitcoin: A Peer-to-Peer Electronic Cash System," https://bitcoin.org/bitcoin.pdf.
[31] L. Literak, "Bitcoin dosáhl parity s dolarem," *AbcLinuxu*, February 22, 2014, www.abclinuxu.cz/zpravicky/bitcoin-dosahl-parity-s-dolarem.
[32] "History of bitcoin," *Wikipedia*, https://en.wikipedia.org/wiki/History_of_bitcoin.

Blockchain refers to a distributed network of computers in which digital transactions are recorded in a public database using cryptography to digitally sign them and connect them to previous transactions. This process creates a chain of grouped transactions, or blocks, that cannot be tampered with or altered. One way to think of this is as if everyone in a network of peers acted as a digital notary. In this network, transactions are notarized by multiple notaries, and notaries publicly broadcast the existence of a record by linking it to an existing and already notarized transaction or document in a public ledger. Among the most interesting attributes of such a system is the fact that trust is not placed in the nodes, but rather in the strength and openness of the network and the science behind the protocol for transactions.

Outside of currency exchange, people can access bitcoins not by labor, but rather by computation. The currency is ephemeral and is not backed by gold or any other representation in the physical space. Its creation is the result of software and hardware computations that have to solve increasingly complex mathematical operations. Once a solution is found, bitcoins are the reward. This process is called "mining." Each one is awarded to the person behind the computation using a unique identifier, also the result of computation, that the user obtains when installing the mining software. Such a key is also referred as a wallet, and the wallet is where the awarded (or purchased) bitcoins are stored. In other words, besides exchanging them directly, as a person could do with any foreign currency, the only way to get them is by solving computational problems.

Blockchain has several human rights applications. It could be used to certify that a video, image, or other type of digital document existed at a given time. This attribute, normally referred as proof of existence, increases the evidentiary weight of a digital asset, like a video or image of human rights abuse that appeared in social media, by increasing the ability of investigators to validate or reject claims of authenticity over the material and map its chain of custody. Preliminary uses of this technology can be seen in projects like Video Vault,[33] a system that I created and maintain, which allows human rights practitioners to preserve digital resources of any kind for later reference. Video Vault facilitates the verification of digital assets by providing an online content sample with a trusted time stamp[34] reflecting the collection time. This time stamp is added as a transaction to the blockchain, where it can be accessed to validate that such asset, picture, video, or web page existed at a particular point in time. Digital assets collected by an individual or organization can in this way be "notarized" and added to the blockchain to create a public ledger, to enhance the verification of media that may contain evidence of human rights abuses.

It is also possible to imagine applications for blockchain technology in other areas of social activity that relate to human rights practice. One example is trade and the

---

[33] Video Vault, www.bravenewtech.org/.
[34] A trusted time stamp is a form of proof of existence that relies on a trusted third party to create and maintain a hash of a file to certify that a particular asset existed at a given time. A hash is a unique alphanumeric string created from the digital file that is time-stamped using cryptography and allows tracking of the creation and modification of a file.

distributed manufacturing or production of goods. Technology like blockchain could be used to create a chain of trust or custody around specific steps of manufacturing, thus increasing the ability to monitor the life cycles of the goods we consume. Such a system could, at least in theory, enhance the ability of agencies, unions, regulators, and civil society to enforce compliance with laws and guidelines that defend the rights of workers, indigenous people, and the environment, to name a few.

This traceability feature is already part of the offerings of companies like Provenance[35] to food producers and supply chain watchdogs. What is learned from this process could benefit its implementation in fact-finding in human rights practice. Provenance is a UK-based company using currencies like Bitcoin and Ethereum, which are implementations of blockchain, to create a public record of the supply chain from the origin of a product to its end consumer. This technology could help consumers learn where their clothes were made or where the fish they are thinking about purchasing for dinner was netted. Perhaps more importantly, it could help consumers understand the environmental and labor conditions where their goods were produced or obtained.

As is often the case with new technologies, a group of forward-looking technologists and entrepreneurs have proposed other creative applications for blockchain, including to increase transparency and reduce corruption in public spending by governments or the use of charitable funds; to create efficient ways to transfer currency to support basic rights, like access to health care and food security, when traditional financial institutions fail in the context of humanitarian crisis; to create alternative and inclusive systems for land registration for migrants; or to provide access to identities in order to prevent discrimination of ex-convicts.

A recently formed e-governance consultancy called Humanitarian Blockchain[36] is attempting to make some of these ideas a reality. Because of its distributed and open nature as well as its reliance on sound mathematical concepts, blockchain is resistant to manipulation. It does not matter if a large government or a local paramilitary organization disagrees with what it carries. Because of its distributed nature, the public ledger will remain unmodified and available to its users. Recently, researchers from the Massachusetts Institute of Technology and Tel Aviv University proposed a decentralized personal data-management system that would ensure that users own and control their data. Such a system would enhance the privacy of sensitive data, including that of human rights practitioners.[37]

Today, blockchain-based systems may be complicated to access and understand by grassroots organizations, but this will rapidly change. It is likely that at this pace,

---

[35] T. Levitt, "Blockchain technology trialled to tackle slavery in the fishing industry," *The Guardian*, September 7, 2016, www.theguardian.com/sustainable-business/2016/sep/07/blockchain-fish-slavery-free-seafood-sustainable-technology.

[36] Humanitarian Blockchain, Facebook, www.facebook.com/HumanitarianBlockchain.

[37] G. Zyskind and O. Nathan, "Decentralizing Privacy: Using Blockchain to Protect Personal Data," in *2015 IEEE Security and Privacy Workshops (SPW)*, Washington, DC, May 21–22, 2015, pp. 180–84.

just as it is beginning to happen with encryption and security mechanisms like Secure Sockets Layer/Transport Layer Security (SSL/TLS), which is used to secure most transactions over the Internet, and end-to-end encryption, which is used to secure communications on tools like WhatsApp and Signal, the benefits of this technology will soon be available in seamless, low-cost ways for practitioners of all kinds.

C Open Hardware, Affordable Sensors, and the Internet of Things

For many years, human rights technology has been limited to software. Software can be written on virtually any computer. There are also numerous well-documented programming languages that, with some patience and basic literacy, anyone can learn. Furthermore, there is no need for a project to start from scratch, because with the growth of open source and free software, many libraries and code bases can help anyone jump-start a project. Such availability and simplicity were, without a doubt, key to the explosion of software products for many disciplines, including human rights practice.

Over the past decade, slowly but incrementally, hardware has followed suit. Similarly to software, the advent of open source hardware has created a vast arena for experimentation and has expanded the toolkit for problem solving that practitioners can access. In 2003, Hernando Barragán, a master's student at the Interaction Design Institute Ivrea in Italy, created Wiring as part of his thesis project. Wiring was aimed at lowering the barrier to accessing prototyping tools for those interested in developing electronics. It consists of the complete tool set needed to develop functional electronic prototypes, from an integrated development environment (IDE) and a simple programming language for microcontrollers to a bootloader to update programs and a well-documented online documentation library. In a controversial move, Barragán's thesis advisors and fellow students copied the project to create Arduino in 2005. Arduino rapidly became the platform of choice for a new generation of open source hardware tinkerers. By 2013, there were 700,000 Arduino devices registered and at least an equal number of clones or copies. The number of prototyping platforms grew, and there are now dozens of different boards and platforms to choose from. The projects enabled by this new generation of hardware range from simple LED controlling projects to sophisticated motor control and sensor management devices. Some of these projects illustrate what we could see at the intersection of human rights practice and open hardware in the near future, especially as issues of environmental justice are increasingly rooted in human rights, including but not limited to nuclear disasters, oil spills, and water safety.

An important example of environmental justice-oriented open hardware involved the creation of sensors to measure radioactivity in the aftermath of the March 11, 2011 earthquake and destructive tsunami that severely damaged the Daiichi nuclear power plant in Fukushima, Japan. The radiation leak that occurred at the power plant was followed by panic and misinformation. Citizens with enough money acquired Geiger counters to measure the scale of the catastrophe, both for personal safety and for the eventual accountability of officials whom they felt were not appropriately responding to the crisis. These devices, which are designed to measure ionizing radiation, became a critical source of reliable information for the affected population. During the early response, the supply of these devices began to decline and prices became too high for many citizens to purchase them. A group of developers, activists, and responders held Skype discussions to brainstorm a possible solution. After a few days, this group met in person at Tokyo Hackerspace. Within a week, they had created the first bGeigie, a DIY Geiger counter that could increase access to reliable data, and they set off for Fukushima. Today, that project has evolved into Safecast, founded by Sean Bonner, Joi Ito, and Pieter Franken as an international, volunteer-centered organization devoted to open citizen science for the environment.[38] A similar story is that of the Public Laboratory for Open Technology, founded in the wake of the April 2010 Deepwater Horizon oil spill in the Gulf of Mexico on the BP-operated Macondo Prospect. During the spill, there was an information blackout for residents of the region. In response, a group of concerned residents, environmental advocates, designers, and social scientists launched DIY kite and balloon aerial photography kits over the spill to collect real-time data about its impact.[39] The success of the mapping effort encouraged the group to found Public Lab as a research and social space for the development of low-cost tools for community-based environmental monitoring and assessment. Among the tools that Public Lab offers is a Desktop Spectrometry Kit, which puts a low-cost, easy-to-use spectroscope or spectrophotometer in the hands of any individual or organization interested in collecting spectra, which are the electromagnetic "fingerprints," or unique identifiers, of materials.[40]

The above examples comprise a small sample of the vibrant community around microcontrollers, sensors, and citizen science. They can help us imagine how the availability of easy-to-use and low-cost sensors and measurement kits may have a transformative effect in the future of human rights. Could we measure the fingerprint of a tear gas canister with sufficient accuracy to point to its origin? Could we allow for communities to directly and reliably collect information about the quality of the water before and after an extractive industry development? Could we take samples with remote equipment of chemical agents used against vulnerable populations? Human rights practitioners need to engage with the vibrant open source hardware community to find answers to questions like this. While the above uses

---

[38] M. Prosser, "How a Crowd Science Geiger Counter Cast Light on The Fukushima Radioactive Fallout Mystery," *Forbes*, March 10, 2016, www.forbes.com/sites/prossermarc/2016/03/10/how-a-crowd-science-geiger-counter-cast-light-on-the-fukushima-radioactive-fallout-mystery/.
[39] Public Lab contributors, "Public Lab: Gulf Coast," https://publiclab.org/wiki/gulf-coast.
[40] Public Lab contributors, "Public Lab: Desktop Spectrometry Kit," https://publiclab.org/wiki/dsk.

may not yet seem related to traditional human rights work, this may change rapidly as environmental issues, like those related to extractive industries or access to water, permeate human rights practice. More importantly, technologies like those discussed above are aligned with the type of technology transfer that Dalindyebo Shabalala calls for in Chapter 3, both because they enable low-cost and broad access, and because they can contribute to the creation of complex monitoring ecosystems that could inform future human rights frameworks.

Hardware is not only sensors and microcontrollers. Over the past ten years, there have been efforts to reduce the cost of, and increase access to, computers. Perhaps the most known example of this is Raspberry Pi. Raspberry Pi is a series of credit card–sized single-board computers developed in the United Kingdom by the Raspberry Pi Foundation to promote the teaching of basic computer science in schools and developing countries.[41] More importantly, it is open source and available anywhere in the world for under $50. The advent of this device has created a great deal of excitement among developers and technologists, as the processing power and the possibilities are immense when compared to a microcontroller like Arduino. This excitement can be seen in its adoption. Since the launch of its first model, the Raspberry Pi 1 Model B, in February 2012, more than ten million have been sold.[42] Enthusiasts and developers have started to create potentially relevant projects for human rights practice. For example, developers have used Raspberry Pi computers to create specialized routers that increase the anonymity of their users. Others have created advanced remote sensor units that can automatically consume data and broadcast it in real time.

The Novena laptop, launched in 2014, was designed for users who care about free software and open source, or who want to modify and extend their hardware. Its creator, Andrew "bunnie" Huang, promoted it as "a laptop with no secrets."[43] It is this claim that makes the Novena interesting for the future of human rights practice. A laptop with nothing but modifiable and open source hardware and software may allow practitioners to access hardware that they can trust to carry out sensitive work and transfer sensitive information. Open source hardware and software are potentially more trustworthy than proprietary technology, as they can be reviewed and audited by anyone who is willing to do so.

The future of Novena is unclear, as it has not yet found commercial success, but its existence has ignited a generation of entrepreneurs willing to compete with large manufacturers to offer options for general users. An important example

---

[41] Raspberry Pi Foundation, "About Us," www.raspberrypi.org/about/.
[42] "Sales Soar and Raspberry Pi British Board Beats Commodore 64," *The MagPi Magazine*, March 16, 2017, www.raspberrypi.org/magpi/raspberry-pi-sales/.
[43] A. Huang and S. Cross, "Novena: A Laptop With No Secrets," *IEEE Spectrum*, October 27, 2015, http://spectrum.ieee.org/consumer-electronics/portable-devices/novena-a-laptop-with-no-secrets.

of this is the Librem 13, a laptop available since 2016 that promises to respect privacy and enhance security in "every chip in the hardware, every line of code in the software."[44] The laptop ships with the option of two operating systems, Purism OS or Qubes OS, which are both well regarded in the security and open source communities as strong and reliable options for those with security and privacy in mind. It also includes hardware kill switches that shut down the microphone, camera, Wi-Fi connection, and Bluetooth. These are important characteristics that practitioners should consider, given the scope of unchecked surveillance by governments exposed to a broad public by the revelations of Edward Snowden and other whistleblowers, as described by Lisl Brunner in Chapter 10.

If these devices survive and evolve, or if they encourage other open and secure products, they will provide valuable tools for human rights practitioners seeking to protect the data of vulnerable populations. As the market for open source or privacy-enhancing hardware is in its early stages of development, it is unclear whether the scale of production will be sufficient to reach human rights practitioners around the globe. Scale will not only impact the affordability of a device, but also determine whether it moves into common usage. If it does not, it could raise red flags for governments when crossing borders or adversarial checkpoints.

It is essential that secure tools are not just available for human rights researchers but are also adopted by wider communities. The general adoption of features by nonspecialized products makes the use of these features by human rights researchers less risky, because they are less identified with behavior the state wants to control. A powerful example of this is the adoption of end-to-end encryption by the popular messaging application WhatsApp. In 2016, WhatsApp announced that it was making end-to-end encryption the communication default for its billion-plus users.[45] The notion of end-to-end encryption, which refers to the use of communications systems in which only the originator and recipient of a message can read its contents, is nothing new to human rights practice. For many years, dozens of human rights and technology advocates have promoted end-to-end encryption as critical for the future of journalistic and human rights work,[46] but it was not until this development that such technology became widely available. If projects like the Novena and Librem 13 laptops successfully compete for a small fraction of the market share of companies like Lenovo and Hewlett-Packard, they could create pressure for other manufacturers to adopt the privacy-enhancing features that distinguish them, and in doing so offer secure computing alternatives for human rights practitioners.

---

[44] Purism, "Discover the Librem 13," https://puri.sm/products/librem-13/.
[45] "End-to-End Encryption," *WhatsApp Blog*, April 5, 2016, https://blog.whatsapp.com/10000618/end-to-end-encryption.
[46] Access Now, "Encryption TK: Securing the Future of Journalism and Human Rights," YouTube, March 20, 2014, www.youtube.com/watch?v=uxidkrhOo-o.

Beyond the expansion of these existing technologies, we are also likely to see innovation around the Internet of Things, or IoT, which references the increased connectivity or networking among devices of all kinds and purposes. The IoT, which allows the devices of smart homes and smart cities to be controlled remotely, and in many cases automatically, is linked directly to the growing availability of open hardware and sensors. From thermostats and refrigerators to wearable devices and new forms of personal and mobile devices, we are likely to see connected devices in virtually every aspect of human life. This will likely create excellent opportunities for new forms of fact-finding and research, but will also likely create new perils for human rights practitioners and general users alike. Perhaps the biggest challenge will come from the ability that governments and organized criminals have developed to access and analyze data stored and in transit. We are only starting to understand what this might mean, for instance in recent analyses of the privacy implications of fitness trackers,[47] for how law enforcement could use our intelligent personal digital assistants in criminal and national security investigations,[48] and how connected home cameras could be infiltrated by organized criminals, governments, and other nefarious actors.[49]

V CONCLUSION

Events of the past five years have significantly shaped the discourse around human rights technology. What has been learned and confirmed after Edward Snowden's revelations of mass and unchecked surveillance by nation-states and corporations has necessarily focused the attention of global civil society on the dire effects of surveillance and the need to counter them.[50] The state of surveillance has cast a dystopian shadow over the future of human rights, as Mark Latonero points out in Chapter 7, where practitioners fear technology will be used for control rather than liberation. The hypersurveillance practices of our times, as well as the role that technology plays in them, are indeed an extensive attack on human rights.[51]

[47] A. Hilts, C. Parsons, and J. Knockel, "Every Step You Fake: A Comparative Analysis of Fitness Tracker Privacy and Security," *Open Effect* (2016).
[48] A. Wang, "Can Alexa help solve a murder? Police think so – but Amazon won't give up her data," *The Washington Post*, December 28, 2016, www.washingtonpost.com/news/the-switch/wp/2016/12/28/can-alexa-help-solve-a-murder-police-think-so-but-amazon-wont-give-up-her-data/.
[49] "Hacked Cameras, DVRs Powered Today's Massive Internet Outage," *Krebs on Security*, October 21, 2016, https://krebsonsecurity.com/2016/10/hacked-cameras-dvrs-powered-todays-massive-internet-outage/.
[50] E. MacAskill et al., "NSA Files: Decoded: What the revelations mean for you," *The Guardian*, November 1, 2013, www.theguardian.com/world/interactive/2013/nov/01/snowden-nsa-files-surveillance-revelations-decoded.
[51] E. Piracés, "From Paranoia to Solidarity: Human Rights Technology in the Age of Hyper-Surveillance," Canada Centre for Global Security Studies, March 28, 2014, www.cyberdialogue.ca/2014/03/from-paranoia-to-solidarity-human-rights-technology-in-the-age-of-hyper-surveillance-by-enrique-piraces/.

However, human rights practitioners should not let that hinder their ability to imagine alternative visions that could guide the intersection of human rights and technology.

The technologies discussed in this chapter do not represent an exhaustive compilation of trends that will shape the future of human rights practice, but rather are a starting point to expand our understanding of what technology could do for us in the near future. Challenging current technology transfer models and expanding the ecosystem of actors around them is key, because in creating a more inclusive, deliberate, and forward-looking interdisciplinary field around human rights technology, we will be creating a better opportunity to advance the larger human rights field.

A change in the dynamics of technology transfer will challenge the traditionally asymmetrical power dynamics between human rights practitioners and their transnational supporters. We can foster this by promoting capacity-building in the Global South, favoring open source software and hardware, and critically evaluating budgetary allotments to technology. In the process, grassroots practitioners will be at the helm of designing and adapting human rights technology. We must be conscious that this will challenge the growth of professional opportunities for Global North practitioners. There are important questions that will be critical for any next step. Can human rights play a role in the governing of technology? What role can the private sector play in advancing human rights technology? Can human rights challenges drive technological innovation? To answer them, we should be open to interdisciplinary conversations like the one taking place in this volume, and encourage an inclusive and participatory multistakeholder ecosystem.

The approach of human rights practitioners to technology will be a determining factor in their ability to advance accountability, transparency, and justice in the years to come. This book is an invitation to imagine the future of the intersection of human rights technology and human rights practice. For this intersection to benefit practitioners, it must adopt a solidarity-based framework for technology transfer.

A solidarity approach requires technologists to understand and respect the cultural context of the environment they are working within. They must reimagine the relationship as bidirectional and characterize their counterparts in technology transfer as active collaborators. Technologists must establish partner relationships with practitioners, from designing solutions that involve technology all the way through to evaluating them. Practitioners should also be able to tinker with and modify the technologies they are using, and technologists should support them in doing so. This commitment should be reflected in the timeline, budget, and conceptualization of the project. Solidarity requires careful consideration of how technology may displace human resources or compete with scarce resources available in the human rights funding landscape. This technology transfer approach prioritizes human capacity and sustainability above technical complexity and sophistication. Finally, technologists must continuously question their own role within larger power structures – are they helping to reduce the burden of inequality and dependency, or are they just recreating it through the deployment of technology? Ultimately, a solidarity approach demands that technologists not contribute to long-term inequalities while working with human rights workers and communities in crisis.

Index

abortion rights, 77, 85
accountability
  algorithms and, 191
  Bitcoin and, 299
  criminal framework, 113–22
  for human rights violations, 7, 9–13
  introduction to, 3
  legal accountability, 44–45
acculturation process, 12
Adaptation Fund, 62–63
adaptive capacity and resilience, 62, 66
affect-laden conception of humanity, 181
Afghanistan, 94
Agreement on Trade-Related Aspects of Intellectual Property Rights (TRIPS), 54–55
agriculture, industrial, 69
Al Qaeda, 102, 110
algorithms
  computer vision algorithms, 180
  in data collection, 171–72
  social media, 202–5
  transparency/accountability problems with, 191
American Association for the Advancement of Science, 76, 174
American Civil Liberties Union (ACLU), 280–81
American Convention on Human Rights, 219, 225
Amnesty International
  data visualization reports, 163, 177
  education in digital literacy, 209
  human rights technology, 291
  interactive data websites, 185
  Twitter communication issues, 201
  user-generated video, 194
  violations by Boko Haram and Nigerian forces, 141–45, 172
anonymity, 157, 196, 216, 225, 293, 304

Anti-Privatization Forum (APF), 26
apartheid, 28
Arab Charter on Human Rights, 219
Arduino project, 302
Arendt, Hannah, 120, 122
armed conflict, 93–96
ART cases. *See* assisted reproductive technologies
artificial intelligence (AI), 1, 161, 296–99
Asia Pacific Economic Cooperation Privacy Framework, 233
assisted reproductive technologies (ART)
  assisted insemination, 72
  introduction to, 72–74
  law/law-making in, 86–90
  margin of appreciation and, 83–86
  overview of, 78–83
  summary of, 90–92
Austria, ban on IVF donors, 79, 84
autonomous weapons
  capability, 93
  criminal accountability, 115–17
  defined, 117
  international humanitarian law, 113–15
  introduction to, 1, 14
  legal implications of deploying, 93
  overview, 98–101, 123
  principles of humanity and, 94
  private contractors and, 122–23
  separation of powers framework, 101–2
  war and, 22

Bali Conference of the Parties (COP), 58
Ball, Patrick, 153, 170–71
bar charts, 181

Bellingcat investigative team. *See* Russian intervention in Ukraine
Belmont Report, 36–38, 41–43
beneficence, 41–42
beneficiaries of action, 68–69
Benetech, 209
bias in big data reporting, 154
bias in data collection, 171
big data
   analytics and, 151, 153, 157
   applications for human rights, 159–61
   collection of, 151–52
   ethics considerations, 173
   human rights violations, 157–59
   introduction to, 7, 125, 127, 149–51
   permissible limitations on privacy, 155–56
   privacy tradeoffs, 153–57
   use of, 151, 153
bioethics and human rights, 2–3, 71–78, 86–90
biosurveillance, 158
Bitcoin, 299
Blockchain, 299–302
Boko Haram, 141–42
The Book of Trees: Visualizing Branches of Knowledge (Lima), 164
Bretton Woods Institutions, 50, 54
Buk missile launcher, 139–41
Bureau of Democracy, Human Rights and Labor (U.S. Department of State), 291
Bureau of Investigative Journalism, 98
bureaucratization of human rights, 13

call detail records, 155
Cambodia's Family Trees, 179
Cameras Everywhere, 130
CameraV app, 172
Cannataci, Joseph, 223. *See also* UN Special Rapporteur on Privacy
carbon emissions, 46
Carnegie Mellon University, 180
   Center for Human Rights Science, 298
Carter Center. *See* Syria Conflict Mapping Project
cell phone data tracking, 270
Center for Applied Legal Studies at the University of Witwatersrand (CALS), 31
Center for Economic and Social Rights (CESR), 176
Center for Strategic and International Studies (CSIS), 201
Charter of Fundamental Rights of the European Union, 219, 233
child soldiers, 142
China, 183, 262
chlorofluorocarbon (CFC) emissions, 56

cholera deaths, mapping of, 165
citizen videos, 9
civic engagement online, 244
civil responsibility in tort, 116
civil rights, 28, 266
civil society organizations, 223
Civilian Joint Task Force (JTF) (Nigeria), 142–45
civilian witnesses, 193
classification of data, 152
clickbait content on social media, 204
climate change
   beneficiaries of action, 68–69
   development approach, 60–69
   impact on human rights, 65–69
   introduction to, 1, 8, 21, 46–48
   modernization and development, 48–55
   summary of, 70
   UNFCCC framework, 55–60
climate change law, 56
Clinton, Bill, U.S. President, 102–6
cluster munitions, 135
Code of Conduct for social media, 16–17
code-sharing websites, 186
Cold War, 49
colonialism, 49–50, 53, 56
Comitato Nazionale per la Bioetica (Italy's National Bioethics Committee), 75
command responsibility, 118
Committee on Economic, Social and Cultural Rights (CESCR), 66–67
common but differentiated responsibilities (CBDR)
   ineffectiveness of, 57, 60
   overview of, 60–63
   solving problems with, 64–69
communication risks, 192–94, 198–200
communication technology. *See also* digital communications
   advocacy communication risks, 198–200
   anonymity rights in communication, 225
   bulk collection of communications data, 217
   fact-finding and advocacy, 190–92
   fact-finding risks, 192–94
   introduction to, 5
   privacy rights in, 221
   surveillance communication rights, 219–20
   transparency and, 10
Communications Decency Act, 250, 260
community values, 2–3
computer vision algorithms, 180
constitutional precepts, 2–3
contact-chaining, 221
contextual integrity frame for privacy, 151
conventional weapons, 94

corporate social responsibility (CSR), 249, 254
corporate surveillance, 158–59
cost-benefit calculations, 47–48
Costa and Pavan v. Italy, 73, 80–81, 89–90
Council of Europe's Committee on Bioethics, 75, 84
Council of Europe's Data Protection Convention, 233
Court of Justice of the European Union (CJEU), 225, 234–35, 239–40
crisis reporting, 154
cryopreserved embryos, 84–85
cultural rights, 71
currency exchange technology, 299–302
cyberlaw, 2, 5–6
cystic fibrosis, 80–81

Daiichi Nuclear Power Plant in Fukushima, Japan, 302–3
data analysis, 167
data-driven policing, 171–72
data journalism, 186
data protection laws, 216, 233–35
data sharing, 153, 176
data visualization
    access and inclusion, 182–83
    brief history of, 163–67
    bulk collection, 217
    challenges to, 171–73
    defined, 162
    distortion techniques in, 175–76
    encoding techniques, 180–81
    interactive, 182
    introduction to, 7, 125–27, 162–63
    mapping of, 183–84
    meanings from technical decisions, 180–82
    mobilization and outreach, 184–85
    qualitative data, 177–80
    quantitative data, 173–77, 182
    summary of, 187
    technical sustainability, 185–86
datafication of human rights, 163, 167–71
Deep Mind, 297
Deepwater Horizon oil spill, 303
deforestation, 67–68
demand-management technology, 25, 28–31, 35
Democratic Republic of Congo (DRC), 178
Detroit Geographic Expedition, 183
Deutscher Ethikrat, 75
developing countries, 49–51, 55–60
development approach to climate change technology, 48–55

Dickson v. United Kingdom, 73, 84
digital communications. *See also* communication technology
    legality, necessity, and proportionality, 225–30
    privacy rights and human rights law, 224–33
    to publics, 200
    surveillance rights, 219–20, 230–31
Digital Millennium Copyright Act (DMCA), 260
digital security, 196
digitization, 10, 19
dignity rights, 26–27, 34, 81–82
discovery, defined, 152
discretion of *demandeur* developing countries, 48
discrimination in targeting of minority groups, 11
disease surveillance, 158
DIY Geiger counters, 302–3
DNA technology in forensics, 295
Doctors' Trial at Nuremberg, 75
documentation technologies, 9
donor gamete ban, 79, 84
Draft Code of Conduct of International Transfer of Technology, 53–54
drone technology
    autonomous functioning of, 99
    international humanitarian law, 113–15
    introduction to, 11–12, 123
    against ISIS, 110–11
    in Kosovo, 103–4
    in Libya, 107–8
    overview of, 96–98
    principles of humanity and, 94–95
    separation of powers framework, 101–2
Dropbox, 294
drought events, 59, 68
Drupal, 294
dual use technologies, 38
due process, 117

e-governance, 301
E-ZPass, 271
Ebola crisis, 154
econometrics, 167
economic rights, 33, 252
economies in transition, 51
emerging economies, 51
encoding techniques, 180–81
encryption technology, 279, 287, 289
energy management software, 69
Engine Room, 209
environmental law, 56
epidemiology data and human rights, 167
equality principle, 3
Ethereum, 301
ethical developments of human fertilization, 89

ethical safeguards, 43
ethics and law of war, 21–22
EU Charter of Fundamental Rights, 225, 234
EU General Data Protection Regulation, 160, 233, 239–42, 259
EU Regulation on Privacy and Electronic Communications, 233
EU-US Umbrella Agreement, 239
European Commission, 16, 75
European Convention for the Protection of Human Rights and Fundamental Freedoms (European Convention), 72, 219, 224
European Convention on Human Rights and Biomedicine, 71
European Court of Human Rights (ECtHR)
 artificial intelligence, 298
 assisted reproductive technologies cases, 78–83, 90, 92
 case law of, 256–57
 communication privacy, 221, 226–27
 digital communication privacy, 224, 235
 extraterritorial right to privacy, 236–37, 242
 margin of appreciation, 83–86
 online domain rights, 253
 overview of, 72–73
 surveillance rights, 220, 232
European Union, 268
Evans v. United Kingdom, 73, 79, 82, 84
Event Labeling through Analytic Media Processing (E-LAMP) project, 180, 298–99
executive branch (president), U.S. government
 allocation of power, 95
 ISIS, 110–12
 Kosovo military action, 102–6
 Libya, 106–10
 separation of powers framework, 101–2
expert systems (ES), 297
extrajudicial killings, 143
Extraordinary African Chamber, 148
extraterritorial right to privacy, 236–42
extreme weather risks, 59

Facebook
 advertising and privacy rights, 265–66
 Code of Conduct with, 16–17
 communication issues, 201
 data collection, 152
 First Amendment protection on, 250
 freedom of expression rights, 243, 261, 264
 gatekeepers of, 251
 human rights impact of, 245–46
 rebel forces against ISIS, 137
 restricting access, 275

Russian intervention in Ukraine, 138
unlawful content processes, 269
user-generated content on conflicts, 134
fact-finding and advocacy. *See also* human rights advocacy
 communication and mediation, 190–92
 communication risks, 192–94, 198–200
 discursive approach, 211–12
 educational approach, 208–9
 inequality in, 205–8
 introduction to, 12, 188–90
 misinterpretation risk with, 194, 196–98
 reflexive approach, 210–11
 risk assemblages, pluralism and inequality, 205–8
 silencing risk double-bind, 208–13
 summary of, 212–14
 surveillance risk, 194–96
 technical approach, 209–10
fairness claims, 52, 55–60
falsified media, 133
First Amendment (U.S.) protection of speech, 250, 268
flooding events, 59
forced marriages, 142
Ford Foundation, 290
Forensic Architecture, 179
Fourth Amendment (U.S.) protection, 276–77, 285
Free Syrian Army (FSA), 137
freedom of expression, 19, 261–66, 268
Freedom of Information, 176
Front Line Defenders and Tactical Technology Collective, 209

G77 membership, 51
G Suite (Google), 294
gamete donation for IVF, 85
gatekeepers, 14, 251
General Agreement on Trade and Tariffs (GATT), 54
geolocation, 138–39
GitHub, 186
Global Network Initiative (GNI)
 Accountability, Policy and Learning Framework, 262
 freedom of expression rights, 261–64
 Implementation Guidelines for Freedom of Expression, 264
 international human rights standards, 246
 internet companies and, 258
 overview of, 249, 261
 soft law frameworks, 267
Global Positioning System (GPS), 177, 271

Global Witness, 179
Google
  de-indexing search results, 240
  European Data Protection Directive, 259
  file management, 294
  First Amendment (U.S.) protection on, 250
  freedom of expression rights, 261–62, 264–65
  G Suite, 294
  human rights impact of, 246
  Jigsaw project, 292
  online domain rights, 243
  restricting access, 275
  unlawful content processes, 269
Google apps, 294
Google Earth, 132, 141
Google Maps, 185
Google Street View, 132
graph theory, 179
graphs/graphics data. *See also* data visualization
  graph theory, 179
  importance of, 173–74
  mapping tools, 174
  for medical treatment, 165–66
  network graphics, 178–79
  racist violence graphs, 166–67
grassroots activists, video use by, 172
Green Climate Fund, 61
greenhouse gas (GHG) emissions, 56, 60, 67
Group Areas Act (South Africa), 28
Guardian Project, 293
guided munitions, 100

hacking technology, 278–79, 287
Harvard Humanitarian Initiative's Signal Code, 160
hate speech, 16–17
Higgins, Eliot, 134–35
High Court of South Africa, 31–33
Hindu nationalists in India, 181
Hippocratic principle, 41
HIV/AIDS, 32, 35, 185
homemade barrel bombs, 135
Human Development Index (HDI), 51
human rights. *See also* online domain and human rights
  beneficence, 41–42
  big data applications for, 159–61
  bioethics and, 71–78, 86–90
  climate change impact on, 65–69
  datafication of, 163, 167–71
  dignity rights, 26–27, 34, 81–82
  financial compensation, 39
  freedom of expression, 261–66
  informed consent, 38–41
  international treaties, 4, 23

  justice and, 42–43
  as moral claims, 4, 13
  practices, 4–5
  rights restricting/enhancing technology, 38
  safeguards, 36
  scrutiny requirements, 36–38
human rights advocacy. *See also* fact-finding and advocacy; user-generated video
  communication channels, 207
  communication risks, 198–200
  democratizing effects for, 126–27
  introduction to, 12, 125–26, 128
  miscalculations and resources, 202–5
  mistakes and reputational integrity, 200–2
Human Rights Center at the University of California, Berkeley, 290
Human Rights Data Analysis Group (HRDAG), 170–71
human rights law. *See also* international humanitarian law
  assisted reproductive technologies, 90
  bioethics and, 72, 74–78
  introduction to, 2–4, 6
  new technologies and, 23
  private actors and online privacy, 253–61
  role of, 17–20
  as state-centric in nature, 267
  violations of, 11
human rights protection, 1, 19–20, 215–16, 218–21. *See also* privacy rights
human rights technology. *See also* technology rights; user-generated video
  artificial intelligence, 1, 161, 296–99
  Blockchain technology, 299–302
  defined, 2, 293–95
  future of, 290–93
  introduction to, 1–2, 289–90
  open hardware, sensors, and the Internet of things, 302–6
  summary of, 306–8
  technological trends in, 295–96
  violations with, 44
human rights violations
  accountability for, 7, 9–13
  big data collection and, 157–59
  effective remedy for violation of, 232–33
  introduction to, 21
  Nigerian military war crimes, 142–45
  with technology, 44
Human Rights Watch (HRW), 163, 177–78, 280–81, 291
Humanitarian Blockchain, 301
Humanity United, 290
hydrological cycle, 61

immigrant detainees, mapping data, 174
in vitro fertilization (IVF)
  Austrian ban on donor gametes, 79, 84
  gamete donation for, 85
  introduction to, 72, 78
  legislation governing, 86–87
inalienability principle, 3
inclusion principle, 3
indexing of data, 152
industrial agriculture, 69
industrialization, 52
information and communication technologies (ICTs). *See also* user-generated video
  advocacy communication risks, 198–200
  communication issues with, 190–92, 201
  fact-finding and, 188
  gatekeepers, 14
  human rights technology, 294
  introduction to, 8
  non-state actors, 16
  summary of, 212
  surveillance risk, 194–96
  threat to privacy, 150
  transnational nature of online platforms, 268
  transparency and, 10
informed consent rights, 38–41
integrated development environment (IDE), 302
intellectual property rights, 8
Inter-American Commission on Human Rights, 225
Inter-American Court of Human Rights, 225
interactive data visualization, 182
interactive data websites, 185
Intergovernmental Panel on Climate Change (IPCC), 66–67
International Bill of Human Rights, 256
International Committee for the Red Cross (ICRC), 98
International Court of Justice (ICJ), 224, 236
International Covenant on Civil and Political Rights (ICCPR), 65, 218, 236, 272–74, 282–85
International Covenant on Economic, Social and Cultural Rights (ICESCR), 26, 65, 174, 286–87
International Criminal Court (ICC), 143, 296
International Energy Agency, 61
international humanitarian law (IHL)
  criminal accountability framework, 113–15
  enforcement strategies from, 4
  introduction to, 1, 3–5
  new technologies and, 22, 245
  overview of, 94
  violations, 4

International Military Tribunal (IMT), 118–19
International Monetary Fund (IMF), 54
international treaties, 4, 23
Internet
  browsing activity tracking, 270
  introduction to, 1, 11, 19
  online hate speech, 16–17
  surveillance concerns, 15
Internet of things (IoT), 306
Internet service providers (ISPs), 249, 276
Investigatory Powers Act, 241
Iraq, 94, 98
Islamic State in Iraq and Syria (ISIS), 98, 102, 110–12, 137
Israeli cyberwarfare company, 196
Israel's Harpies, 93, 122
Italy, ban on pre-implantation genetic diagnosis, 80–81

Jigsaw project (Google), 292
Johannesburg Water, 26, 28, 30–33, 37, 39–40
justice and technology, 6–9
justice claims in developing countries, 55–60

Knecht v. Romania, 73, 84, 89
knowledge-based systems (KBS), 297
knowledge creation, 5
Korean War, 112
Kosovo military action, 102–6
Kuziminsky camp, 139
Kyoto Protocol (1997), 47, 58, 62

La Rue, Frank, 248. *See also* UN Special Rapporteur on the Promotion and Protection of the Right to Freedom of Opinion and Expression
laws of Caliphate, 142
least-developed countries (LDCs), 50, 55
legal implications of human fertilisation, 89
lethal force, 12
Libya, 106–10
Libya, no-fly zone, 107
line charts, 181
LinkedIn, 250
Live Leak, 134

MacArthur Foundation, 290
Malaysia Airlines Flight 17 (MH17), 138
mapping data visualization, 183–84
mapping tools, 174
marginalized communities, 48, 170, 183
Martus project, 293

massacres and video evidence, 132
McNaboe, Chris, 135
medically-assisted procreation, 90
Meier, Patrick, 154
metadata, 172, 197–98, 277
micro-grid applications, 69
Microsoft, 241
middle income countries, 51
misappropriation of personal data, 235
mobile phone technology, 7, 9–10, 154–56
mobilization, 184–85
modalities of regulation, 6
modernization, 48–55
monsoon events, 59
moral claims, as human rights, 4, 13
multilateral environmental treaties (MEAs), 48, 53
multinational corporations (MNCs), 235

National Adaptation Plans (NAPs), 62
national security protection, 226
nationally determined contributions (NDCs), 58–59
natural disaster responses, 154
Nazi Germany, 220
network graphics, 178–79
New International Economic Order, 48, 63
New Tactics in Human Rights, 209
new technologies. *See also* big data; communication technology; drone technology; human rights technology; information and communication technologies; technology rights
  accountability challenges, 9–13
  advocacy and, 200, 205
  cyberlaw and, 5–6
  doctor/patient relationship, 77
  fact-finding and, 194, 205
  human rights and, 36, 125–28, 159
  human rights law and, 18
  impact of, 102, 164, 216
  introduction to, 1–2, 8–9, 15, 289, 301
  normative approaches, 21–24
  private authority and, 13
  resistance to, 278
  of warfare, 104, 112, 209
New York University's Center for Human Rights and Global Justice, 163
Nigerian military, 142–45
Nightingale, Florence, 165–66
non-discrimination principle, 3
non-governmental organizations (NGOs), 130, 168, 188, 193, 199, 206

non-state actors, 14–16, 20, 215
nonpublic forums, 249–50
North Atlantic Treaty Organization (NATO), 103
Northrop Grumman contractors, 100
Nuffield Council on Bioethics (U.K.), 75
Nuremberg trials of Nazi war criminals, 114, 117, 119–21
Nuremburg Code, 36

Oak Foundation, 290
Obama, Barack, U.S. President, 107–9, 111–12, 287–88
Office of the Prosecutor at the International Criminal Court, 143
online domain and human rights
  challenges to, 266–69
  gatekeepers, 251
  human rights within, 261–66
  introduction to, 243–47
  private actors and human rights law, 253–61
  role of internet companies in, 247–53
  types of forums, 249–50
  user expressions, 251–53
Ontario's Human Rights Commission, 157
open data sharing, 176
open hardware, 302–6
Open Net Initiative, 253
Open Society Foundations, 290
open source intelligence, 146, 296
Open Technology Fund, 291
OpenEvSys project, 293
OpenStreetMap, 184
Orange Order in Northern Ireland, 181
Orange Revolution in Ukraine, 181
Organization for Security and Co-operation in Europe (OSCE), 49, 56–57, 139, 254
overseas development assistance (ODA), 63

Panama Papers, 179
parent, in reproduction technologies, 19–20
parent-child relations, 89
Paris Agreement (2015), 58–59
Parrillo v. Italy, 73, 82–85
participation principle, 3
pie charts, 181
pluralism in fact-finding advocacy, 205–8
poverty and water, 27
power and technology, 6–9
pre-implantation genetic diagnosis (PGD), 72, 80–81
Predator drones, 95
prepaid water meter technology
  Belmont Report, 37–38
  overview of, 25

prepaid water meter technology (cont.)
　　South Africa, 28–29, 31
　　United Kingdom, 32
PRISM Program, 221–22
privacy rights. *See also* big data and privacy rights
　　bulk collection privacy, 227–30
　　data protection laws, 216, 233–35
　　digital communication, 224–33
　　effective remedy for violation of, 232–33
　　extraterritorial right to privacy, 236–42
　　Facebook users against advertisers, 265–66
　　introduction to, 19, 22, 217–18
　　legality, necessity, and proportionality, 225–30
　　permissible limitations on privacy, 155–56
　　protection of, 218–21
　　Snowden, Edward, impact on, 150, 217, 221–24
　　summary of, 242
　　unlawful interference with, 272–73
Privacy Shield agreement, 240
private actors and human rights law, 253–61
private contractors
　　autonomous weaponry and, 100–1, 122–23
　　criminal accountability, 117–22
　　drones, 97–98
　　international humanitarian law, 113–15
　　against ISIS, 111
　　overview of, 28
　　separation of powers framework, 101–2
　　summary of, 123
　　use in Kosovo, 104
　　use in Libya, 108
privatization of war, 22, 95
procreative tourism, 88
proof of existence attribute, 300
Provenance company, 301
public ethics, 75
Public Laboratory for Open Technology, 303
Putin, Vladimir, 138

qualitative data visualization, 177–80
quantitative analysis, 7
quantitative data visualization, 173–77, 182
Qubes OS, 305

Radio Free Asia, 291
Raspberry Pi technology, 304
Reaper drones, 95
reproduction technologies, 1, 5, 14, 19, 91
Responsible Data Forum (2016), 173
retention of data, 152
right to dignity, 26–27, 34
right to enjoy the benefits of scientific progress (REBSP), 26–27, 34–36
right to food, 65–68

right to free expression, 19
right to health, 65–67
right to life, 65–66
right to privacy. *See* privacy rights
right to water, 65–67
rights-based approach, 64
　　human rights-based approach, 2–4, 6, 18, 24, 64, 73, 78, 80–81, 90
rights restricting/enhancing technology, 38
Rio Declaration on Environment and Development, 53
Royal Statistical Society, 165
RQ-4 Global Hawk (Northrup Grumman), 97
Ruggie, John, 254–55, 257–58
rule of law principle, 3
Russian intervention in Ukraine, 138–41

S. and Marper v. United Kingdom, 78, 234–35
safe motherhood and abortion, 77
satellite imagery, 9, 132, 138, 178, 296
science and technology studies (STS), 2–3, 5–7
scraping, defined, 152
sea-level rises, 61
search, defined, 152
second-hand witnessing, 133
secondary trauma, 146–47
Security in-a-Box, 209
security training for digital technology, 281
semi-autonomous offensive systems, 99–100
separation of powers, 101–2
sexual violence, 142
S.H. and Others v. Austria, 73, 79, 84
shame/shaming, 4, 11–12
Silverman, Craig, 129
SITU Research, 296
Skype, 278, 294
smartphones, 168, 193
Smith v. Maryland, 276–77
Snowden, Edward, 150, 217, 221–24
social developments of human fertilization, 89
social justice mapping, 183–84
social media
　　advocacy messages via, 198–200
　　clickbait content on, 204
　　Code of Conduct, 16–17
　　content culture, 204
　　human rights violations and, 10
　　introduction to, 1
　　miscalculations and resources, 202–5
　　privacy concerns, 284
　　proofing of, 200–2
　　quantification methods, 199
　　rebel forces against ISIS, 137
　　user-generated video, 134

Index

sociotechnical processes in technology, 151
soft law frameworks, 267
South Africa
   apartheid, 28
   free water allowance, 29–30
   litigation, 21
   prepaid meter technology, 28–29, 31
South African Constitution, 26
South African Constitutional Court, 26, 43
special and differential (S&D) treatment, 51–52, 55–60
spyware, 195
SSL/TLS security, 302
Stasi secret police in East Germany, 220
state obligation in human rights protection
   individual data protection, 286–88
   introduction to, 215
   setting privacy limits, 263
Statistical Breviary (Playfair), 164
Statistical Society of London, 165
storing data, 152
surrogacy concerns, 81–82
surveillance
   communication surveillance rights, 219–20, 230–31
   extraterritorial right to privacy, 236–39
   government surveillance, 160
   risks of, 194–96
   Snowden, Edward, impact on, 150, 217, 221–24
surveillant anxiety, 157–58
Syria Conflict Mapping Project (Carter Center)
   introduction to, 134–37
   network dynamics, 136–37
   overview of, 134
   weapons tracking, 135–36

Taliban, 102
technological solutionism, 13
technology. *See also* communication technology; human rights technology
   challenges for non-state actors, 20
   conclusions about, 282–88
   demand-management technology, 25, 28–31, 35
   encryption technology, 279, 287, 289
   enforcement of, 125–28
   hacking technology, 278–79, 287
   ICCPR waiver, 283–85
   individual choice and, 271–77
   introduction to, 270–71
   new technologies, 21–24
   non-state actors and, 14–16, 20, 215
   power and justice, 6–9
   third-party doctrine, 276–77, 285
   voluntariness of insecure use of, 277–82

technology-enabled surrogacy, 81
technology entrepreneurs, 18
technology of distance, 168
Telecommunications Industry Dialogue, 262
Telegram network, 134, 294
telephony metadata, 234
terrorism, 93–94
third-party doctrine, 276–77, 285
traditional public forums, 249–50
transnational nature of online platforms, 268
transnational solidarity, 293
transparency concerns, 10, 191, 299
truth-claims, manipulation of, 197, 206
Twitter
   communication issues with, 191, 201
   content removal by, 259
   freedom of expression rights, 261
   human rights impact of, 245–46
   introduction to, 16–17
   overview of, 134
   rebel forces against ISIS, 137
   Russian intervention in Ukraine, 138

U.K. Investigatory Powers Act, 224
UN Committee on Development Policy List, 52
UN Conference on Trade and Development (UNCTAD), 50, 53
UN Educational, Scientific and Cultural Organization (UNESCO), 75
UN Framework Convention on Climate Change (UNFCCC), 47–48, 51, 55–60
UN General Assembly, 50–51
UN Global Compact, 254
UN Global Pulse, 160
UN Guiding Principles on Business and Human Rights, 14–16, 247, 255–58, 264, 267
UN Human Rights Committee, 220–21, 225, 236, 271, 273
UN Human Rights Council (UNHRC), 150–51, 223, 248, 255, 271
UN Office of the High Commissioner for Human Rights (OHCHR), 157, 223, 225, 238–39
UN Security Council, 107
UN Special Rapporteur on Privacy, 159. *See also* Cannataci, Joseph
UN Special Rapporteur on the Promotion and Protection of the Right to Freedom of Opinion and Expression, 243, 248. *See also* La Rue, Frank (former)
UN World Summit on the Information Society, 247–48
Unionism in Northern Ireland, 181
United Arab Emirates, 195
United Nations (UN), 14, 26, 34, 114

United Nations Declaration for a New International Economic Order (NIEO), 49, 52–53
United States v. Miller, 275–77
Universal Declaration on Bioethics and Human Rights, 36, 39, 44, 71
Universal Declaration on Human Rights (UDHR), 26, 65, 218–19
universality principle, 3
unmanned aerial vehicles (UAVs), 96, 103, 294–96. See also drone technology
U.S. Central Command, 110
U.S. Congress, 95, 101–12
U.S. Department of Justice Office of Legal Counsel (OLC), 105–6, 109
U.S. Foreign Intelligence Surveillance Court, 230
U.S. House of Representatives, 104, 108–9
U.S. National Security Agency (NSA), 172, 195, 281
U.S. Presidential Commission for the Study of Bioethical Issues, 75
U.S Constitution, 101–2
use of force, 20
user-generated video
  Boko Haram, 141–42
  case studies, 133
  challenges in law/advocacy, 132–33
  discussion, 145–48
  graphic nature of images, 203
  introduction to, 125–26, 129
  Nigerian military war crimes, 142–45
  role in human rights investigations, 130–32, 194
  Russian intervention in Ukraine, 138–41
  Syria case study, 134–37
  video evidence of crimes, 131
  weapons tracking, 135–36
user welfare and technology, 38

Verification Handbook for Investigative Reporting (Silverman, Tsubaki), 129
video evidence. *See* user-generated video
Vienna Convention on the Law of Treaties, 282
Vienna Declaration and Platform of Action
  indivisibility principle, 3–4
  inter-dependence principle, 3–4
  inter-relatedness principle, 3–4
Vietnam War, 102–3

vision-based video analysis system, 298–99
voluntariness of insecure technology use, 277–82
vulnerable populations, 6

War on Terror, 274
War Powers Resolution, 96, 105
Washington Consensus, 28
water
  beneficence, 41–42
  demand management devices, 28–31, 35
  free water allowance, 29–30
  Free Water Policy, 33
  informed consent and, 38–41
  introduction to, 25–28
  justice and, 42–43
  Mazibuko Litigation, 31–33
  right to science, 34–36
  scrutiny requirements with, 36–38
  technologies, 14
  tokens for, 29
  weapons tracking, 135–36
WhatsApp, 197, 209, 294, 305
whistleblowers, 231
WikiLeaks, 179
Wikipedia, 251
willful blindness, 120–21
Wiring project, 302
WITNESS (NGO), 130
Wordpress, 294
World Bank, 54
World Intellectual Property Organization (WIPO), 53–54
World Medical Association, 36
World Trade Organization (WTO), 50, 52, 54

YouTube
  communication issues with, 191
  freedom of expression rights, 264
  introduction to, 16–17
  metadata doctoring, 197–98
  number of monthly visitors, 243
  overview of, 134
  Russian intervention in Ukraine, 138
Yugoslavian weapons stockpiling, 135

Zuckerberg, Mark, 261